Ecosystem Management

About Island Press

Island Press is the only nonprofit organization in the United States whose principal purpose is the publication of books on environmental issues and natural resource management. We provide solutions-oriented information to professionals, public officials, business and community leaders, and concerned citizens who are shaping responses to environmental problems.

In 2002, Island Press celebrates its eighteenth anniversary as the leading provider of timely and practical books that take a multidisciplinary approach to critical environmental concerns. Our growing list of titles reflects our commitment to bringing the best of an expanding body of literature to the environmental community throughout North America and the world.

Support for Island Press is provided by The Nathan Cummings Foundation, Geraldine R. Dodge Foundation, Doris Duke Charitable Foundation, Educational Foundation of America, The Charles Engelhard Foundation, The Ford Foundation, The George Gund Foundation, The Vira I. Heinz Endowment, The William and Flora Hewlett Foundation, Henry Luce Foundation, The John D. and Catherine T. MacArthur Foundation, The Andrew W. Mellon Foundation, The Moriah Fund, The Curtis and Edith Munson Foundation, National Fish and Wildlife Foundation, The New-Land Foundation, Oak Foundation, The Overbrook Foundation, The David and Lucile Packard Foundation, The Pew Charitable Trusts, The Rockefeller Foundation, The Winslow Foundation, and other generous donors.

The opinions expressed in this book are those of the author(s) and do not necessarily reflect the views of these foundations.

Ecosystem Management

Adaptive, Community-Based Conservation

Gary K. Meffe

Larry A. Nielsen

Richard L. Knight

Dennis A. Schenborn

ISLAND PRESS
Washington • Covelo • London

Library of Congress Cataloging-in-Publication Data.

Ecosystem Management : adaptive, community-based conservation / by Gary K. Meffe ... [et al.].
p. cm.
ISBN 1-55963-824-9 (cloth : alk. paper)
1. Ecosystem management. I. Meffe, Gary K.
QH75 .E327 2002
333.95—dc21 2002007408

British Cataloguing-in-Publication Data available.

Printed on recycled, acid-free paper

Manufactured in the United States of America

09 08 07 06 05 04 03 02 8 7 6 5 4 3 2 1

To our wives—
Nancy Meffe, Sharon Nielsen, Heather Knight, and Elaine Schenborn—
for their love, patience, and support, and for reminding us of
what is truly important in life.

Contents

Preface

The world is changing quickly, and our models of learning, communicating, and acting must change accordingly. Throughout society we must rethink basic notions of how we define and accomplish our goals in a complex and changing environment, including how we prepare ourselves for professional careers. This book is an effort in that direction for students in natural resource management. It is a response to some of the challenges we perceive students will face in the early twenty-first century, and it is a practice field on which they may begin to develop their skills.

This is a different kind of textbook for a different kind of course. It is based on the proposition that college education in general, and education in the natural resources in particular, must be active and engage participants in the collaborative venture of learning. We've created a text consistent with that proposition, partly because we believe people learn more effectively in an active mode, but especially because it is necessary that all of us become more capable problem solvers in a complicated world. Hence, we like to think of the people using this text as participants rather than readers, as collaborators rather than students.

Learning technical information is important, of course, to any field of endeavor, and it is the foundation upon which professionalism is built. Understanding the basic theories and empirical bases that constitute fields such as ecology, economics, fisheries and wildlife management, specialized taxonomic studies, physiology, genetics, sociology, and so forth is necessary to function as a professional—*necessary*, but not sufficient. It has been our experience that much more is needed to be a professional, and that individuals planning careers in natural resource management need more preparation in and experience with other skills and ideas beyond the technical aspects of our work.

In particular, we need to understand that good scientific and technical knowledge is not enough, by itself, to succeed in natural resource management, because science is only one component of a complex world of decision making. Environmental policy and management decisions are set within a much larger socioeconomic and institutional context, one that can swamp and effectively neutralize the best scientific information if those who represent science do not know how to work effectively with decision makers and diverse stakeholders. This context can be a challenging place in which to work, but work within it we must if the science is to be used to guide environmental policy and management. We hope that this book, and our approach, helps make participants more effective within that arena.

Our approach in this text is to engage participants early and often in active problem solving, using realistic and complex landscape scenarios. Although we cover the technical scientific information important to natural resource management ranging from genes to landscapes, the real progress will be made by integrating this information into the human context. We believe this approach will better prepare participants for the very complex and challenging world beyond the safe confines of a university.

This book grew out of a training course we developed and have presented for 7 years for the U.S. Fish and Wildlife Service through their National Conservation Training Center in Shepherdstown, West Virginia. In 20 offerings to more than 600 natural resource professionals, we learned a great deal about what is needed to function successfully at that level. Participants from numerous federal and state agencies, nongovernmental organizations, the U.S. military, and industry took our course and educated us about their worlds; to them we owe a great deal of gratitude. They have provided the insights and wherewithal to help future practitioners in their profession travel a smoother road.

We thank Rick Lemmon (Director of NCTC), Chris Horsch (Head of the Aquatic Resources Training Branch), and especially June McIlwain, the leader of our course, *An Ecosystem Approach to Conservation.* It has been a pleasure to work with and learn from such a fine group of professionals. The folks at Island Press have been outstanding to work with and very professional. We are indebted to Barbara Dean for encouraging us to attempt this book in the first place and for keeping the project moving along. Her insights, instincts, and intellect are something to behold. Barbara Youngblood, Cecilia González, Amelia Durand, and copyeditor Betsy Dilernia all performed their jobs with grace and determination. We thank our respective institutions for their logistic support throughout this process. James Gibbs, Steven Yaffee, and an anonymous reviewer provided excellent comments on an early draft, and even where we did not heed their advice we certainly appreciated and learned from it. GKM is indebted to Margaret Flagg and Ellen Main of the editorial staff of *Conservation Biology.* Their professionalism, dedication, levels of excellence, and good humor not only made it possible to write this book on top of my "real job," but they make it fun to go to work every day. RLK wishes to thank his colleagues and former students at CSU for expanding his thinking regarding natural resource management. Part of this work was written while on a sabbatical leave granted by Colorado State University. LAN would like to thank Danielle Young-Kocovsky, who assisted him in many ways. DAS wishes to thank the men and women of the Wisconsin Department of Natural Resources for their dedication to resource management with their many stakeholders. *"Res non verba"*—deeds, not words.

Finally, and most importantly, our families not only have shown infinite patience and much guidance, but have offered the greatest of gifts—understanding and a solid foundation of love and support from which to work and to believe in ourselves. We owe you one.

About the Authors

Gary K. Meffe is an Adjunct Professor in the Department of Wildlife Ecology and Conservation at the University of Florida. He is senior author of the widely used college textbook *Principles of Conservation Biology*, coauthor of *Biodiversity on Military Lands: A Handbook*, and coeditor of *Ecology and Evolution of Livebearing Fishes*. Since 1997, he has served as Editor of the international journal *Conservation Biology*.

Larry A. Nielsen is Dean of the College of Natural Resources at North Carolina State University in Raleigh. A fisheries biologist by training, he has been honored for his teaching while on the faculty at Virginia Tech and for enhancing diversity while at Pennsylvania State University. He is sought widely as a speaker on topics related to ecosystem management, the future of natural resources, and community-based conservation.

Richard L. Knight is a Professor of wildlife conservation at Colorado State University. His interests deal with the interdependency of healthy human and natural communities. He has edited several books, including *A New Century for Natural Resources Management*, *Stewardship Across Boundaries*, and *Aldo Leopold and the Ecological Conscience*.

Dennis A. Schenborn is Chief of Planning and Budget for the Bureau of Fisheries Management and Habitat Protection of the Wisconsin State Department of Natural Resources. For more than 25 years, he has organized, led, and taught public involvement and organizational management for natural resource management agencies throughout North America. He is past President of the Organization of Wildlife Planners and formerly served as a research biologist with the United States Antarctic Research Program on two Antarctic expeditions.

Essay Contributors

Roger L. Banks is the Field Supervisor for the Ecological Services Field Office of the U.S. Fish and Wildlife Service in Charleston, S.C. He devoted most of his career to protecting wetlands and other important habitats through traditional regulatory means. For the past 10 years he has pursued private and public partnership efforts in South Carolina geared toward promoting proactive, long-term habitat protection on a landscape scale.

Mark W. Brunson is an Associate Professor in the Department of Environment & Society at Utah State University. He studies social and psychological aspects of rangeland and forest management to better understand what makes people accept or take part in activities that contribute to environmental sustainability.

Heather A. L. Knight, a native of Australia, is The Nature Conservancy Program Manager for the Phantom Canyon Preserve in the Laramie Foothills. This 300,000-acre site is one of the last remaining ranching communities along Colorado's northern Front Range. and is a community-based conservation program supported by diverse public and private partners.

Bill McDonald is a fifth-generation rancher on the Sycamore Ranch in southeastern Arizona. He is a long-time supervisor of the Whitewater Draw Natural Resource Conservation District and a past recipient of the Arizona Game and Fish Commission's award for Outstanding Wildlife Habitat Stewardship. He helped form, and is the Executive Director of, the Malpai Borderlands Group, for which he was awarded a MacArthur Genius Fellowship in 1998.

Michael O'Connell is Managing Director of the California South Coast Ecoregion for The Nature Conservancy. His involvement in conservation planning dates to the time when only four Habitat Conservation Plans existed; over 400 are now in preparation or are implemented. For 5 years, Mike directed The Nature Conservancy's involvement in the Southern California NCCP program.

Riki Ott has dedicated her academic training in marine biology and toxicology to help the public understand the effects of oil, mining, and timber industry activities on water quality and marine and aquatic ecosystems. She has helped citizens use this knowledge to redefine business practices and government accountability to improve the quality of life through environmental protection, social justice, and economic stability.

Kristin Smith is Executive Director of the Copper River Watershed Project. For ten years she has worked in the public sector on low-income housing and community-development projects. She holds a Master of Public Policy degree from Harvard University.

George N. Wallace is an Associate Professor at Colorado State University in the College of Natural Resources, where he teaches and does research in the areas of land use and protected area management. He also directs CSU's Center for Protected Area Management, serves as a Larimer County Planning Commissioner, and owns and operates a farm that has won several stewardship awards.

Steven L. Yaffee is a Professor of Natural Resource and Environmental Policy in the School of Natural Resources and Environment at the University of Michigan, where he directs the school's Ecosystem Management Initiative, a center focused on imagining, evaluating, and promoting innovative approaches for sustainable natural resource management. Dr. Yaffee's research focuses on collaboration and adaptive management in ecosystem management and public policies that affect biodiversity conservation.

Introduction: New Approaches for a New Millennium

Human activity over the past several hundred years has left a significant and growing footprint on planet Earth. In no period of human history has our species had a greater impact on the biophysical world. Ozone holes at the poles and microcontaminants in virtually every living organism attest to the far-reaching effects of human activities on every ecosystem. We build roads and log the hot zone of equatorial Africa and then carry emergent viruses across oceans. We burn neotropical rain forests to make way for grazing and farming on land that can sustain those practices for only a few short years. We mine ancient aquifers to make the deserts bloom, while other land-use practices expand the deserts of North Africa. We dam rivers for irrigation that wither the Caspian and Aral Seas. We harvest the world's oceans until the catch is depleted and then move on to a new place or to another trophic level. We develop and use land with only the barest knowledge of the consequences of our actions on the complex food webs and the bioenergetics of oceanic and terrestrial systems too vast to understand, yet so vulnerable that we have altered them in irrevocable ways.

In the United States, we reduce timber harvest on our public forests, while we increase our consumption of wood products and decry the cutting of boreal and tropical forests. We build cities in the desert and let them sprawl with far-flung subdivisions, while we ponder the politics and technology necessary to move water from the Great Lakes to the Southwest. We construct subdivisions over rich mesic farmlands of the Midwest, while building elaborate irrigation systems to grow crops in green circles on arid, short-grass prairie. We let chance plan our cities while we meticulously bioengineer transgenic crops and other species to solve problems we have created but do not understand. We cannot, however, effectively engineer what we do not understand, and what we largely do not understand is our impact on the ecosystems upon which we depend.

During the twentieth century, we became detached from the land that supports us and often lost sight of its complexity. Although it is crucial that we consider the long-term consequences of

Figure I.1. *Three examples of the application of ecosystem management principles in large landscapes.*

(a)

(a) A Florida panther habitat in a mixture of agricultural fields and forests in Collier County, south Florida. (Photo by David Maehr.)

(b)

(b) The Malpai borderlands region of southeastern Arizona and southwestern New Mexico, the site of innovative ecosystem approaches to sustaining ranching lifestyles and maintaining diverse and healthy native biota. (Photo by Charles Curtin.)

(c)

(c) Volunteers in the Chicago Wilderness initiative help maintain some 200,000 acres of forests, prairies, and wetlands in and around the heavily populated Chicago region. (Photo by Carol Freeman.)

our actions on the ecological systems that support life on Earth, we typically fail to do so. Now, as we measure ozone holes at the poles, shrinking ice caps, rising seas, and the lack of fresh water, a growing knowledge about the impact of humans on the environment compels changes in our business-as-usual attitude. We are more than 6 billion people who have crossed into a new millennium, and we have a clear choice: to continue our destructive relationship with the ecological world or to diverge from the path taken for the last several hundred years.

This book is about the application of the sciences of ecology and conservation biology to real-world problem solving. Emphasizing the complex ecological, socioeconomic, and institutional matrix in which natural resource management functions, it will illustrate how we can be more effective in that challenging arena. This book is also about people in communities of interest and communities of place—people who care so much about their quality of life today and in the future that they have chosen to work with others to improve the places where they live, work, and play, while restoring the land. It is about the interface of science, people, and their governments as they struggle to understand their collective impacts on ecosystems and change their approaches. It is about people who believe that, because their actions affect ecosystems in profound ways, they must learn to live more gently on the land. Here are a few examples (Figure I.1):

- In south Florida, a coalition of public agencies, environmental groups, and private citizens are restoring and protecting critical habitat for the Florida panther and many other species on a million acres of private land.
- A group of ecological and social scientists in China is trying to influence their government's population control, emigration, and economic policies to better balance the needs of the local human community with the habitat needs of the last giant pandas living in the wild.
- Amidst large pressures from development interests, nearly 1 million acres of native Arizona and New Mexican grasslands and forests are cooperatively managed by the Malpai Borderlands Group to maintain ranching lifestyles and restore the natural processes that sustain a healthy, unfragmented landscape.
- Along the Blackfoot River of Montana, ranchers and other private landowners are working together to restore the river while maintaining the rural working character of the landscape. They have restored 100 miles of the river, recreated 2100 acres of wetlands, and placed 45,000 acres in conservation easements.
- More than 135 private and public organizations have created the Chicago Wilderness initiative to protect, restore, and manage native prairie, wetlands, and other natural communities in and around the city of Chicago. Their mission includes community outreach and education efforts.
- In the Applegate Valley of Oregon, environmentalists, loggers, and government officials set aside their differences and found a common ground centered on managing for a healthy forest with natural and economic values.

These and many other examples of community-based approaches to ecosystem management affirm that we can, as Aldo Leopold wrote in 1938, "learn to live on a piece of land without spoiling it." The people who are making progress toward resolving natural resource issues at the ecosystem level do so by avoiding prolonged court battles and win-lose situations. Success comes from rational discussion among groups with different viewpoints and the development of common goals. Sharing scientific information is, of course, essential to mutual understanding of the natural processes affected by human activity, but it is not, by itself, enough. The success stories come not only from understanding scientific information, but also from the motivations of various people and their ways of facilitating dialog and consensus to reach common goals.

The Appearance of Ecosystem Management

In the 1990s, natural resource management in the United States underwent a major change in philosophy and direction. As past efforts using top-down, government-mandated, expert-driven approaches to managing natural resources failed or met with public resistance and resentment, new ideas came into play that took a different approach. For the first time in conservation history, shared decision making, cooperation rather than confrontation, and grass-roots, community-based involvement at the local level began to replace or supplement government-mandated programs imposed on landscapes from the outside. These efforts are also focusing on large natural systems (such as watersheds), rather than staying within artificial and ecologically meaningless straight lines on a map. Known variously as ecosystem management, community-based conservation, adaptive management, or landscape-level conservation, these efforts are not only working, but are sweeping through natural resource management agencies, nongovernmental organizations, the private sector, and even industry as a more reasonable way to conduct land and resource management. There is no going back at this point, and as we move further into the twenty-first century with an expanding human population and shrinking resource base, the demand for ecosystem management as an appropriate problem-solving mechanism will only increase.

This book is intended to address this change in approach and to prepare you—today's students of natural resource management and conservation—for the many challenges that await you as professionals. The book is based on three fundamental premises:

1. The effective and efficient use and management of Earth's natural resources are critically important to both human welfare and the continuance of functional ecosystems and biological diversity on this planet.
2. The management of such use is an increasingly difficult challenge, as each year witnesses more people chasing fewer resources in a more contentious way.
3. Traditional university curricula may not fully prepare students in natural resource management programs for grappling with the extraordinarily complex, uncertain, multidimensional, and often contentious arena in which these challenges are played out.

Our intention is to directly address the third premise: enabling you as students to more effectively deal with the second premise when you become professionals, so that the first premise can ultimately be achieved. But first we must ask whether these premises are true, or at least reasonable approximations. Let's examine them in turn.

First premise. Obviously, all of the resources humanity uses come from Earth, driven by the energy source of the sun. So in a trivial sense, at least, the first premise is true: We only have materials from Earth with which to prosper as a species. But in a less trivial vein, scientific evidence continues to show that functional ecosystems provide humanity with many and diverse services that we could not live without: oxygen production, purification of fresh water, erosion control, fertile soil production and retention, climate control and temperature amelioriation, food production, crop pollination, waste decomposition and detoxification, mitigation of floods and droughts, and so forth. And biologically diverse systems seem better than impoverished systems at providing these services. The loss of such functions cannot help but be harmful to humanity. At minimum, they are prohibitively costly to replace through technological means, and they obviously are harmful to the diversity of life on Earth.

Second premise. Our experiences and those of other scientists and managers unequivocally indicate that natural resource management increasingly faces complex challenges. Special interests, ideologically driven politics, competition for limited space and resources by a growing human population, and an increasing disconnection from the land by Americans and others have combined to exploit natural systems in a degradative and unsustainable manner and offer challenging management dilemmas.

Third premise. A trend in the second half of the twentieth century toward specialization and highly focused, disciplinary training in university curricula means that you may be ill-prepared to meet the challenges of a complex, contentious atmosphere that requires skills well beyond technical, scientific knowledge. We have repeatedly heard from natural resource professionals that their university training did not come close to preparing them for the non-scientific aspects of their jobs—the "people" parts of their work that often dominate their days. To better prepare you as professionals, we wish to introduce you to such a world in a "safe setting" where scenarios may be played out, experience gained, and new skills developed.

This book is intended to actively engage you in problem solving by melding the scientific principles of conservation biology with the complex human dimensions that prevail in everyday life, with the goal of equipping you to address real issues in conservation and management. This problem-solving, inquiry-based mode of learning is, we think, more effective for professional development than the traditional lecture-and-listen mode, because conservation and resource management are dynamic activities that require active, engaged people able to adapt to changing circumstances. It is also vastly more exciting, as it enables you to grapple with problems and develop your own solutions through direct experience and participation, and apply your technical knowledge to real issues. Thus, this book and its accompanying course will likely be different from traditional classes you have had and books you have used. We note in particular that this is not a comprehensive textbook in conservation biology; there are other books that serve that purpose. Rather, we use basic principles of conservation biology, integrated with practical aspects of the human dimensions, to pursue and forge problem solving for real landscapes.

How to Use This Book

This book is structured around three main parts. Part I (Chapters 1–4) provides the conceptual toolbox of, and sets the stage for, ecosystem management. These chapters present the basic models and concepts that will be followed throughout. Part II (Chapters 5–9) provides the biological and ecological background necessary to conduct effective ecosystem management by discussing levels of biological organization from genes progressively up through landscapes. This will be a review for many, and new material for others, and will get everyone on the same playing field. Part III (Chapters 10–12) uses the various human dimensions to implement the technical, ecological knowledge that you have within contemporary socioeconomic and institutional settings.

In Chapters 2–12, problem-solving exercises directly engage you with the material at hand. These exercises are perhaps the most critical aspect of this approach, as they will challenge you to use the materials in an applied, hands-on manner and often will give you the opportunity to discuss the materials (sometimes in a heated fashion!) with your fellow students. In places, the material is also complemented with "boxes," or supplementary material having some bearing on the subject at hand. These should also enrich your experiences and stimulate further thinking on the topic.

There are eight essays—"Experiences in Ecosystem Management"—presented at the end of selected chapters. These are firsthand accounts of ecosystem management and community-based conservation, written by the people who were there and are trying to make this approach work on the ground in real places. Although the essays do not necessarily correspond to the specific contents of the chapters, they are good examples of the challenges that are occuring in many places of finding innovative ways to reduce conflict and live better upon the land. You should use these essays as guides to applying this approach to real-world situations. Each essay ends with several questions that offer fodder for further discussion on that particular ecosystem experience. These will prove very useful, as they derive from real situations that professionals have had to grapple with.

What really sets this course apart from others, and, we think, makes it quite exciting, is that it is built around hypothetical but realistic and complex landscape scenarios that you will work with to make decisions and recommendations in an ecosystem management framework. Three scenarios are included in Chapter 1: one represents the northeastern or midwestern part of the United States (The ROLE Model); a second represents the intermountain west (SnowPACT); and a third reflects the humid, lowland southeast (PDQ

Revival). Individually and as a class you will become intimately familiar with one or more of these scenarios, including their geographic settings; ecological features (such as major habitat types, prevalent species, hydrology, climate, and management issues); and the human landscape, including the socioeconomic features, political scene, and major players. The scenarios will be used throughout the course to address natural resource management problems and issues that represent those likely to arise in such settings. We recommend that you address most of these problems and issues as interactive groups or an entire class and seek solutions collectively via discussion and careful planning.

The problems and issues you will face are intentionally complex, sloppy, difficult—and maybe even frustrating at times. But they are realistic and reflect the problems that professionals in this field face nearly every day. The goal is to come to grips with the realities of the world so that you may be more competent as a professional to confront such challenges when it really counts.

You will find that often there is no one correct solution—or many possible solutions—for a given problem. And you will not know whether the responses you develop will actually work. There are no patently right or wrong answers to most exercises, and no "answer guide" is provided to check your results. This approach reflects how the world actually operates; you must learn to deal with vagueness and uncertainty, and make decisions with incomplete information and conflicting pressures. You also must learn to work with people outside your profession who represent different value systems, hold perspectives that may be unfamiliar to you, and have the power to do things that you cannot do or cannot stop. The approach we lay out here will provide an important opportunity to experience a realistic professional setting before you find yourself in such a situation where you eventually work—when it really counts, and when the future of the natural and human communities may be at stake.

Have fun with the scenarios! Embrace them—learn the players, begin to "inhabit" the places. Feel free to "think outside the box" to develop innovative solutions to very complex problems. Use this as an opportunity to apply what you already know about science and human behavior, combined with new materials and skills you will need to learn and develop, to address very practical and applied issues. Do not feel bound by convention, though you will be bound by laws and community standards of behavior. Always remember to act with integrity, conviction, and attention to detail. Regardless of your personal feelings about an issue, they should not cloud your objectivity as a scientist or your ethical obligations as a citizen. You may find yourself making recommendations you are uncomfortable with or would prefer not to do. This again reflects the challenges that professionals must deal with every day.

The scenarios, and the book as a whole, focus on the United States. We do this for two reasons: (1) Our collective experiences are in the U.S., and it is best to write what you know about; and (2) many of the ideas and approaches used here have been developed by and used in resource management in the U.S. Regardless, this approach and the ideas behind it fundamentally are without political boundaries. The scenarios are useful anywhere, or they can be modified to fit the special needs of any locality.

AN OVERVIEW AND THE FLOW OF THE TEXT

Our approach will be practical, will orient you toward active problem solving, and will center on realistic land management problems and issues. Chapter 1 provides the landscape scenarios that you will use to address management problems throughout the remainder of the book. Of course, this chapter should be read first (perhaps more than once) and the scenario information carefully absorbed. Chapter 2 formally defines ecosystem management and examines how successful action at an ecosystem scale requires the involvement and long-term commitment of ecological and social scientists, human communities, and government. Although they can produce short-term results, single-species oriented and unilateral, top-down approaches to managing natural resources often fail when applied to larger-scale systems or over longer time frames. Chapter 3 considers why uncertainty and variation dominate natural and human systems, pointing out how simple solutions focusing on only one or a few parameters do not take into account the complexity of most ecosystems. In Chapter 4, we see that natural resource policies and actions can be viewed as experiments that provide opportunities to learn about ecosystems, rather than prescriptions to be faithfully followed. The inherent complexity and uncertainty of ecosystems mandate the monitoring and evaluation of management actions, and modifications of our adaptive management approach as needed.

Chapters 5–9 present advances in the fields of genetics, population ecology, and landscape ecology that have given us new concepts and scientific tools with which to better understand ecosystems. These chapters collectively form a "primer" of conservation biology. They may be a review for some students and new material for others; regardless, they will lay the foundation for bringing science onto a firm footing with socioeconomic and institutional considerations in good management.

In Chapter 10, we discuss why natural resources cannot be managed effectively without public support. Court battles over ecosystem or environmental issues have not solved ecological problems. Lawsuits are time-consuming and expensive, and they produce losers as well as winners. Dialog between scientists and public interest groups can result in mutual goals that meet both ecological and human needs. People protect what they learn to value and fail to protect what they do not know how to value. Clearly, natural resources cannot be managed effectively without the application of objective, science-based ecological understanding, yet they also cannot be managed successfully without public support.

The scale of ecosystem management necessitates cooperation across multiple government jurisdictions and on both public and private lands. Piecemeal actions do not work as well as coordinated actions that focus on common objectives. A systematic and explicit process is essential to sustaining action and evaluating progress. Chapters 11 and 12 are concerned with the process of strategic thinking: deciding what the objectives should be, how to achieve them, and how to measure success.

You may notice that, contrary to many scientific textbooks, we generally do not include source citations for information within the text (other than to attribute direct quotes or ideas). Rather than break up the message with reference support for every point made, we conclude each chapter with appropriate references and suggested readings on that topic. We encourage you to pursue these writings as authoritative sources for the topics; they will provide more specialized information on each topic than we can offer here. Also, we move back and forth between metric and nonmetric measurements. Virtually all scientific work is done using metric units (meters, hectares, and so forth), but much conversation in the "real world" uses nonmetric measures (yards, acres, and so on). We retained both of these approaches to reflect the complexities of the world and to illustrate the flexibility needed by professionals to work within both scientific and nonscientific circles. Finally, note that terms in boldface are important enough to be formally defined in a glossary toward the back of the book.

PART I

The Conceptual Toolbox

The Landscape Scenarios

YOUR EXPERIENCE WITH THIS BOOK AND THE SUCCESS of this course largely will revolve around and depend upon the landscape scenarios. These are where you will work on many of the problems and questions embedded in the chapters, to help you work through and "experience" the materials presented. Get to know your scenario thoroughly in every aspect: ecologically, socioeconomically, politically, and geographically.

Three landscape scenarios follow. All of them are equally challenging, and they all address the same basic problems.

- The ROLE Model is set in a midwestern/northeastern landscape of mixed industrial and agricultural land use.
- SnowPACT is set in the intermountain West, with large private and public ownerships and associated conflicts of changing uses.
- PDQ Revival is set in the humid Southeast, is influenced by a major military base, and captures the changing sociopolitical climate of that region.

Your instructor will inform you which scenario(s) to use. As you read the assigned scenario, begin to digest its richness and complexities. Study the maps, look at the photographs, and get a good feel for the landscape. Begin to "inhabit" the place and become part it. You will refer to the scenario throughout the course and use it as a reference source for detailed information. In the chapters that follow, you will use your growing scientific knowledge base, combined with processes and techniques we will cover, to address and explore many challenging questions and issues to be addressed in this place. Dive in and have fun!

Note that each scenario contains names of individuals who play various roles in those systems. All names are fictitious, and any resemblance to persons living or dead is purely coincidental.

THE ROLE MODEL

Just 6 months ago, an unprecedented event occurred in the area known as Round Lake (Figure 1.1). Representatives of communities, agencies, and interest groups stood together before a press conference and read the following statement:

An old adage says, "Today is the first day of the rest of your life." We can paraphrase by saying today is the first day of the rest of the Round Lake Ecosystem's life. We are here today to sign an agreement that dedicates the people, agencies, and resources of our area to a new style of managing our natural resources and environment. We pledge to work together to assure that the qualities we love and need—clean water, clean air, abundant and diverse wildlife and fish, healthy land, and productive farms and forests—will continue and prosper through time and space.

We have chosen to call this initiative the Round Lake Ecosystem Model—or ROLE Model—because we believe this effort can truly be a model for ourselves and the rest of the nation. We know that the ways of the past, which have fragmented land and communities and have pitted neighbor against neighbor, cannot continue. We all have too much to lose by those behaviors. And we have so much to gain by working together, using reason, and seeking win-win solutions to issues.

We often talk about being role models. We know that our children will behave as they see us behave, so we try to be honest, just, and forgiving within our families. We know that as responsible members of the public community, we must establish rules and procedures that are fair, open, and respectful of others. To these roles and role models, today we add the necessity of treating the land and its resources with the same care and respect that we extend to other humans. We recognize that we depend on the health and productivity of our lands to provide us the essentials of life—air, water, soil, plants, and animals—and also the beauty and comfort that nurtures our character.

Today we begin a long, difficult, and expensive journey, but a journey that we know will take us where we want to go. We are confident the people of the Round Lake ecosystem want to take this journey. We are proud that our citizens, businesses, agencies, and community groups are leading themselves and the nation in becoming the ROLE Model!

THE ROLE MODEL AGREEMENT

The Round Lake Ecosystem Model Agreement is a simple document with profound implications. Most importantly, it establishes the Round Lake Ecosystem Team as a broadly based coalition of representatives of all groups that wish to join. It has an initial 10-year charter, with the expectation that it will be renewed continuously and become a leading focus for community planning and action.

The state Department of Natural Resources (DNR), through its secretary, and the U.S. Fish and Wildlife Service, through its regional director, have committed their resources to provide the base operations for the team. Each agency has agreed to assign one professional to coordinate the team's work for the next 5 years—commitments that were considered essential (and inspirational). In addition, each agency has agreed to assign its most senior local staff person to serve on the team. These are the DNR's District Director, Margaret Staples, and the Bingham National Wildlife Refuge manager, Oliver Adams. And, of course, these agencies have pledged the support of their staff and physical resources to help along the way.

All signatories to the agreement are automatically members of the team, and a subset has been elected by the members to comprise the Steering Committee. The list is impressive (Steering Committee members are noted by an asterisk):

ROLE Model Members

Benson City Council*
Bingham National Wildlife Refuge*

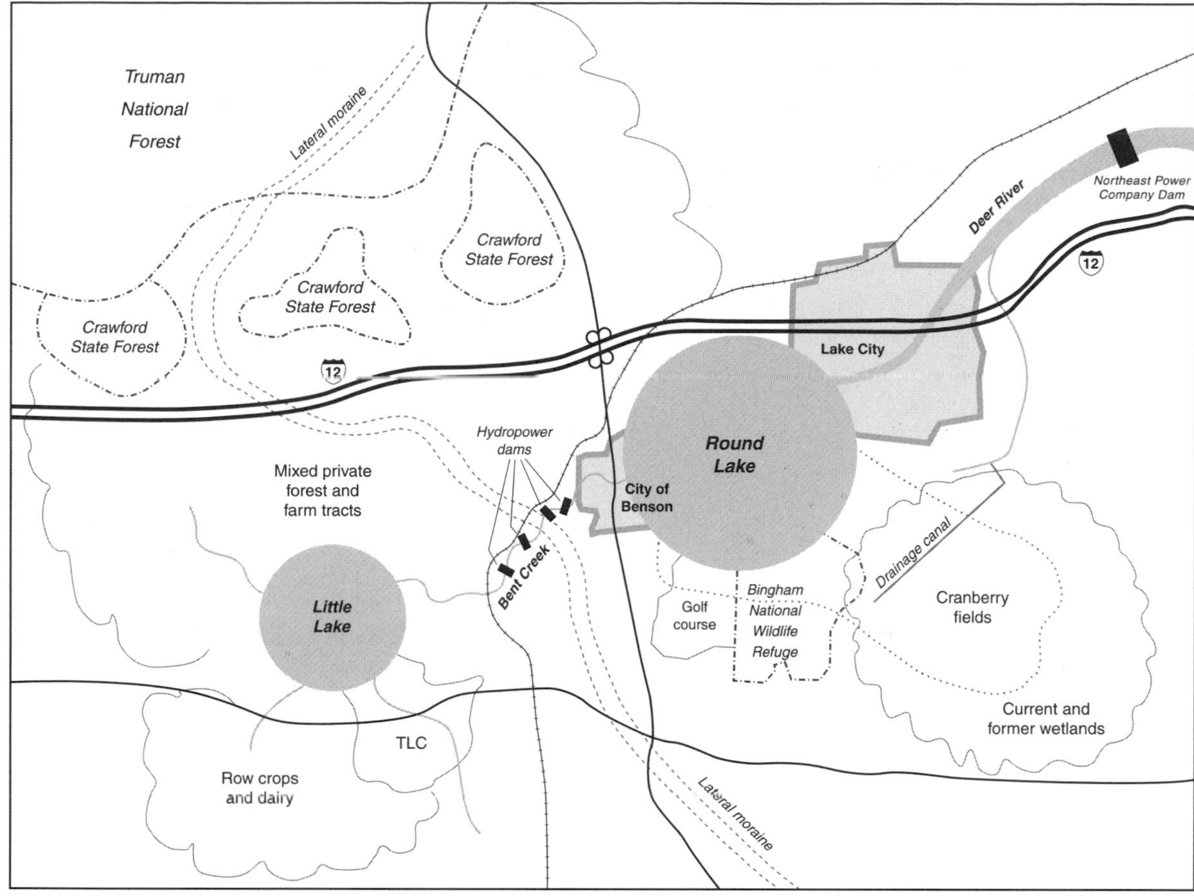

Figure 1.1. *A map of the Round Lake ecosystem.*

Crawford County Planning Commission*
Cranberry Growers' Association*
Cranberry Marsh Audubon Society
Crawford County Grange*
Department of Natural Resources*
Friends of Round Lake
Hardwood Lumber Manufacturers' Association
Hunters for Waterfowl*
Lake City Council*
League of Women Voters
Little Lake Shoreline Association
Mid-State Outdoor Writers Association
Northeast Power Company*
Penowa Indian Nation*
Round Lake Area Chamber of Commerce*
Round Lake Forest Landowners Association*
Society for North American Plants (SNAP)
Truman National Forest*
Trust for Land Conservation (TLC)*
U.S. Army Corps of Engineers*
U.S. Natural Resource Conservation Service
Walleyes for Tomorrow

The agreement lists several core values the group chose as guidelines for their long-term operation:

The ROLE Model's Core Values

We seek to create a place that meets the needs of ourselves and future residents. We seek to do this in a way that will be a model of civility, common sense, rationality, and efficiency. We pledge ourselves to be guided by the following principles:

- We will be inclusive, rather than exclusive, inviting all people and viewpoints; but we will not tolerate attempts to delay or derail our efforts.

- We will use all the expert knowledge we can get to help guide decisions, including that of scientists, economists, and sociologists; but we will not shrink from decisions or actions because of "insufficient data."
- We will supplement the maps of ownership and jurisdiction with maps of natural features and functions.
- We will work with all decision-making groups, from county commissioners to national agencies, to bring the ideas and goodwill of our citizens forward.
- We will find win-win situations, so that no individuals lose in decisions that bring gains to all of us.
- We will seek voluntary cooperation rather than rules, regulations, and laws.
- We will set our vision on the long-term and will be prepared to discuss openly the short-term costs of such a vision.
- We will be realistic, recognizing that we start from *here* and that our first steps probably will be small.
- We will succeed!

THE ROUND LAKE ECOSYSTEM

The Round Lake ecosystem is a large watershed that drains into the Great Lakes–St. Lawrence system. The major water system is the Little Lake–Bent Creek–Round Lake–Deer River drainage, which generally flows northeast. Most of the area was glaciated in the Wisconsinian era, but fingers of unglaciated lands intrude from the south. The elevation is about 800 feet, with flat to rolling terrain. The glaciated areas support rich farms and relatively productive forestlands in a mixed patchwork that reminds people of a calendar photograph (Figure 1.2).

The Round Lake ecosystem is split into two primary physiographic regions by a lateral moraine that runs north to south just west of Round Lake (see Figure 1.1). The soils to the east of the moraine are a mix of silty loams overlaying a complex geology of glacial till that forms the flat outwash plain to the east. This creates a shallow aquifer with high transmissibility through the sand and gravel outwash with scattered clay lenses. Groundwater flows in a general northeastern direction, but flow varies from location to location because of the clay formations. West of the moraine, sandstone and limestone formations underlay the thinner soils of what once was contiguous forest.

The area is dominated by Round Lake, a 40,000-acre natural lake named for its nearly circular shape. Round Lake is relatively shallow (maximum depth 85 feet; average depth 30 feet) and has expansive littoral areas, some rocky and some silty;

Figure 1.2. *The Round Lake ecosystem has mixed land uses of field crops interspersed with deciduous forests and natural lakes. (Photo by Larry A. Nielsen.)*

it stratifies in summer and is ice-covered in most winters. The lake is roughly divided into two basins, separated by a relatively shallow section that runs west-east across the lower third of the lake. The southern basin has relatively slow water turnover rates because the main flow of water through the lake occurs in the larger, northern basin. The lake holds a typical fauna of warm-water and cool-water fishes, including largemouth and smallmouth bass, various panfishes, walleyes, carp, and suckers; 33 fish species were recorded in the most recent biological surveys. Round Lake is used extensively for recreational boating, served by substantial marinas in the towns of Benson and Lake City.

The southern shore of Round Lake was once connected to a wetland system almost as large as Round Lake itself. Much of the wetland area was drained for farming. Today most of the farms in the wetland region grow cranberries; the Round Lake region supplies about 30% of the nation's industrial cranberries (i.e., those that go into food processing). A portion of the wetlands is protected via the Bingham National Wildlife Refuge, a 9000-acre refuge created in the 1940s. The refuge is named after the nineteenth-century artist George Caleb Bingham, who did an extensive set of paintings depicting pioneer and Native American life along the southern shore (many of those paintings are on display in the Lake City Art Museum). Other parts of the wetland have been drained for golf course developments; other wetlands, especially those close to the lakeshore, are privately owned.

Little Lake is a smaller version of Round Lake, about 10 miles upstream, linked to Round Lake via Bent Creek. Little Lake, 12,000 acres in area, has a similar limnological profile to Round Lake and a similar fauna. However, walleyes are uncommon in the lake, prohibited from upstream movements by the series of low-head power dams on Bent Creek. The land around Little Lake was once owned entirely by Howard Brown, who invented the movable carriage for the typewriter. Brown, who was somewhat eccentric, wanted to be able to stand on the shore of the lake and own everything he could see. He succeeded, but his family was not as fortunate financially, and after his death in 1944, they sold off the land bit by bit. In the 1970s, the family regained its feet financially, realized that they had lost most of what had been a tremendous resource, and gave the remaining parcel, about 5000 acres and 1 mile of shoreline, to the Trust for Land Conservation (TLC).

Truman National Forest, in the northwestern portion of the watershed, is named after President Harry Truman. The land had been in federal ownership since the 1920s, after it had been logged, farmed, and abandoned. Truman issued an executive order making it a National Forest in 1948, along with several others in the eastern U.S. It contains a largely even-aged forest, with most stands 80–100 years old. Major stands are white oak, red oak, sugar maple, hemlock, and white pine. Truman National Forest has been one of the very few national forests that conduct profitable timber sales, on approximately 200,000 of its 300,000 acres. Truman is a true multiple-use forest, with major recreational uses and extensive interests in developing old-growth forests from the 5000 acres of old growth remaining. The lands that fall within the Round Lake area include a mixture of mature forests and about half of the old-growth tracts.

Managers at Truman National Forest have looked at their old growth in the Round Lake region and decided that they should develop a management plan that eventually will link their old-growth remnants into a continuous band. They are particularly interested in using this base for linking with other old-growth and mature forests on state and private lands.

Crawford State Forest is a three-unit forest in the region. Just like the Truman, it has mostly mature stands of mixed hardwoods with occasional stands of hemlock. Each unit of the forest is about 5000 acres, the minimum size the state will accept or keep as state forest. The state forest is surrounded by mixed farmland and private forestland; most private tracts are small (averaging 55 acres), and the owners have other jobs that provide their major income.

Real estate values are soaring anywhere near the Crawford and Truman forests. Larger tracks of private agricultural land are being divided into 3-to-10-acre home sites and sold to people who want to be close to nature while remaining within commuting distance of Lake City.

THE SOCIAL AND ECONOMIC SETTING

Lake City is a city of 100,000 residents on the northeastern shore of Round Lake (Figure 1.3). Like most cities of its size, it grew rapidly around 1900, spurred by the Industrial Revolution. It profited greatly from its proximity to larger midwestern cities. A rail line put Lake City on the path of agricultural products moving north and east and manufactured products moving south and west. Lake City developed a diversified economic base, which continues now. Always a civic-minded city, Lake City has built a reputation for being a good place to live. Annual surveys place it about halfway down the list of the "100 Best Places to Live in America." Lake City's long-time mayor, Tom Morning, is the perfect representative of the town. He is down-to-earth, action-oriented, trusting of people, suspicious of government, ambitious, and hard working. Although the mayor was slow to warm to the idea of the ROLE Model, once he became convinced that it could be the way to move Lake City up the list of the Best 100, he got behind it fully.

Because of its civic character, Lake City is alive with groups that work on its behalf. The group known as Friends of Round Lake works constantly to keep the water clean and the lake accessible to all citizens. They annually sponsor shoreline cleanups, coordinate boating safety classes, and sponsor an annual Aquatic Envirothon. They have pledged their membership to being active in the ROLE Model idea, suggesting especially that they would love to stage community events that would get people involved—and might generate money. A series of other similar organizations feel and act the same way, although they sometimes tend to be a bit more narrow in their interests. Friends of Round Lake has clearly become the leading environmental/civic group in the area. For example, the group's part-time executive director, Chris Gallagher, has just been asked by the governor to become co-chairman of his new Commission for the 21st Century Environment.

Figure 1.3. *Lake City is a prosperous community surrounding the outlet of Round Lake into the Deer River. (Photo by Larry A. Nielsen.)*

Benson is across Round Lake from Lake City, where Bent Creek enters the lake. Benson has 20,000 residents, down from its highest population of nearly 50,000. Benson has not been as fortunate as Lake City. It thrived on heavy industries, which were located along the rail-line in the town and down Bent Creek toward Little Lake. The rusting of the industries, starting in the early 1960s, took its toll on the economics of the community. When Interstate 12 was completed in 1967, continuing west from Lake City, rather than following the rail-line south, Benson went into an economic downturn. Many of the heavy industries closed up, leaving old facilities that have now become environmental problems. Long considered a rival of Lake City, the city of Benson has recently realized that it should look to Lake City as a partner. Nonetheless, Benson and its mayor, Nancy Lyons, remain fairly traditional. Although signatories to the ROLE Model Agreement, they enter the team fairly skeptical and certainly cautious.

Benson has two bright spots economically. The leading citizens of Benson always lived on the south side of town, along the lakefront. In recent years, several residential developments have begun popping up on the south side of town and beyond. Folks from Lake City and surrounding areas have begun buying property on 5-to-10-acre sites, building upscale houses and generally raising the prestige of the area. Several major real estate investment trusts, searching for relatively cheap sites for development in proximity of desirable communities, have taken options on tracts of several thousand acres each.

The second bright spot has been the growth of golf in the region. Prominent Benson citizens funded and built a 9-hole country club golf course in the 1920s, south of the city and adjacent to the Bingham Wildlife Refuge. In the post-war boom time of the early 1950s, they expanded the course to 18 holes. The quality of the course, along with its beautiful setting (holes go from forest to wetland to lakeshore settings), attracted increasing interest. In conjunction with real estate developers, the club went semipublic in the 1970s, selling condominiums along the fairways. In 1990, a second golf course was built, along with a medium-sized conference hotel complex. Today, plans are under way for a major golf resort, with two 18-hole championship courses, 200 fairway condominiums, and a 200-room hotel; the resort will be called Sandhill at Bent Creek.

The Round Lake region is also strongly linked to the Penowa People, a Native American tribe. The tribe lived around Round and Little Lakes when the first pioneers came to the area. They fished the lakes, gathered wild rice in the marsh of Round Lake, and eventually practiced some farming. They fished the spring runs of walleyes that came up Deer River and Bent Creek, spearing spawning walleyes. They smoked and dried the walleye meat. The Penowans chose the wrong side in the war of 1812, however, fighting with the British against the United States. After the war, they were convinced to cede their lands to the U.S. government, but they retained their rights to hunt, fish, and gather wild rice for both their own use and trade.

Today about 1000 Penowans live in the Round Lake area. Tribal members have continued to gather wild rice through time and sell it through local outlets at a high price. Known for its large, meaty grains and earthy flavor, Penowan wild rice is considered a delicacy throughout the region and has been tapped by organic and health food restaurants in the East. The business thrives today. Tribal members also have been increasingly interested in the reestablishment of their walleye fishing traditions and the possibilities of developing a highly profitable business along the lines of their wild rice ventures. They are increasingly interested in reestablishing runs of native walleyes, but only if legitimate stocks will be used.

SPECIAL RESOURCES

The following species illustrate some of the leading "players" in the biological and ecological issues that need to be addressed with the ROLE Model ecosystem.

WALLEYES. The walleye population in Round Lake is quite vigorous, supported by a DNR walleye hatchery at the mouth of Bent Creek that annually catches and strips thousands of adults, raises the eggs to fry, and stocks them in the lake. Through time, the hatchery has supplemented its catches with eggs brought in from Lake Erie and the Ohio River. The walleye fishery is closely monitored by Walleyes for Tomorrow, a group whose members have many ideas for improving the fishery. Walleyes for Tomorrow employs their own biologist, Jacek Wajda, who reviews agency plans and participates in all technical and citizen task forces. The group is interested in stocking various strains. They also want a fishway around the Northeast Power Company Dam, to get back the walleye runs their grandparents talk about (Figure 1.4), and are pushing to introduce zander, the European equivalent of the walleye. (One reason they hired Wajda is his previous experience working with zander in Poland.)

Walleyes in Little Lake are another story. Walleyes disappeared from Little Lake in the 1920s, when a series of low-head hydropower dams were built on Bent Creek; four dams still exist and still block upstream migration. In 1970, when the DNR unexpectedly caught 2 ripe females and 20 ripe males in routine spring netting in Little Lake, they immediately spawned them and kept them separately in the hatchery. They restocked the fry in Little Lake, and a small, unstable population has developed. The Little Lake walleye population has grown and shrunk repeatedly over time, but it has never grown to a size to support a fishery.

Recently, Walleyes for Tomorrow has demanded

Figure 1.4. *The Northeast Power Company Dam is an aging hydroelectric facility on the Deer River downstream from Lake City. (Photo by Larry A. Nielsen.)*

that the DNR increase the stocking of walleyes in Round Lake in response to several weak year classes. The Penowan tribe, the DNR, and Walleyes for Tomorrow also have proposed developing Little Lake as a trophy-only walleye lake, using the walleye population already in that lake. They wish to develop a sport fishery to rival Lake Erie within 5 years. They also wish to directly compete for an annual booking on the North American Walleye Tournament that will bring more than $3 million to the local community if successful.

THE SHINERS. Maps of the state distribution of fishes always show many dots for Little Lake. It has been an ichthyologist's delight for a century. Early surveys by David Starr Jordan remarked about the unusual diversity of shiners. Later surveys confirmed the reality: Little Lake held an unusual diversity of cyprinids (minnows) and percids (darters and perch), including five species of a shiner genus known nowhere else except Little Lake and Lake Erie. The shiners are quite abundant within the lake, but are carefully watched because of their uniqueness.

The shiners are particularly interesting because they are hosts for the glochidia (larval form) of a species of freshwater mussel, the radiant mussel, that lives along the shores of Little Lake and nowhere else. The radiant mussel is also unusual because it is a lake mussel, living on the rocky shoreline areas of Little Lake, which includes TLC property.

The shiners also have a special significance to fishing-tackle buffs. They were the models for the original five colors and patterns used by the American Tackle Company for their "Looks-Alive lures" in the 1930s.

BOG TURTLE. The wetlands south of Round Lake are home to the bog turtle, a widely spread but uncommon turtle found east of the Mississippi, from New York to South Carolina (Figure 1.5). The bog turtle lives in freshwater marshes and clear, slow-moving streams with muddy bottoms. It is a small turtle, seldom growing larger than 4 inches across. It is an omnivore, but dearly loves mussels and crayfish. It grows slowly and becomes sexually mature at 8 years. It is a secretive animal, except when it suns on rocks or logs. Rather like a sea turtle, it leaves the water to lay eggs (1–6 per nest) in June, moving to more upland areas as much as a half-mile away from water. Young turtles hatch in August or September and take the reverse route back to the water.

The most successful reproduction occurs in three areas: the Bingham National Wildlife Refuge,

Figure 1.5. *Bog turtles are found in the southern regions of Round Lake and in the swamps and uplands adjacent to the lake. (Photo by R.G. Tuck, Jr.)*

the fringes of ponds on the back-nine of the original golf course, and in an undeveloped wetland just past the last cranberry field. From these areas, postreproductive and young turtles spread out to at least a dozen known habitat areas along the southern edge of Round Lake.

Bog turtles have been declining gradually over time, partly due to illegal sales (a pair can sell for $2000 in Japan and Europe), but mostly for a suite of reasons that have not been clearly defined. Although not currently listed as a protected species, the U.S. Fish and Wildlife Service is closely watching its decline, for a judgment of listing as threatened. Recent studies have also found another interesting fact: Bog turtles have high concentrations of heavy metals in their shells and pesticides in their soft tissues.

A 1964 research report by a local herpetologist stated: "The seasonal flooding and natural summer draining of the wetlands that surround the upland nesting sites are necessary to the bog turtles' survival. Either too much or too little water at the wrong time of year can suppress nesting success to near zero." A recent census of the three known ROLE nesting areas (A, B, and C on Figure 1.6) found the following:

	A	Bingham	C
Number of nests/area	18	6	31
Mean % of eggs hatched/nest	80%	35%	70%
Mean clutch size	5	5	6

The Cerulean Warbler and Other Neotropical Migrants. The cerulean warbler has declined by nearly 50% in the past decade in the area, as shown by annual breeding bird counts. The warbler lives in the highest branches of dominant and codominant trees in mature forests. It is found in relatively high densities in the southern area of Truman National Forest, but almost always in old-growth stands and the surrounding mature stands. It is especially fond of hemlock stands. The cerulean warbler is also found occasionally in private forestlands around the national forest, and in late summer, immature birds are often found in the three tracts of Crawford State Forest. Rarely, however, are nests of cerulean warblers found in the state forest; when nests are found, they are always in the largest, eastern tract.

A college student recently analyzed the data from annual breeding bird surveys, from the late 1960s to the present, for the three tracts of Crawford State Forest. He found an interesting pattern of distribution and changes in abundance:

Present in Latest Survey

Species	East	Central	West	Status
Cerulean warbler		x		Decreasing
Acadian flycatcher		x	x	Stable
Blackburnian warbler		x		Increasing
Olive-sided flycatcher		x		Decreasing
Wood thrush		x		Decreasing
Great crested flycatcher		x	x	Stable
Eastern woodland peewee	x		x	Decreasing
Hooded warbler	x	x	x	Increasing
Ovenbird		x		Stable
Yellow-billed cuckoo	x			Decreasing
Red-eyed vireo	x			Stable
Ruby-throated hummingbird	x	x	x	Increasing
Scarlet tanager	x			Stable
Yellow-throated vireo	x	x		Increasing

The Hemlock Woolly Adelgid. The hemlock woolly adelgid is a small aphidlike insect that feeds on several species of hemlock. Infestations are recognizable by the white, woolly looking material that the insects produce. Native to Asia, the

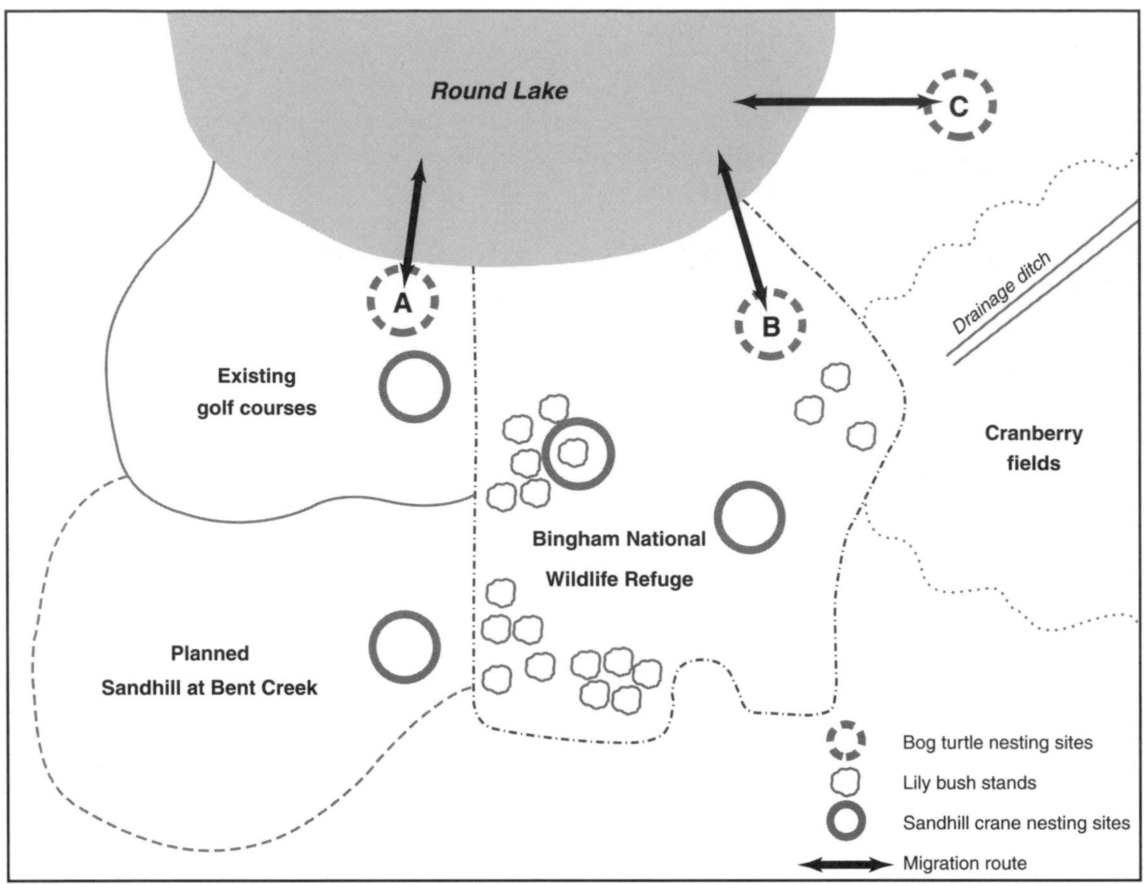

Figure 1.6. *A detailed map of the southern shoreline of Round Lake.*

insect was introduced into the United States accidentally and has been spreading slowly across the country. It has been relatively harmless on hemlocks and other conifers in the western U.S., but has been very destructive on eastern hemlocks. It attacks the young branches of hemlock trees, sucking out the sap, destroying the tissue in the process, and killing an infected tree in 1–4 years.

Serious infestations are now present about 100 miles east of the Round Lake watershed, moving westward about 10 miles per year. A large number of studies are being conducted to develop control methods, including mechanical removal, planting genetically engineered resistant species (using genes from western hemlocks), and pesticide control. Studies of two natural predators from Japan, the oribatid mite and the ladybird beetle, have indicated that biological control may be possible.

A major agent of dispersal is contact with animals that move through the trees, including cerulean warblers, squirrels, porcupines, and deer. Clear-cutting of infested stands is often recommended for these reasons and for the effective salvaging of damaged trees.

SANDHILL CRANES. Sandhill cranes return annually to the Round Lake area, taking up residence south of the lake (Figure 1.7). They nest in selected areas in the Bingham Wildlife Refuge, on the existing golf courses, and in the area where the new golf development is planned; they build nests in large mounds of grass or uprooted plants, usually on slightly elevated hills or drier ground. They feed in marshes or on prairielike abandoned and active farm fields. For as long as anyone can remember, bird watchers have come to the area to watch the

Figure 1.7. *Sandhill cranes are common and popular in the Round Lake ecosystem. (Photo by John and Karen Hollingsworth, U.S. Fish & Wildlife Service.)*

unusual and active courtship process of the sandhill cranes. The wild dancing and strutting is such a sight that PBS (the Public Broadcasting Service) has filmed the activities and featured them in several nature documentaries for television.

Sandhills do not currently nest in the TLC property (which has yet to receive a name), but several sites appear to be ideal for nesting, with raised hillocks located at appropriate intervals near the edge of Little Lake. The TLC folks are eager to attract sandhills to their property, but they are at a loss as to how. They have suggested that trapping some adults and/or young from the refuge and transferring them to their property would be good, but they cannot seem to get by an imposing set of state and federal regulations.

The Trust for Land Conservation is also interested in making their holding on Little Lake as significant as possible. Therefore, they are open to the idea of expanding their holdings, trading lands for other more significant lands, or operating within a larger context of adjoining lands that have various conservation measures (e.g., easements). They have hired a new part-time staff biologist to work on the property and its possibilities. Shondra Jefferson has taken the position as her first job, right out of Tuskegee University's conservation biology program.

Freshwater Mussels. In addition to the radiant mussel, which lives in Little Lake, Bent Creek holds a suite of freshwater mussels. Although beds once occupied the whole length of stream, now they are found only within the first half-mile or so below each of the low-head hydropower dams. As with most freshwater mussels, these populations seem to be troubled. Size distributions show that the populations are getting increasingly older, with few intermediate-aged individuals present. Concern has been growing because zebra mussels are moving up the Deer River from the Great Lakes; however, no zebra mussels have been found upstream of the Northeast Power Company Dam.

Lily Bush. Bingham Refuge contains several self-sustaining stands of the federally threatened lily bush, a medium-sized evergreen bush that grows in relatively open woodlands with well-drained soils. It is found on the refuge's higher elevations, atop the glacial drumlins that dot the property, and in a few locations off the refuge along the edge of the lateral moraine.

The plant is called the lily bush because it is usually associated with a complex of wildflowers from the lily family. Wildflower enthusiasts look for the distinctive patch of lily bush (a low shrub with dark bark and bright green, elongated leaves that flutter in the breeze) and know that they are also likely to find eastern troutlily, midland Camaslily, and wood lily. The complex also includes a characteristic set of other wildflowers from the lily-of-the-valley and iris families.

Though generally hardy and relatively free from insect problems, the lily bush is susceptible to gypsy moths and particularly prone to overbrowsing by deer in late winter. Although not a preferred food item of deer, their late-winter browsing will kill whole stands of the shrub. Ecologists have estimated that deer densities in excess of 15 per square mile will jeopardize the lily bush.

In addition to the known populations of the lily

bush south of Round Lake, scattered stands are found along the moraine north of Interstate 14. A few stands have been mapped on drumlins in the Crawford State Forest. These stands have become increasingly isolated and surrounded by maturing timber stands in the forest.

A 1935 botanical survey that helped establish the boundary of the federal forest reported that many species of the lily family can be found on the drumlins immediately south of the proposed forest boundary. It is likely that more stands exist on private lands surrounding Crawford. However, the drumlins that dot the private landholdings have not been surveyed recently.

WHITE-TAILED DEER AND ELK. White-tailed deer are the basis of a thriving hunting industry—as well as having become the area's biggest pest (Figure 1.8). According to the DNR, the carrying capacity for deer in the area is 22 per square mile; the current density is 31 per square mile. The consequence of deer overpopulation is the elimination of almost all forest regeneration, as well as the loss of most native wildflowers. Deer consume virtually everything that the timber industry likes (they leave striped maple, a low-value timber crop, and hay-scented ferns, which crowd out other vegetation). For successful forest regeneration, harvested areas must be fenced to keep out deer for 5 years after harvest. The mess is spilling over into other places as well, with deer and people coming into more contact via vehicle collisions, the eating of landscaping plants, depredating crops, and even roaming down city streets. Lyme disease is on the increase as well, with three cases reported last year, all in children on hikes with the county nature program.

For a decade, the DNR has been trying to reduce the herd, and these efforts are working. In 1980, the density was 38 deer per square mile. Recently, however, the hunting lobby has been getting restless, because hunting is becoming more difficult. Deer hunters want the density to go back up.

Deer on the Bingham Refuge also have become a problem. Cross-country skiing on the golf courses and snowmobile activity in and around the Bingham Refuge have concentrated deer onto the refuge during much of the winter. Winter deer densities on the refuge are estimated at twice the region's average. Human activities in the area around Bingham have made hunter access to the area increasingly difficult. Antihunting sentiment and demonstrations on the roadways leading to and from Bingham have made hunters uncomfortable. Fewer people hunt the refuge each year, and many have decided to hunt on the agricultural lands north of the interstate highway instead.

Figure 1.8. *White-tailed deer have become so abundant in the region that most people consider them pests. The problem is especially acute in the Bingham National Wildlife Refuge. (Photo by Larry A. Nielsen.)*

The DNR also has another idea: to reintroduce elk. Elk were native in the watershed, but were eliminated during the 1850s. Sportsmen have learned of the successful establishment of elk herds in a number of places around the Midwest and East, and they want them back here as well. The Penowans like the idea, too. They are interested in leasing land to breed elk and slaughter them for sale in restaurants and specialty markets. Farmers are less excited, because elk are known to carry a variety of respiratory diseases that can be transmitted to cattle—and the area's number 1 agricultural crop is milk. The forestry community is not happy either, because they see elk as large deer, capable of eating more and higher than even deer can.

SPECIAL INTERESTS AND ISSUES

Several issues and special interests are and will be important drivers of decision making within the ROLE Model ecosystem.

The Cranberry Industry. The cranberry industry is growing at an annual rate of 8%, buoyed by the news that cranberry juice helps prevent kidney stones and bladder infections, an increasing concern of aging baby boomers. The industry wishes to expand its operations into the remaining unprotected wetlands, right up to the southern shore of Round Lake. Consequently, the Cranberry Growers' Association, a national organization headquartered in Lake City, has hired a marketing firm to develop ideas and begin building local support for expansion in the region, as well as developing national markets for cranberries. The firm has assigned Marsha Kwan to the task, and she and a team are operating out of the Cranberry Growers' Association's offices. One of her first ideas has been to work with the Penowans to develop paired products centered around Penowan wild rice and Round Lake cranberries.

However, expansion of cranberry farming in the region faces some obstacles because of toxic chemicals in fishes and turtles in Round Lake. The area south and east of Round Lake is a groundwater maze. Cranberry farmers claim that all their runoff flows north into their cooperative drainage ditch and then into a tributary of Deer River. Their soil and hydrologic consultant insists that there is an impervious clay lens that separates the commercial fields from the Bingham Refuge, so none of the environmental problems in the refuge or lake are of their doing. He also says that studies have shown no environmental damage from the chemical or management practices of the cranberry farmers, as long as they follow manufacturer's label instructions and recognized Best Management Practices (BMPs)—which they do.

The Round-About Trail. The Lake City and Benson economic communities have recognized that they have a valuable tourism resource, and they have decided, based on a consultant's report, to target cyclists. The mayors have proposed a bikeway to go all the way around Round Lake (the Round-About Trail), which would have various stopping places where cyclists could view natural features of the ecosystem, visit cultural resources, and refresh themselves. The Crawford County Commissioners are excited about the idea, and they are considering modifications to their master plans to accommodate the bikeway. Around the southern end of the lake, they have proposed that the 12-foot-wide asphalt trail be elevated above the wetlands. The Round Lake Chamber of Commerce is very enthusiastic, and their director, Jesse Stern, has applied for and received a state tourism development grant to conduct an in-depth feasibility study.

Toxic Contaminants in the Round Lake Ecosystem. Toxic sediments exist in Bent Creek below the second low-head hydropower dam. The sediments contain heavy metals and other early industrial chemicals. There is no identifiable responsible party for the toxic chemical—so it is everyone's problem now. A major flood, or the removal of the dam, could cause the suspension of these toxicants in the water column and their transportation downstream to Round Lake and beyond.

A recent public health series on the local PBS television station highlighted contaminants in soils and water and their accumulation in fish and wildlife. This prompted a local university study of fishes found in Round Lake and Deer River. Although still preliminary, the results show that walleyes in the lake and carp in the river have measurable amounts of mercury, halogenated polycyclic aromatic hydrocarbons (PAHs), and pesticides. The local community is split between users who scoff at the potential health risk (notably Walleyes for Tomorrow) and those who think the fisheries should be shut down as a public health risk (including many members of Friends of Round Lake). The public health community continues to debate risk assessment and relative risks, arriving at no conclusion and everyone's confusion.

Canada Geese on the Golf Courses. The Bingham National Wildlife Refuge was established primarily to protect and enhance the habitat of ducks and geese. Efforts have been successful. The movement of waterfowl through the refuge each spring and fall is a sight to behold, and many people travel to watch the annual migrations. Along with the migratory waterfowl, the refuge has also attracted a large population of resident giant Canada geese. The geese are abundant on the refuge, but also on the adjacent golf courses. They interfere with golfers directly—and indirectly due to their droppings on the greens. This is a major topic of letters to the editor in the local paper. The Sandhill at Bent Creek developers are eager to work to avoid this problem on their resort. They are also eager to get some sandhill cranes onto the property. They are willing to talk about planning their golf course to attract cranes and avoid geese. They have learned that golf courses can now be certified by a national conservation organization, and they are interested in pursuing this opportunity. The Sandhill at Bent Creek project manager, Alice "Berty" Bertrand, wants to be active in the ROLE Model initiative.

NORTHEAST POWER COMPANY DAM. The Northeast Power Company has petitioned the U.S. Army Corps of Engineers to take over ownership of their nearly century-old dam before it fails. By agreements written when the dam was authorized in 1911, ownership of the dam reverts to the government if the dam is no longer used for power generation. The Northeast Power Company has found the cost of renovating the dam for modern hydropower generation to be too high, and they are planning to abandon it as a power-producing facility. The Corps has also received a petition from the Eastern States Boaters Coalition, located in Boston, to remove the old dam and open navigation downstream. With the dam removed, boats would have access to the Deer River and eventually the Intra-Coastal Waterway.

SnowPACT

Excitement grew as the first printing *of The SnowPACT Way* hit the mail. The people who received the first few copies could hardly believe what had happened in the past year. From a situation in which individuals and groups fought about everything related to natural resources in their community, they had come together to work collaboratively on their future. And a major symbol of their success was the publishing of their newsletter, *The SnowPACT Way*.

The first issue of *The SnowPACT Way* carried a full-page description of the ideals of the group. It is reproduced here:

The SnowPACT Way

INAUGURAL ISSUE, PAGE 1

Distributed quarterly to all Snow River residents without charge.
Subscriptions available at cost outside the Snow River watershed.

The SnowPACT Way is the newsletter of the Snow River Ecosystem Compact, popularly known as SnowPACT. We formed SnowPACT as concerned citizens of the Snow River watershed, people who wanted to ensure that our economic prosperity and quality of life would continue forever.

The kinds of ideas that we are developing have many labels in today's world—ecosystem management, conservation biology, sustainable development, and community-based management. We don't really worry about what these terms mean, because we know what we mean, in our minds and our hearts.

We mean that we want to live, work, play, and worship in this place—this wonderful place—harmoniously, with our families, our neighbors, the resources that inspire and support us, and the ecological systems that ensure our continued existence through time.

We mean that we've watched the destructive arguing among ourselves—about how this acre will be used, about who has access and who doesn't, about whose "rights" are being lost. And we've learned that such arguing gets us nowhere. It focuses on the wrong questions—how to divide a shrinking pie—rather than focusing on how we can make the pie bigger and better.

We mean that we pledge to work together to make the present and the future better for all of us—and for each of us. We came here because it was a beautiful and prosperous place. We have invested heavily through time to make our people safe, healthy, and learned. To complete the journey, we are now also investing to make our use of our lands just as safe, healthy, and intelligent.

Over the past year, as the members of SnowPACT have met and begun to chart our future, we've learned more than we ever imagined and have grown together in ways that have been extremely rewarding. We also know that more of us need to become involved in the land-use decisions being made in our county and communities. Today we invite everyone in the Snow River watershed to join this effort.

And we don't care what you want to call it—as long as you call it successful!

—The SnowPACT Community Circle

The inside of the newsletter documented the way that SnowPACT got started and how it works. Essentially, SnowPACT is a broadly based group of representatives of government agencies, private organizations, and citizens from throughout the Snow River watershed; any group can join, as long as they are committed to the statement of principles that guides the assemblage. The agreement itself is a simple document that pledges the member groups to work together for the benefit of the people and resources of the Snow River watershed. Although it does not constrain the members or the member organizations to certain decisions and actions within their organizations, it does commit them to "come to the table, over and over again" as a way of making the best of every situation. As such, it encourages group solutions, made voluntarily and consensually, rather than seeking administrative or judicial decisions among contending parties.

The initial charter of the group is for 10 years, but the clear intent is that this will become a permanent organizing basis for community planning and action.

SnowPACT Principles of Intent and Operation

We seek to create a place to live that meets the needs of ourselves and future residents. We seek to do this in a way that is based on community, civility, common sense, rationality, and efficiency. We pledge ourselves to be guided by the following principles:

1. We will be inclusive, rather than exclusive, inviting all people and viewpoints; but we will not tolerate attempts to delay, derail, or fractionate our efforts.
2. We will use all the expert knowledge we can get to help guide decisions, including that of natural scientists, social scientists, Native Americans, and experienced community members; but we will not shrink from plans, decisions, or actions because of "insufficient data."
3. We will seek a sustainable level of economic and recreational activity, as part of a sustainable ecosystem that conserves biological diversity and ecosystem processes.
4. We will recognize and work with the natural cycles of events, such as fire, game population levels, and economic activity, rather than trying to control them.
5. We will emphasize maps of natural features and functions, rather than maps of ownership and jurisdiction; but we will respect and work with the legal authorities that have created property and jurisdictional lines.
6. We will recognize that ecosystem boundaries are vague and that the area of interest will vary depending on a specific resource, function, or use.
7. We will find win-win solutions, based on considering many options rather than just two sides, so that no individuals lose in decisions that bring gains to us collectively.
8. We will seek voluntary cooperation among our members rather than rules, regulations, and laws; but we will respect the spirit and letter of those laws to which we are subject.
9. We will set our vision on the long-term and will be prepared to discuss openly and accept the short-term costs of such a vision; and we will provide nonpartisan input to improve our local decision makers.
10. We will be realistic, recognizing that we start from here and that our first steps may be small.
11. We will be guided not only by the head, but also by the heart and soul of community and resource stewardship.
12. We will succeed!

The state governor, through the Green Government Council and the Department of Natural Resources (DNR), and the U.S. Fish and Wildlife Service, through its regional director, have committed their resources to provide the base operations for SnowPACT. The governor's Green Government Council has assigned one of its regional coordinators a primary task of ensuring easy access to all state government agencies and programs, directly through the governor's office. The state DNR has assigned a planner from its central office half-time to facilitate activities of SnowPACT and has assigned a full-time biologist from its regional staff for the next 5 years to help with technical information and activities. The U.S. Fish and Wildlife Service has assigned one professional resource manager from the Kachina Arch Resource Management Area, also for 5 years. In addition, all state and federal resource agencies with lands or other responsibilities in the watershed and region have pledged the support of their staff and physical resources to help along the way.

The state's senators and representatives from both parties have taken an intense interest in SnowPACT as a unique approach and a possible model for the entire nation. It is also supported by one independent representative, Bill Hamilton, who is in his third term and has run each time on the idea of community-based decisions as best for the country. During the past year, legislators have authorized $2,000,000 annually for the work of SnowPACT for 10 years and have funded the appropriation for the first year and pledge they will get it every year. Moreover, they made the work of SnowPACT exempt from the rules of the Federal Advisory Committee Act (FACA) and have assigned the U.S. Fish and Wildlife Service as the lead agency for all federal resource management and environmental responsibilities in the watershed.

Planning staff in Repose County, where most of the SnowPACT lands are located, have agreed to work with SnowPACT representatives via task forces that will be revising the county master plan. In addition, both Repose County and the town of Altavista have added SnowPACT to the entities asked to review subdivision, rezoning, and development applications.

All signatories to the agreement are automatically members of the Snow River Ecosystem Compact, which entitles them to send a representative to all meetings as "speaking members" (Figure 1.9). They are also voting members, if the need arises to hold formal votes.

The core work of SnowPACT is carried out by the Community Circle, a set of 10–15 representatives of member organizations who are endorsed by a majority of the signatories to the agreement. The Community Circle does the hard work of SnowPACT—organizing, overseeing, making decisions, seeking funds, and generally ensuring that the principles of the agreement are maintained and enhanced.

Figure 1.9. *Community collaboration, like that shown here, is a standard way for working on common interests in the Snow River ecosystem. (Photo by Larry A. Nielsen.)*

SnowPACT Community Circle

- Altavista Mayor (Wayne Orr)
- Bluestone River Cattleman's Association (Sam Henry III)
- Department of Natural Resources regional planner
- Green Government Council
- NCTC Bluff Canyon (Kristin Bagley)
- Red Cliff Association (Eleanor Sanchez)
- Repose County Commissioners (Dutch Markson)
- Representative Bill Hamilton
- ROCin' (Jacques Moreau)
- Semak Council of Elders (Howard Two Feathers)
- U.S. Fish and Wildlife Service, Kachina Arch Resource Management Area
- Westfir CEO (Katherine Slater)

THE SNOW RIVER ECOSYSTEM

The Snow River ecosystem is a medium-sized watershed located in the U.S. intermountain West (Figure 1.10). The watershed is approximately 300,000 acres in extent and is well-defined by low mountains on the west and north, by undulating low ridges on the east, and by the Bluestone River on the south. Most of the watershed is drained by the Snow River, which flows generally southward. The Snow River enters the Bluestone River, flowing from west to east, in the middle of the town of Altavista.

The Snow River has three main tributaries, called South, Middle, and North Creeks. The South and Middle Creeks both drain into Pine Lake, an impoundment on the Snow River, but North Creek enters the river below the impoundment. Smaller streams enter the Bluestone River throughout the watershed, but they carry relatively little water and are often dry in the summer and fall.

The land south of the Bluestone River is a relatively narrow strip of undulating hills that end in a continuous ridge that parallels the river. The land on the south side of the river is 1–2 miles wide and is widest within the vicinity of Altavista.

The watershed is bisected by a continuous escarpment, called Red Cliff, which runs northeast to southwest (Figure 1.11). Red Cliff varies in height from 50 to 300 feet at various places along its face.

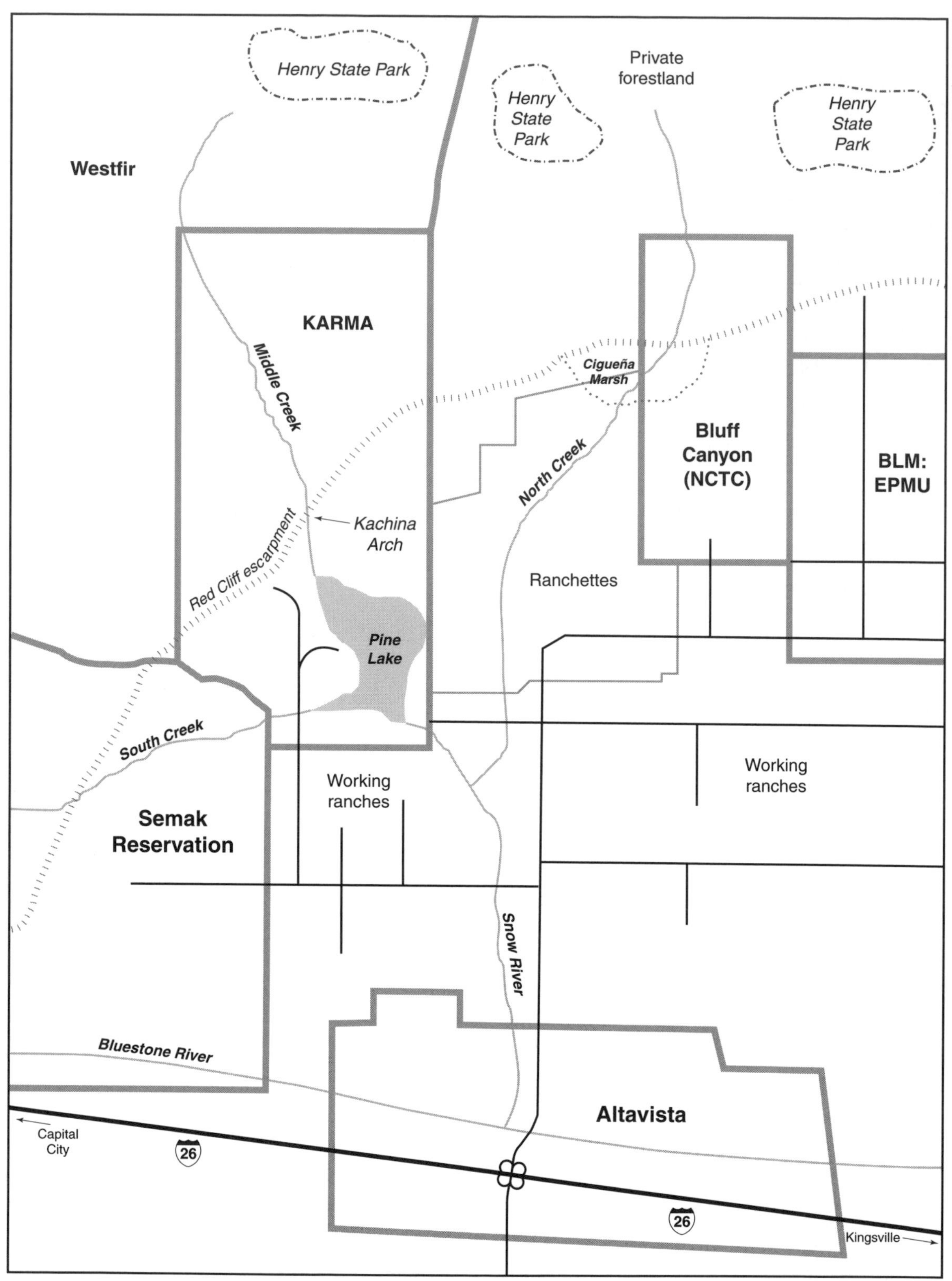

Figure 1.10. *A map of the Snow River ecosystem.*

Figure 1.11. *The defining feature of the Snow River ecosystem is Red Cliff, a stunning geological escarpment that traverses the watershed. (Photo by Larry A. Nielsen.)*

As it nears the Bluestone River, Red Cliff gradually falls to river level, producing something like a wide gateway to the lower watershed on the west. Red Cliff also slowly declines in height toward the northeast. The presence of Red Cliff contributes many special qualities to the watershed, related to ecology, recreation, waterfowl, spirituality, and scenic beauty. Red Cliff is a common photograph subject on landscape calendars, and although most U.S. citizens could not name the place, they would know it from seeing its picture hanging on their kitchen wall.

Red Cliff creates two physiographic and ecological zones within the watershed. The upper watershed, above Red Cliff, is a high-elevation plateau, starting at about 5000 feet and rising to the mountain ridges. It is primarily granitic, vegetated with a mixed coniferous forest dominated by ponderosa pine. The upper watershed is fed by runoff from snowmelt and rainfall, which is concentrated in the spring and early summer. The area gets a total of about 25 inches of precipitation annually.

The watershed below Red Cliff slopes gently to the Bluestone River, beginning at the base of the cliff at elevations of 4800–4500 feet and ending at about 4000 feet at the mouth of the Snow River. Vegetation in the lower watershed is primarily dryland shrubs and grasses. Very little rain actually falls on the lower watershed, except in spring, but groundwater is recharged from winter snows and spring melting.

Red Cliff itself is an important part of the ecosystem. It is primarily red, hard sandstone with interspersed limestone deposits. The limestone areas have eroded over time, creating unique areas. Many springs emerge from the face of Red Cliff, running through the limestone deposits; they run all year long and feed the South, Middle, and North Creeks and a series of moist meadows along the base of the cliff. The limestone has also created many caves that are important ecological habitats for cave-dwelling animals, especially bats.

Several major limestone areas have eroded into deep ravines. The largest of these contains a dramatic sandstone arch, called Kachina Arch. It is an important landmark (also common on calendars), but its primary importance is as a sacred and symbolic area of the Semak Nation of Native Americans.

THE SOCIAL AND ECONOMIC SETTING

Originally, the lands of the Snow River ecosystem were inhabited by Native Americans of the Semak

Figure 1.12. *Members of the Semak Nation have lived in this region for as long as anyone can remember; signs of their ancestors are depicted on rocks throughout the area. (Photo by Larry A. Nielsen.)*

Nation (Figure 1.12). This seminomadic tribe used the region as a winter home to take advantage of the sheltered lands at the base of the Red Cliff escarpment and the large herd of elk that migrated into the area each fall. The deep ravines along the face of Red Cliff are sacred to the Semaks because they are held to be the home of the Buffalo Calf Woman and other beneficent spiritual beings, called kachinas.

In 1840, an itinerant trapper reported the discovery of gold in the Snow River. This event caused a minor land rush by people who surmised that the Red Cliff escarpment and the riverbeds below it contained large gold deposits. The town of Altavista originated at the confluence of the Snow and Bluestone Rivers. Gold speculators convinced the federal government to seek a treaty with the Semak Nation. The treaty was completed in 1843. Although the Semaks ceded most of the land in the Snow River watershed to the U.S. government, they kept their 40,000-acre reservation; retained the rights to use the remainder of their traditionally used lands for hunting, fishing, and other customary purposes; and specifically retained rights for using the sacred sites for traditional and ceremonial purposes forever. By 1849, when the real gold rush began farther west, interest in the Snow River declined.

Although the gold rush subsided, mining interests continued intermittently in the watershed. Among the most stable was a mining operation above Red Cliff, within the watershed of North Creek. A small and shallow seam of coal runs through the rock formations near North Creek, and it was surface-mined for the first three decades of the twentieth century by a local family. In the 1930s, when the Great Depression hit, the family walked away from the land and the mining operation, leaving open seams, piles of tailings, and rusting equipment. In the spring and occasionally after heavy rains, North Creek still takes on odd colors and odors.

Federal lands in the Snow River watershed were assigned to the Federal Lands Office, and many acres were sold or given to private citizens. Eventually, the federal holdings were reassigned to the Bureau of Land Management (BLM) and used for a variety of purposes, such as timber harvesting, cattle grazing, and an occasional mining frenzy. In 1932, major segments of the remaining federal lands were sold to private individuals, speculating on the probability that west-bound farmers and ranchers would settle in the area.

During the 1940s, pressure among the ranchers in the lower watershed led to the construction of a minor Bureau of Reclamation dam and reservoir, 3000-acre Pine Lake, downstream of Red Cliff. For about 10 years, Pine Lake provided water for ranching operations. However, increasing needs and the development of high-capacity wells caused ranchers to install their own wells and begin pumping groundwater for irrigation and cattle. Consequently, Pine Lake has become primarily a recreational site.

After World War II, the state began an aggressive system of developing state parks, under the leadership of Sam Henry, born and raised in Altavista. His idea was that every state resident should be able to reach some state park for day use from anywhere in the state. He was Secretary of the Department of Natural Resources for 20 years, and he achieved his vision. In the Snow River watershed, his agency purchased three parcels of forested land above and bordering on Red Cliff.

Interstate 26 was completed in 1974, linking Altavista to Capital City (the state capital) to the west and to Kingsville, the state's major metropolitan area, to the east. Altavista sits about equally distant from each, within a 2-hour drive. The interstate highway connection was a shot of development adrenaline for Altavista, making the town accessible as a weekend retreat for urbanites and a recreational destination for local and long-distance travelers.

Because of the floodplain cut by the Bluestone River, the riparian area has always been a human travel corridor west. A minor branch of westbound settlement trails (called the Ingel's Path) followed the river and was active through the gold rush and subsequent two decades. Later the railroad came through and continues to carry freight traffic through the region. Finally, I-26 was created in the same corridor.

A 1981 study of potentially significant western watersheds, conducted for the Western Governors' Association, gave the Snow River watershed a B+ for scenic beauty, a B+ for resource development potential, and (a bit unfortunately) a C for civic harmony. In all, though, of nearly 100 western watersheds rated, the Snow River watershed was among the 15 most significant for human, resource, and ecological potential.

Until developing its first master plan in 1995, Repose County was quite vulnerable to unplanned development because it had no zoning or land-use code. Uncontrolled ranchette development highlighted the need to improve the planning process (Figure 1.13). Development pressure continues to increase, as it has in nearly all parts of the intermountain West next to public lands. Most recently, an issue of *Living Well* magazine published an article entitled "How the West Will Be Re-Won," which highlighted 21 places for the twenty-first century, with short profiles. The Snow River was among them, and it was described this way:

Figure 1.13. *Ranchette development, including both upscale homes like this one and more traditional homesteads, is a new feature of the Snow River ecosystem. (Photo by Larry A. Nielsen.)*

> SNOW RIVER—A treat for the eyes and the soul, the Snow River watershed seems to be every person's quest. The area includes virtually everything, from the scenic Red Cliff to mature forests, bubbling streams, rolling hills, and plenty of public lands amid the private holdings. Access is great, along the east-west corridor. Altavista provides a fine gateway to the watershed, including all the recreational facilities one would want. An aggressive development community is bringing health care, education, and cultural resources to the area, based on the promise that the Snow River will be the place to be in the future. The watershed is still a working landscape, with ranches, farms, and logging. We expect that land values will increase at twice the national average, owing partly to an increasingly active sense by current residents that some limits on growth are needed. We believe Snow River will become a "closed-end fund" relatively soon—so get in now, and let the good times and good profits roll!

PEOPLE, PLACES, AND INTERESTS

The diversity of landforms, cultures, and activities within the Snow River watershed are impressive, as illustrated by the following summaries.

KARMA. A central feature in the Snow River ecosystem is the Kachina Arch Resource Management Area, or KARMA. KARMA is a 20,000-acre property owned by the U.S. Fish and Wildlife Service (USFWS) located in the middle of the Snow River watershed, incorporating parts of Red Cliff, Pine Lake, and the range of habitats that exist throughout the watershed.

KARMA is a landholding that is part of a new concept for the USFWS. Resource management areas are just beginning to be created, from appropriate holdings in land management agencies throughout the U.S. Department of the Interior. Incorporating natural, cultural, commodity, amenity, and recreational values, they can and should be used for a combination of purposes.

In entrusting these lands to the USFWS, the Secretary of the Interior designated them as lands to be managed according to adaptive management—using the decisions as experiments from which we all can learn. Consequently, resource management areas are designed to be living laboratories, for both technical management and policies.

KARMA is one of the first resource management areas to be designated, so many people are watching the progress closely. In the initial instructions for managing KARMA, the Director of the USFWS said that it should be managed fully under the principles of ecosystem management. In this case, the directive means that KARMA will be a central part of the Snow River Ecosystem Compact.

HENRY MEMORIAL STATE PARK. Sam Henry set up a series of state parks across the state during his term as Secretary of Natural Resources. Within the Snow River watershed, a series of three state park units lies along the top of Red Cliff; together they are called the Henry Memorial State Park.

"The Henrys," as they are known locally, are tracts of 3000, 5000, and 8000 acres. Each tract is a contiguous area, with mostly mature ponderosa pine forests. The easternmost tract, adjacent to the New Century Trust for Conservation (NCTC) Bluff Canyon property, has a large old-growth area, covering the northern half of the park.

Throughout the Henrys, the tracts are highly developed with campgrounds, roads, trails, picnic areas, and interpretive areas. The park system has been criticized recently for its lack of strategic approaches to developing and maintaining their lands, especially the lack of attention to ecological aspects. In response, the park system is considering major changes to some parks to reverse or better plan development. The Henrys are candidates for such changes.

BLUFF CANYON, AN NCTC PROPERTY. Bluff Canyon is owned in fee-simple by the New Century Trust for Conservation. It is a beautiful slice of the Snow River watershed, incorporating elements of the high-elevation forest, Red Cliff and one of its deep ravines (Bluff Canyon), mountain meadows at the base of Red Cliff, and part of Cigueña Marsh.

The property was acquired in several stages by NCTC, beginning with a large bequest by a former forestry operator. NCTC sold most of the upper portion of the property to Westfir, thereby raising the funds to purchase other parcels from private landowners.

Bluff Canyon is noteworthy because it holds one of the bachelor caves used by big-eared bats and several other cave sites that might attract bats. It also straddles a series of very popular rock-climbing sites on the escarpment, with one within its eastern border.

Bluff Canyon and the general NCTC interests in the region are overseen by a half-time professional naturalist, educator, and communication specialist, Kristin Bagley. Kristin moved to the area several years ago with her husband and young family. Her husband is associate pastor of the largest Protestant Church in Altavista.

Like many NCTC properties, Bluff Canyon will eventually be turned over to a public agency for ownership and management. For the present, however, NCTC is eager to be involved in the idea of ecosystem management in the Snow River watershed—to ensure that it is done well and to get credit for having pulled it off.

NCTC is also very concerned about the fragmentation of the landscape surrounding Bluff Canyon and about preserving more of Cigueña Marsh. It has a strong interest in acquiring additional lands or helping develop conservation easements that will prevent the carving up of the relatively large tracts of open space that remain.

COMMERCIAL FORESTLAND. Much of the area above Red Cliff is commercial forestland. To the north and east, the lands are mostly owned in small tracts by individuals or family trusts. Although they have a variety of management objectives for their lands, harvesting commercial timber is a major objective (Figure 1.14). Many of these landowners have developed cooperative relationships with

Figure 1.14. *Forest management in the ecosystem includes replanting harvested sites with evergreens that are native to the region. (Photo by Larry A. Nielsen.)*

large timber companies, which help them manage their land and have a first option to buy timber at a small premium over market price.

The major forestland owner in the northwestern part of the watershed is Westfir, a private family business that has held this property for a century. They maintain the forest in even-aged stands of ponderosa pine, which they harvest via clearcutting on an 80-year rotation. The family owns more than 500,000 acres of forest throughout the region, operating their entire holdings as one dispersed forest. They manage and harvest primarily for sawtimber, which they mill, dry, and cut for their home construction business in Colorado. The CEO of Westfir, Katherine Slater, was educated at a western forestry school as a forest engineer, and she maintains a very active role in the national forestry community.

Katherine Slater and Westfir have been very supportive of the principles of forest sustainability. Westfir harvests their timber according to Best Management Practices (BMPs) and complies with the principles of their Sustainable Forestry Initiative. Slater is on the board of several national programs dealing with sustainability, with an interest in both the market value of "certified wood" and the ecological importance of sustainable management. She has been a leader in pushing for market-driven improvements in forest management, including certification of her company's lands. Consequently, Westfir is now working to develop a management approach for its lands that matches the certification standards of various third-party groups. She is eager to make Westfir's lands part of the SnowPACT program, providing an example and a stimulus for others.

Westfir works closely with state and federal land management agencies and with local communities. For example, it funded the Altavista Youth Recreation Center and was a major contributor to the nonprofit Bluestone Cancer Hospice. The company employs many local teenagers in its planting operations and has been nationally recognized for a work-release program for jailed teens. CEO Slater and other family members are among the social elite throughout the state, and the governor would not dream of doing anything ecologically or socially without including her.

Westfir believes strongly that the nation's first mission is to keep the land in working status, rather than in preserves. Along with its willingness to be good stewards of its land, it is adamant against excessive government regulation and additional land purchase. Here is an excerpt from the company's most recent Annual Report:

> This nation is like no other in the world—built on the premise that free people can and will act not only in their own interest, but in the interest of their fellow citizens. We have proven over and over again that the best solution to our problems is to look to our families, friends, and neighbors. This is no less true for a community of landowners, who can prosper now and be stewards for the future. When government takes over, it saps our strength, drives out our good will, and punishes our innovation. Westfir believes that the answer to all problems is the goodwill of the people, and we commit our time, creativity, and fortune to that belief.

Consequently, Westfir has been a very willing partner in the new concept for the Snow River watershed. However, it will participate only if it believes the work is founded fundamentally in its own volunteer contributions and those of the other community members.

Commercial Ranching. The southeastern portion of the Snow River watershed is mostly in medium-sized working ranches, as is the land to the south of I-26, almost to the state line. Like Westfir, the ranches have remained largely in the hands of private families for many years. Over time, however, they have been divided into smaller parcels, now mostly 3000–5000 acres. With increasing development pressure, escalating land values, and marginal cattle prices, more ranchers are thinking about selling and are being approached by developers from outside the region who are well financed. State laws allow them to subdivide into parcels of 35 acres or larger without going through any county development review.

The ranches run beef cattle in fenced pastures throughout the year, supplementing grazing with raw hay and silage during the winter when the snow accumulations are deep. Consequently, about 50% of any ranch is unimproved pasture and 50% is irrigated hay fields. Some ranchers run beef cattle in fenced pastures on their own lands, whereas others depend on summer grazing permits on KARMA and BLM lands to supplement their base properties.

The limiting time for forage production is mid to late summer, when rainfall is sparse and the local water table has dropped to its annual low. To keep the forage in good condition, ranchers begin irrigating their fields in mid-May and continue through September. Irrigation water comes from two sources: Pine Lake and high-capacity wells. Pine Lake was the historical source for irrigation water, but it is now used as mostly an auxiliary source, providing water by gravity flow when pumps break down or other problems occur.

High-capacity wells pump from an underground aquifer that also feeds the marshlands in and adjacent to the NCTC Bluff Canyon site. Irrigation pumping is now running at approximately 10% over the annual recharge of the aquifer, but the ranchers are not worried. They anticipate that more ranches will be taken out of production soon, following the lead of the ranchette developers in the northern part of the watershed, thereby reducing the demands on the aquifer.

The ranchers *are* worried about a proposal they have been hearing to reintroduce American bison to KARMA and the Semak Reservation. They contend that the likelihood is strong that bison will carry a variety of diseases, such as brucellosis and bovine tuberculosis, that will infect their cattle, thereby preventing their transportation across state lines to primary markets. And this concern may be real: Wildlife biologists know that bison can and do carry several significant diseases.

Ranchers also believe that reintroduced bison will invade their range—tearing down fences, eating the most desirable forage, and generally messing up the place. They believe that once the bison enter their land, the ranchers will be subject to several other laws and certainly to public ridicule if they try to remove the bison. They have been watching the events around Yellowstone National Park closely, and they do not like what they have seen and heard.

The ranching community is represented informally by the Bluestone River Cattlemen's Association, headed by Sam Henry III, grandson of Sam Henry. Sam Henry is a moderate man, but his constituents occasionally get excited, and he represents their interests aggressively. His connection with the legendary Sam Henry gives him access to state lawmakers, and he enjoys a certain deference even within the environmental community.

BLM: Eastern Plains Management Unit. The Bureau of Land Management is the other major federal land management agency with holdings in the watershed. The BLM administers a large tract of land in the eastern portion of the watershed and extending in various parcels across several watersheds farther to the east. The holdings are managed as a unit from an office in Kingsville; together they are called the Eastern Plains Management Unit (EPMU).

The EPMU leases parcels to private ranchers for cattle grazing. Over the years, the number of cattle being grazed has gradually risen, while the number of individual leasees has dropped. The rangeland in EPMU is in generally satisfactory condition, with good production of cattle, low calf mortality, and little erosion along riparian areas.

There is often pressure to sell off parts of the EPMU to a variety of landholders, especially in the Snow River watershed. The western border of EPMU lands is adjacent to the NCTC Bluff Canyon, and the northern border runs along Red Cliff—both desirable private property locations.

Ranchette Development. In the 1960s, a ranch owner subdivided his land into a series of small properties designed as "ranchettes." This land lies between KARMA and NCTC Bluff Canyon. The ranchettes are 25–35 acres, served by deep wells, septic systems, and all-weather asphalt roads.

The ranchettes were bought quickly, mostly as investment properties. Since the mid-1980s, however, more people have moved into the area on a permanent basis, since the completion of the Altavista interchange on I-26. Most of the properties are held by professionals who like the idea of owning some land. Almost all residents keep horses or a few other animals, plant large gardens, and consider themselves "ranchers."

They have formed the Red Cliff Association (RCA), an active group in civic affairs and land management. The goal of RCA is to "create a place where people live in beauty, safety, health, serenity, and happiness." Toward that end, their Nature Committee plays a major role in the development of individual properties and common areas. They have built a walking trail through Cigueña Marsh and an observation tower for viewing the American avocet. This large wading bird stops in the wetlands, and around Pine Lake, semiannually on journeys between breeding sites in southwestern Canada and overwintering sites in Guatemala. RCA residents are especially protective of Cigueña Marsh because it has historically had one of the largest migratory flights of American avocets in the United States.

The leadership of RCA is also eager to reintroduce bison to the Snow River watershed. As enthusiasts for both native wildlife and the preservation of earlier western lifestyles, they believe that the bison is a perfect symbol. Moreover, they have learned that the bison is considered a "keystone species," one that has substantial effects on other aspects of the ecosystem. Thus, they contend that reintroducing the bison will help re-create an ecosystem that is like the earlier one they admire—and therefore will be a perfect example of ecosystem management. They have teamed with the Semak tribe in plans to buy and import bison for the Semak Reservation, and they hope to introduce it into KARMA.

However, RCA members are also purists. The long-time chair of the Nature Committee, Eleanor Sanchez, serves on the state's Biodiversity Task Force, where she has become a proponent of restoring native species and populations. Through her, RCA has become aware of genetic diversity, and they do not want just any bison returned to their area. They are eager to have the original genetic strain of bison returned to their ecosystem.

Several blocks of land are held by real estate interests and remain undeveloped. Several development proposals are in the making; one is reported to be an expensive gated community and golf course development. Recently, a small environmental group in the ecosystem has added rural sprawl and open-space protection to its list of priority concerns in its 5-year action plan.

The Semak Nation. Members of the Semak Nation reside in the southwestern portion of the Snow River watershed, with about 10,000 acres within the watershed and about 30,000 acres in watersheds to the west. The Semak tribe is a fully recognized sovereign nation of Native Americans. Most Semaks still live on the reservation, but growing interest in the wealth of the nation has caused many nonreservation members to rediscover their roots. These members live mostly in Capital City, Kingsville, and smaller cities along the Bluestone River–Interstate 26 corridor. About 2000 Semaks live on the reservation itself.

The Semaks are large-scale cattle ranchers, with land management patterns very similar to their non-native neighbors. They have followed excellent advice to develop their cattle business so that they are competitive with any other operation in the region.

However, they are also interested in further ways to develop their business. One idea they are exploring is the reintroduction of bison as a semi-domesticated animal for ranching. They have begun experimental ranching in isolated parts of their reservation, and the results have been encouraging. Their market analysis shows very strong interest in "buffalo meat" throughout the Pacific Rim, especially if the meat is grown by native peoples. Their husbandry also has been successful, but they are concerned about the behavior of animals when constrained in fenced ranges, as well as the potential for disease. Nonetheless, they are strongly committed to trying buffalo ranching on a pilot scale.

As committed as they are about their business, the Semaks are just as committed about preserving their cultural heritage. They work hard to keep traditional customs, including leadership by a council of elders. The current spokesperson for the Council of Elders is Howard Two Feathers. He represents a perfect combination of devotion to the old ways and of seeking opportunities to move the Semak Nation's economy into the future.

The Snow River watershed, the land that includes KARMA, and the Red Cliff escarpment are very important to the Semaks. Over time, each spring they have traveled along Red Cliff to the ravine that contains the natural arch now called Kachina Arch (Figure 1.15). At the arch, they conduct ceremonial dances and reenactments of their creation story. Traditionally, young men have gone to the arch by themselves for several days, returning with a sacred red rock that they carve into a ceremonial pipe. After this rite of passage, they are considered adults and full members of the tribe.

Figure 1.15. *Kachina Arch is the signature landscape feature for both the Kachina Arch Resource Management Area and the Semak Nation. (Photo by Larry A. Nielsen.)*

The Semak elders have been vocal in their concern about the sacredness of Kachina Arch and the other ravines along Red Cliff. The increasing use of the area by rock climbers interferes with their traditional uses and their belief in the sanctity of the rocks. They fear that the watershed development and its associated plans will reduce freedom of access and their use of traditional places along Red Cliff and especially in Kachina Arch itself.

PINE LAKE AND DOUBLE-A. Pine Lake is a 3000-acre impoundment built as an irrigation reservoir and completed in 1947. For some time, Pine Lake was nearly emptied every year for late-summer irrigation, but it now serves principally for emergency and occasional irrigation. The lake is a very popular fishing site for local residents and for weekend visitors from the entire I-26 corridor. Pine Lake is stocked with rainbow trout, which feed on abundant amphipods, giving the trout a very red color and a delicious taste. Rainbows grow well in the lake, but they do not reproduce well. Occasional lake reproduction is seen along the rockier shorelines. Rainbows do migrate up the tributary streams (South and Middle Creeks) of the Snow River in spring and summer, where spawning occurs. Survival of the young is relatively poor, however, requiring the constant replenishment of Pine Lake with hatchery stocks.

Altavista has an active angling group known as the Altavista Anglers, or Double-A. Double-A is a very civic minded group. As part of administering the town's Big Brothers and Big Sisters activities, they sponsor several Pine Lake and riverfront fishing events. They pioneered the Double-A Program, in which they take all middle-school students with at least two grades of A on their report cards on an annual early summer camping-hiking-fishing weekend. They also operate a senior citizens' fishing program and have built several handicapped-access areas along the lower Snow River and Bluestone River.

Double-A has actively assisted with fish management. It supported the development of an auxiliary hatchery to try to increase natural reproduction and supplement natural reproduction in the meantime. It has also paid for and installed in-stream structures to add riffles and pools to Snow River and its tributaries. It collects fish samples for the ongoing pesticide-monitoring programs run by the U.S. Environmental Protection Agency. And, members have listened patiently to proposals for removing rainbow trout from Pine Lake in favor of native species. They have listened, but they have not bought into the idea.

To help keep them informed and involved in decision making, Double-A employs a part-time lobbyist, Thomas "Fins" Polansky. Fins is a statewide personality who hosts an outdoors show on public television that airs weekly. Fins is also a very effective spokesman for the fishing community. Consequently, Double-A is always present and loud at Fish and Wildlife Commission meetings to discuss management plans for Snow River and Pine Lake. And it generally wins.

Representative Bill Hamilton has presented the Community Circle with an unexpected issue: removing the dam on Pine Lake and restoring the Snow River as a free-flowing stream. The proposal to remove the dam has come as part of a national study of small-scale impoundments on federal lands, conducted by Restoring America's Rivers. In its report, it cites 135 small impoundments that no longer serve their intended purpose; the report highlights 13 impoundments/rivers as priorities, and Pine Lake/Snow River is among them. Representative Hamilton is not necessarily a proponent of removal, but he thinks SnowPACT needs to be proactive about addressing the report.

COMMERCIAL AND RECREATIONAL ROCK CLIMBING. Long before rock climbing became a national craze, Red Cliff escarpment was a favorite for dedicated rock climbers. Since the 1950s, Red Cliff has been a prime destination of western rock climbers, with thousands visiting the area annually, especially in early summer. The escarpment offers a variety of sites, ranging from novice to expert. The most exhilarating climb is up the inner walls of the ravine holding Kachina Arch, ending with a walk across the arch itself—a natural high, according to the climbers. With the growing popularity of rock climbing, Red Cliff has become noteworthy internationally. Today, rock climbers come from all over the world to add Kachina Arch to their "life list."

Rock Climbers International (ROCin', as they call themselves), a major interest group for both professional and amateur rock climbers, moved to Altavista in 1984. ROCin' sponsors a well-attended climbing competition up Red Cliff each spring, as the kickoff for the western climbing season. ROCin' also teaches courses for professional rock-climbing guides and instructors throughout the summer and fall. In 1996, it introduced a line of outdoor clothing called Red Cliff, which has been doing very well and is now under consideration for acquisition and national marketing by at least three large clothing retailers.

Altavista, consequently, has developed a thriving industry serving rock climbers. The combination of tourist services, ROCin' operations, and various equipment companies is responsible for about 400 direct jobs in Altavista and many more indirect jobs. ROCin', and its Executive Director Jacques Moreau, are eager to see Red Cliff develop into a highly intensive rock-climbing destination. They anticipate that complete tourism packages could be developed that take advantage not only of the Red Cliff escarpment, but also the other natural features of the entire watershed.

Rock climbing has become an issue recently, after studies demonstrated differences in the ecological communities associated with climbing sites. The Red Cliff escarpment supports 12 species of plants and 8 species of birds found only on cliffs of this rock type. A recent study found that frequently climbed cliffs support few of these plant and bird species; however, they do support other types. For example, starlings and brown-headed cowbirds are found only in association with popular climbing areas, but they are not found at unclimbed cliffs. Likewise, the non-native, invasive cheatgrass is common at the base of climbed cliffs and is not found at unclimbed cliffs.

ALTAVISTA. Altavista, a city with a population of approximately 14,000, has traditionally relied on logging and ranching for its economic base. More recently, tourism has entered the economic scene, providing for a growing service economy.

The mayor of Altavista, Wayne Orr, is very popular: he has been reelected every 2 years since 1988, and it looks like he could go on forever. Mayor Orr has kept his eye on the economic welfare of the Altavista area. He has served as a very effective go-between for ranchers, loggers, ranchette owners, and federal land management agencies. Recognizing the value of the tourist industry, he carefully nurtures the rock-climbing interests.

The mayor also knows one other thing—that people like Altavista and the Snow River area because it is beautiful. And he means to keep it that way. In 1988, he introduced the Altavista Greenway Gateway, a plan to make the Bluestone River and Snow River corridors inviting to travelers on I-26 and to tourists stopping by for a short visit. He has worked in every possible way to establish a greenway along the rivers; he purchased some land with city revenues, created conservation easements for private landowners, gave tax incentives for companies to beautify their streamside properties, and made a walking-jogging-biking trail along the riverbanks. By the end of last year, he had achieved 75% of the goal of a continuous greenway on both banks of the Bluestone and Snow Rivers within city limits; he will get the other 25% over the next 5 years.

Mayor Orr is working with the local chamber of commerce to promote rock climbing and the river project. He is trying to convince the Repose County Commissioners, led by Dutch Markson, to enter into an intergovernmental agreement (IGA) to amend both master plans to include a Bluestone River Special Area Plan. County commissioners are more conservative and are being pressured by development interests to be wary of any protection scheme along this high-value river corridor and its benchlands, which are ripe for more home development.

SPECIAL RESOURCES

A variety of special resources define the biological and ecological systems within the SnowPACT region, as exemplified by the following.

THE AMERICAN MARTEN. The American marten is a medium-sized member of the weasel family that inhabits mature coniferous/mixed hardwood forest across the northern half of the United States and most of Canada. It is a "special interest" species in the U.S. and a threatened species in Canada. In the western U.S., the marten was heavily trapped throughout the 1800s and early 1900s, reaching near extirpation by the mid-1930s. Since then, the fragmentation of mature forest has further reduced its range.

A typical marten pair requires a home range of 2–3 square miles. Martens avoid clear-cuts and other open patches. Studies have shown that a minimum patch size of about 800 acres is necessary for maintenance of a marten pair and that the animals are edge-sensitive. Martens do not disperse across open ground readily.

Historical records show that martens were abundant in the Snow River watershed above Red Cliff. Today, the extent of marten populations in the area includes a known colony of four reproducing pairs in NCTC Bluff Canyon, informal reports of occasional martens seen by logging crews on Westfir property, and the possibility that some exist on KARMA. A geographic information systems (GIS) study of possible habitat areas for martens, conducted by a graduate student at Western State University, demonstrated that large areas of the Henrys are suitable for martens. Approximately 50% of the eastern tract, 25% of the center tract, and 10% of the western tract are suitable marten habitat.

The Big-Eared Bat. The big-eared bat is a federally threatened species that is found in isolated caves along Red Cliff. Enormous flights of big-eared bats were recorded in local newspapers from the early twentieth century, but now bats are found in only three caves along the escarpment, two within Kachina Arch and one in Bluff Canyon.

Big-eared bats have a complex life history, tied to a complex use of various caves. The fundamental habitat unit is a communal cave used by adult males and females during mating (which occurs from April through June). After mating, the males leave this so-called maternal cave, returning to "bachelor caves," inhabited by pre- and postreproductively mature males. Females gestate, give birth, and nurse in the maternal cave; immature females also remain in the maternal caves. Females stay close to the maternal cave during the summer, but males make large flights every evening, sometimes traveling tens of miles in search of insects. Males do not necessarily return to the caves each night but may take up temporary residence in trees and other structures, especially under bridges and porches in Altavista.

At present, the three caves used by big-eared bats include a known maternal cave inside Kachina Arch and two bachelor caves, one in Kachina Arch and the other in Bluff Canyon. A survey of these caves and an expert assessment by specialists from the Alliance for Bat Conservation (ABC) suggest that the use of the maternal cave is dropping regularly. The reasons for the decline are not known, but excessive visitation by naturalists and disturbance by rock climbers are suspected. ABC has sponsored research revealing that 65% of female bats return to the maternal cave in which they were born. An early report by a local mammmalogist indicated that up to ten caves along Red Cliff were used as maternal caves by bats in the 1950s. A survey of numerous other caves along Red Cliff has revealed that several other caves have been used in the past as maternal and bachelor caves, including two maternal caves on the Semak Reservation and one maternal cave in the privately owned ranchette area. A whole series of caves on the KARMA section of the Red Cliff may be useful as big-eared bat sites.

The Snow River Cutter. The Snow River and its tributaries are home to a stock of native cutthroat trout, locally called the Snow River cutter (Figure 1.16). The cutter has been studied electrophoretically and found to be a distinct subspecies of the Bluestone cutthroat, which ranges along the entire length of the Bluestone River. Isolation of the cutter from the more pandemic Bluestone cutthroat apparently occurs because of the entry of a warm spring just above the town of Altavista that keeps a 1-mile section of the Snow River too warm for migration by cutthroat either up or down the river. It is currently abundant in the upper reaches of the South and Middle Creeks above Pine Lake; a vigorous fishery for the cutter occurs in both creeks, restricted to artificial lures only and with a limit of one fish over 14 inches per day. The limit is not really important because most anglers release all the fish they catch. Although occasional cutters are found in Pine Lake, the somewhat higher water temperature in the lake generally restricts cutter use.

The cutter populations in South and Middle Creeks may be affected by introduced rainbow

Figure 1.16. *The Snow River cutthroat trout lives in rivers upstream of Pine Lake. (Photo by Lloyd Hazzard.)*

trout in Pine Lake. In the interest of establishing a more productive year-round fishery, state biologists introduced and continue to stock rainbows from a state hatchery population, which turned out to be a California rainbow that had been sent to hatcheries in Pennsylvania in the 1880s, maintained there since, and used to stock lakes throughout the eastern United States.

The Snow River cutter in North Creek is another story. Two distinct populations seem to exist. A very small population of cutters lives in North Creek above the Red Cliff escarpment. The fish have been up there as long as anyone knows, and genetic studies have shown that they have lost several rare alleles that are present in fishes in the other locations. A population of cutters also exists in North Creek below Red Cliff. The population is stable and fished under the same regulations as in the other streams.

THE MOUNTAIN MEADOW ECOSYSTEM. An unusual ecosystem, the mountain meadow, exists broadly along the base of the Red Cliff escarpment and at various sites within KARMA. Although not unique, the particular conditions within the KARMA meadows are fairly unusual. Because the area is fed by groundwater flow from the escarpment that is fairly continuous except during winter, the meadows retain much of their character throughout the year. However, the relatively open landscape allows for more productivity and a broader set of plant and animals species to be present. Consequently, diversity is quite high within the various meadows along the escarpment.

The nature of water flow and landform along the base of the escarpment also produces a relatively continuous meadow environment along the entire base of the escarpment within KARMA. This unbroken meadow is quite unusual, because most habitats of this kind are small and usually separated by ridges. Additional patches of meadows occur in the NCTC Bluff Canyon and the Semak Reservation.

Surveys by USFWS personnel have shown that the meadow environment within KARMA is home to more than 45 species of plants that are moist-soil obligates (dominated by species in the parsley family, primarily cow parsnip, swamp whiteheads, and Queen Anne's lace); seven amphibians (including an endemic frog, the mountain red-legged frog, and an uncommon population of the yellow-legged frog); a dozen reptiles; and several mammals that prefer moist soils (meadow mouse, pocket gopher, Belding ground squirrel, and mole).

THE CIGUEÑA MARSH. Cigueña Marsh is an expansive wetland area that occurs partly in the NCTC Bluff Canyon and partly in privately owned lands between Bluff Canyon and KARMA. Cigueña Marsh is part of the extensive network of marsh meadows along the base of Red Cliff, but the wetlands expand in the area known as Cigueña Marsh. Cigueña Marsh is not clearly defined, but the areas of interspersed wet meadows, marsh areas with emergent vegetation, open water areas, and isolated upland islands comprise 800–1200 acres, depending on what definition of "wetland" is used. The marsh occurs generally half in and half outside the TNC property, in an irregular oval shape.

As described elsewhere, Cigueña Marsh is home to migrating groups of American avocets. It also supports a regular complement of waterfowl, marsh birds, amphibians, and reptiles, similar to that described for the mountain meadow habitat. However, the marsh may be significant in that it appears to contain a complex of amphibians and marsh plants that are unique. Although various species of the complex occur in other places, the entire assemblage is found nowhere else, presumably because higher levels of human activity, in the form of ranching, home ownership, or recreation, take place.

THE PALE SWALLOWTAIL. The pale swallowtail is an uncommon butterfly that feeds primarily on flowers in the parsley family; it is particularly fond of swamp whiteheads (Figure 1.17). Biologists became interested in the pale swallowtail some years ago when it was discovered that pupae were toxic to starlings. The pale swallowtail was the focus of a number of research projects to identify the mechanism of toxicity and determine whether the toxicity could be managed for use as a repellant or toxicant for nuisance starling populations. Although the work is still under way, the intense study re-

Figure 1.17. *The pale swallowtail butterfly is abundant in the mountain meadow ecosystem. (Photo by John and Karen Hollingsworth, U.S. Fish & Wildlife Service.)*

vealed a great deal about the pale swallowtail's life history.

The butterfly is found in isolated patches throughout the intermountain West, but is now slowly declining. It appears to disperse broadly in the adult stage, but reproduction is not successful unless sufficient patches of parsley species are present as food for the larvae. Consequently, reproduction tends to occur in isolated habitats that act as source populations for surrounding areas. Pale swallowtails of all life stages are quite common in the meadows of KARMA, but larvae are not found in the Bluff Canyon or Semak Reservation meadows, nor are they common in other meadow areas in adjacent watersheds. When adults disperse in the late summer, the migrations are quite spectacular—an event similar to the migratory flights of monarch butterflies, but not as dramatic or directed.

Federal and state authorities have recently focused their attention on the pale swallowtail. Consequently, they have asked SnowPACT to develop something like a Habitat Conservation Plan (but not necessarily with all the requirements and constraints of an official HCP) for use as a model for pale swallowtail conservation throughout the intermountain west.

PDQ Revival

The southeastern United States is a region that has seen unprecedented growth and prosperity in recent decades. Many cities have grown from small manufacturing towns to world-class high-technology and service centers, attracting people from throughout the nation and world. Coastal areas have been converted to major recreational and industrial centers. Along the way, many areas have seen their distinctive landscapes and lifestyles change, with the dilution of southern culture and rural communities. In a few places, communities and individuals have come together to conserve their traditions and their lands. One such place is the PDQ ecosystem.

Few residents of PDQ knew when they began talking about conserving their homes and heritage that they were creating a personalized approach to ecosystem management. Today, many might even reject the idea, not being prone to jump on the bandwagon for some leading-edge social or scientific theory. Instead, they would continue to work locally to make sure that they got what they wanted. The story of the PDQ ecosystem, therefore, is a bit different than other examples of "ecosystem management," and it is a story that is best introduced by one of the local leaders of the effort, Mr. Sonny Tymes.

In the mid-1980s, Sonny Tymes was a household name throughout the United States. He was the nation's best college quarterback, leading his team to the SEC championship 3 years in a row and breaking all SEC records for passing and total yards (he was quite a scrambler on broken plays). He went on to a short career as a starting NFL quarterback, ending when his scrambling led to a knee-shattering tackle in the 1992 playoffs. Although no longer a national figure, he remains a popular sports broadcaster, covering SEC football games on regional television and hosting a syndicated weekly outdoors show, *It's Sonny Outside.*

Sonny Tymes is also a PDQ hero. Born and raised in New Scotland, Sonny became known to people in town who stopped by his parents' small candy store in New Scotland's Riverfront Old Market. Sonny starred in all sports in high school, leading the New Scotland Tartans to state championships in football, basketball, and track. Although he went on to national fame, he has retained his close relationships with the community and the landscape. He and his family still live in New Scotland. He funded the New Scotland Youth Center (known locally as Good Tymes), where local children can go for recreation, but where poorer children have special opportunities; he uses his success as an inspiration for local African American children who, like him, came from modest means.

He has made his most important mark, however, in his efforts to preserve the lifestyle and landscape of his home area. As he is fond of saying (in a recent newspaper interview):

> This place is heaven to me. It gave me what I needed to succeed, and my family has loved living here for generations. The people and the land always nurtured me, and they still do every day. What a great place to make a living and raise a family!

Growing up as a sportsman, Sonny spent his free time as a child in the fields and woods and on the waters of the PDQ basin. As an adult, although he has the opportunity to hunt, fish, boat, and hike across the world, he still most likes spending time at home. He owns a 20,000-acre tract of land that he calls High Tymes. High Tymes is a mixed landscape of pine plantations, farm fields, and natural habitats, straddling the Paumassee River and sharing a border with Camp Fraser.

Sensing that his beloved PDQ might be one of the next places to be taken over by large-scale development, Sonny began to talk with his neighbors, community leaders, and others about 10 years ago. He began exploring ways that they could develop the so-called win-win strategies that might keep their community prosperous, but also preserve the rural lifestyles and natural features of the landscape. His leadership has been crucial to the development of the community-based set of efforts that are known locally as PDQ Revival, or PDQR. As is characteristic of the area, PDQR shies away from formal procedures, programs, and recognition. Instead, it works through the community in a variety of ways.

As Sonny Tymes has often said (in a statement made before the U.S. House of Representatives at a hearing about coastal-zone management):

> I can only speak for the folks of the PDQ region. We know why we came to PDQ and why we stayed, and we want to keep it that way. We want a place that is easy to live in; a place where we can work in good jobs with our neighbors; a place where we can enjoy nature's gifts, whether that be hunting or fishing or wildflower watching; a place where our kids and uncles and grandparents and neighbors know each other and watch out for each other. It isn't an accident that our little efforts are known as PDQ Revival. The revival is part of our heritage as well; it nurtures our souls and keeps us strong with our neighbors. The PDQ Revival is the same idea, because by keeping and renewing our lands, we continue to revive our communities.
>
> Now, we appreciate your ideas and especially your money. But, we really want you to let us use it in ways we know are best for us—and for our kids and their kids. And we're pretty sure that what we decide is best for us will also be best for the rest of the country. So, kick us the ball, and we'll put the scoring drive together—no fumbles, no penalties, and no incompletes!

This is the sort of thing that Sonny Tymes can say, and everyone smiles and applauds. Based on that testimony, Congress agreed to provide funds for 10 years, via various federal programs, to help the PDQ Revival project. It also wrote into the appropriations language that it expected the agencies to provide the help the community needed and that it wanted a report back at 2-year intervals about what the community was doing.

THE PDQ ECOSYSTEM

The PDQ ecosystem gets its name from the Paumaussee, Dee, and Queen Rivers, a set of parallel rivers that run from west to east, flowing into the Atlantic Ocean in the southeastern U.S. (Figure 1.18). The three rivers drain an area of over 1.5 million acres, joining near the coast to form Paumaussee Sound. The Paumaussee River is the largest of the three and the central river; it runs for over 150 miles, originating in the higher reaches of the piedmont region. It is a highly productive river; in the area generally considered the PDQ ecosystem, it is a lowland river, with warm water, slow flow, and often high turbidity. The Dee and Queen Rivers, which flank the Paumaussee on the north and south, respectively, are shorter rivers, originating in swamps in the coastal plain and flowing gently to Paumaussee Sound. Both are blackwater rivers, with low productivity and the dark-stained waters typical of blackwater streams.

The ecosystem itself is a humid subtropical ecological zone, characterized by cool winters and hot, wet summers. Most of the ecosystem is in the coastal plain, but the upper reaches rise up to the lower piedmont. (Local folks, however, generally think of the coastal plain area as the upper limits of their area of concern.) The dominant natural community is the longleaf pine savanna, heavily dependent on a regime of frequent, cool fires for the maintenance of an open understory (Figure 1.19). The region has been used extensively since colonial times, with substantial changes in the vegetation. Pine plantations and croplands cover about half the area, but many of the croplands

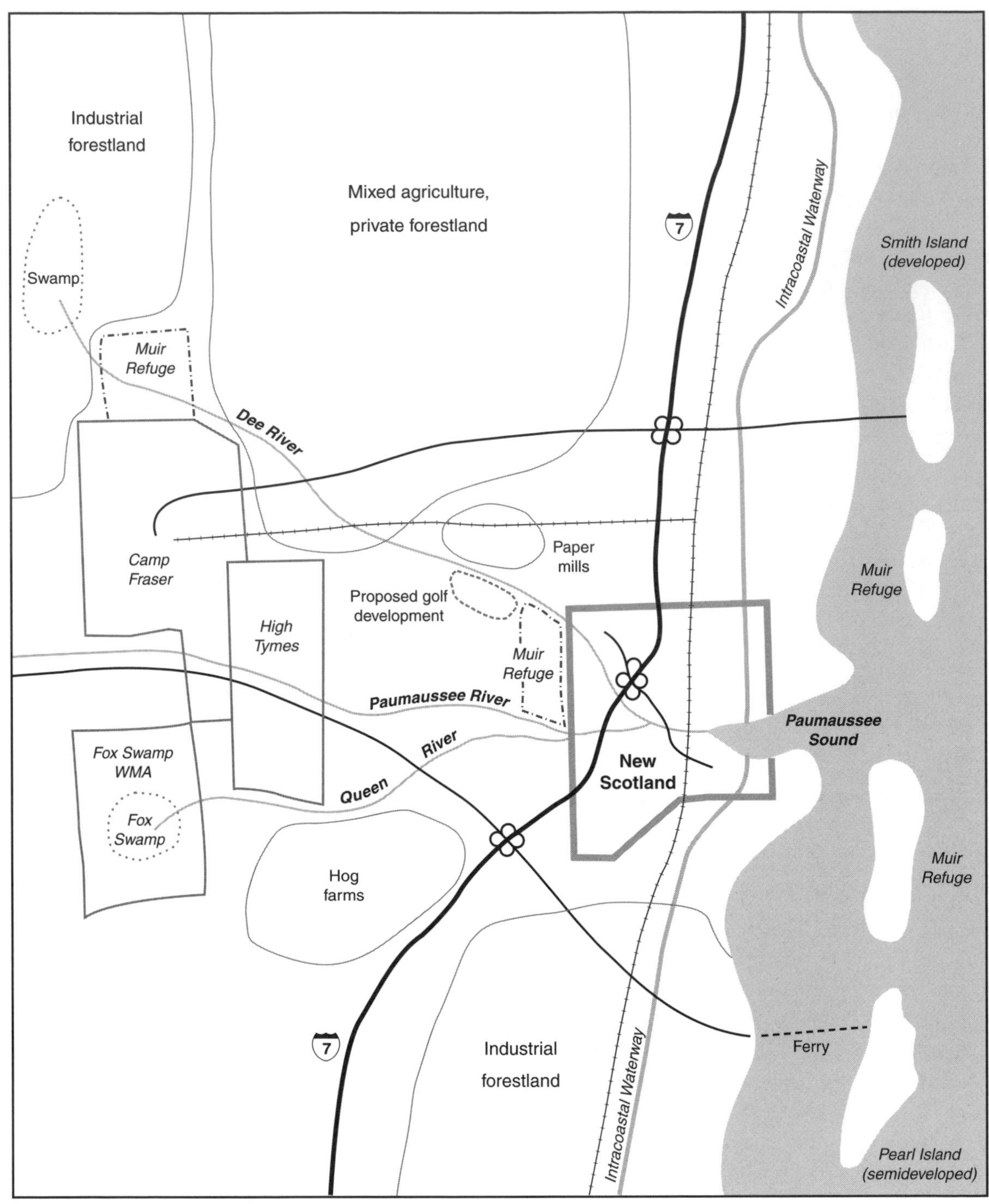

Figure 1.18. *A map of the PDQ ecosystem.*

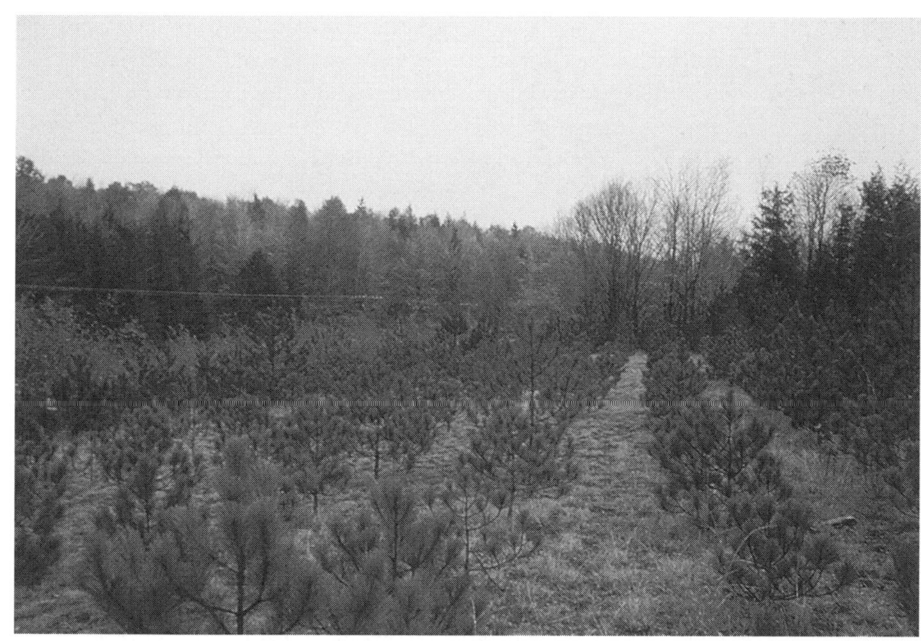

Figure 1.19. *Much of the PDQ region is planted in southern pines or contains natural stands of longleaf pine. (Photo by Larry A. Nielsen.)*

have been idle for years. The coastal region has a series of barrier islands, some of which are developed for recreation and some of which remain natural. In recent decades, the islands have been relatively stable (for barrier islands), with a small amount of erosion along the southern ends of islands and sand accretion along the northern ends.

THE PDQ REGION

The PDQ ecosystem encompasses the area generally known as PDQ. The name became popular after World War II, following the wartime slang usage of PDQ to mean "pretty darn quick." Since then, many local businesses and civic organizations have adopted the PDQ name.

Since colonial times, the three rivers and their surrounding lands have been a relatively single societal unit. Paumaussee Sound was a reasonably good natural harbor that had much ship traffic from Great Britain during later colonial times. The region was originally settled by immigrants from Scotland; hence, the major community in the region, at the mouth of the rivers, is known as New Scotland. The entire region falls within two county governments (Queen Mary and Horatio Counties). The settlers were primarily farmers, converting as much of the land as possible to cropland and eventually to pine plantations. As with most southern farming communities, PDQ originally depended on an enslaved African American population to sustain major farms and plantations.

The years after World War II saw many changes in the PDQ region. A large federal landholding was created in the 1930s, accumulated from abandoned farmlands. After the war, these lands became known as Camp Fraser, a U.S. Army base that today covers approximately 100,000 acres. Camp Fraser became a training center for soldiers in the Korean War, and it grew rapidly in use under the threat of the Cold War. Today, Camp Fraser continues to be one of the most intensively used military training centers in the nation, serving all branches of the military. About 2000 people are assigned to Camp Fraser permanently, and nearly 5000 trainees are present at any one time. General James Aberdeen is the commander of Camp Fraser. General Jim, as he is known locally, sought this assignment (after a distinguished overseas career) because he grew up nearby and has a strong affinity for the region. He has dedicated Camp Fraser to the PDQ Revival idea, to the extent possible. Here is his mission statement for the camp:

PDQ and Camp Fraser: A Mission Statement

BY GENERAL JAMES ABERDEEN, CAMP COMMANDANT

No institution knows better than the military that change is inevitable and must be addressed proactively if we are to safeguard our future. Therefore, I am pleased to commit the people and resources of Camp Fraser to the concept of the PDQ Revival. This initiative will work to continue and enhance the strength of our region and our nation—strength in defense, strength in economy, and strength in our great natural resources.

By law and by reason, Camp Fraser has two missions. First, we are obligated to provide the best training for our military forces so they will be ready to defend our nation in times of crisis. For generations, our country has looked to Camp Fraser as the leader in both training itself and the development of training technology. We will continue this mission without compromise. To do anything less would be a betrayal to the American people.

Second, and no less important, we realize that Camp Fraser is the home for a very special natural resource—the ecosystems and wildlife of our region. We have learned over recent years that Camp Fraser is fortunate to include unique habitats and animals, primarily the longleaf pine savanna and the red-cockaded woodpecker. This area also provides some of the nation's best fishing and hunting, a great tradition and a defining aspect of the American personality.

The experiences of the past half-century—toxic pollution, the loss of valuable species, increasing costs of waste disposal, both in this country and in the ill-fated communist nations—have shown us that our security as a nation is just as dependent on a healthy environment as it is on a strong military. Therefore, we have raised our mission of ecosystem sustainability to an equal status with that of military training. We will do everything we can to make Camp Fraser the standard for resource sustainability.

We also realize that at Camp Fraser we cannot do this alone. Although Camp Fraser is an important part of the PDQ ecosystem, we know that its sustainability also depends on how the rest of the region is treated. Military leaders know that a battle plan drawn on a map bears little relation to the actual battle; nature is the same—it pays no attention to the boundaries humans draw on a map.

As citizens of the PDQ region, as well as stewards of Camp Fraser, we want the best over the long term for this place that we love and depend on. Therefore, we join willingly with others in PDQ to make our region a sustainable ecosystem for humans and nature indefinitely. We encourage other community members to join with us in this effort, and we pledge our resources—people, talent, equipment, and funds—to help make the PDQ Revival a success, both on and off Camp Fraser.

General Jim clearly is progressive, but he rose to his command level because he knows that excelling in military training is his first job. Many people have heard him say, when the cameras and tape recorders were not running, "I understand that we manage for two shades of green these days—the olive drab of military operations and the 'green' of the environment. But one shade always comes first. Do I make my point?"

The town of New Scotland is the major urban area in the PDQ region. It grew early as a port for shipping agricultural crops. The commercial activity helped create a substantial riverfront area where merchants bought and sold, loaded and unloaded goods. Along with this growth came an accompanying service area. After the Civil War, a substantial African American community developed adjacent to the commercial establishments. Together these areas are known today as the Riverfront Old Market. Eventually, however, when ships became larger, Paumaussee Sound was abandoned as a major commercial port. Today it remains important for recreational boating, some commercial fishing, and tour boats (Figure 1.20).

After shipping died away, New Scotland became a rail terminus for transporting cotton, logs, and farm crops. After World War II, however, the expansion of Camp Fraser converted New Scotland to a thriving service center. The completion of the Intracoastal Waterway in the 1950s brought renewed boating interest to the region; today, Paumaussee Sound is a major nexus on the Intracoastal Waterway because boaters can enter or leave the waterway through the sound. Recreational growth has also continued in the area, with the development of several barrier islands as beach resorts as well as extensive golf courses.

Although the area does not rival places like

Figure 1.20. *The economy of the PDQ region has added fishing and nature-based tourism to the traditional farming and forestry industries. (Photo by Larry A. Nielsen.)*

Hilton Head or Myrtle Beach, some developers anticipate such a future. Today, the permanent population of New Scotland is 50,000, and it is growing at a steady pace. New Scotland's business and community leaders understand that their present success rests on the continued presence of Camp Fraser, but they also understand that tourism provides economic diversity and a long-term expanding opportunity. They want both.

Harold Smith, New Scotland's mayor and the son of a beachfront developer (Smith Island is named for his family), knows that the strength of the economy depends on diversifying the base. He was the driving force for New Scotland being awarded one of the state's first "Main Street" grants, whereby major investments were made in the Riverfront Old Market, both preserving its 1900s-era features and building it into a thriving artists' colony. He created the PDQ Tourism Council, which develops and underwrites various activities, and he hand-picked the Executive Director, Joe Danley, who was responsible for tourism promotion at St. Augustine, Florida. Recently the council developed an annual decoy and wildlife print show and an independent southern filmmakers' festival. Mayor Smith, who is an African American, has also worked hard to preserve the African American heritage in New Scotland and the PDQ region by designating the riverfront area as a "black historical region." He also participates in a tri-state governors' initiative to create a system for telling and preserving African American history.

Several other smaller communities lie within the PDQ region, none of which is larger than a few thousand residents. Together, the population of the region totals about 90,000 permanent residents.

PDQ LANDS AND LAND USES

A variety of land uses pervade the PDQ region, as demonstrated by the following.

FARMING. About half the lands in PDQ are agricultural, owned in small tracts by individual farmers and families. Crops vary, with a mixture of corn, soybeans, peanuts, tobacco, cotton, and other field vegetables. Until recently, very little animal agriculture was conducted. Within the past decade, however, several intensive hog production facilities have been located in the region. Each of the facilities is owned by a private investment group, but all are contractually tied to a national pork-processing company, Wisdom Meats, that buys the entire annual production of the facilities. The hog facilities are all located along the Queen River. Each has advanced treatment facilities, including a series of retention ponds; the final stage, however, is the discharge of the ponded water to the Queen River. Agricultural interests in the region are represented by the Farm Bureau, led by Wilbur Boyd, a fourth-generation farmer and major investor in the hog production facilities.

FORESTRY. Another major land use is forestland. As with farmland, much of the land is owned by individuals, who generally plant loblolly pine. However, about one-third of the private small-tract forestland is not managed, being left in longleaf pine or other natural forest types by the owners. Private forestland owners and farmers are served by the extension service, which has a regional center in New Scotland, staffed by more than a dozen agricultural, forest, environmental, home, and 4-H agents.

Several major paper firms own large tracts of land. Total holdings per company are in the range of 50,000–200,000 acres, and collectively these holdings comprise 20% of the land in the PDQ region. These commercial woodlands are almost all in loblolly pine plantations, managed on short rotations as prescribed by growth and yield models developed by researchers at the state university. Consequently, the companies invest heavily in genetic improvement (including biotechnology as well as traditional hybridization), fertilization, site preparation, and intermediate thinning. The industry is rapidly consolidating, so that the companies are always in flux through mergers and sales of lands.

The large companies have all joined the industry-sponsored Sustainable Forestry Initiative, which aims to see forests used for extractive purposes, but with minimal impact and complete regeneration. At the state level, they have created a program in sustainable forestry, led by Barbara Ladd, a professional forester with a previous career in public relations for one of the paper companies. Barbara Ladd often writes opinion pieces for the newspapers, including national outlets; a recent piece included the following description of the industry's view:

> Without question, private forestland is one of our region's great successes. We employ over 20,000 people statewide, and our companies have invested hundreds of millions in our new Dee River Paper Mill (with the best environmental technology!) based on the current and future availability of a reliable supply of pine and hardwood pulp. The lands on which we grow trees provide habitats for many wild species and free access for our region's hunters. We have identified uniquely valuable sites on our lands for biodiversity, and we are protecting those. We're also putting out less phosphorus pollution than 5 years ago, and we're producing more paper. That's good for the environment and good for jobs, and we'd like to expand more in the future.
>
> We strongly believe that the investments we've made are indicative of what industry wants to do when provided with an environment in which we can use our capabilities to do what is best for our land, our workers, our communities, and our shareholders. We do not need government overregulation. We only need clarity about what informed and well-meaning people expect us to do. That is why we continue to sponsor community roundtables to discuss our work and your interests.

The John Muir National Wildlife Refuge. The Muir Refuge is a series of dispersed holdings of the U.S. Fish and Wildlife Service that extend for 87 miles along the coastal region (Figure 1.21). The refuge is approximately centered around New

Figure 1.21. *The John Muir National Wildlife Refuge includes wetland sites like this one, along with upland, riparian sites. (Photo by Larry A. Nielsen.)*

Scotland, where its offices and an educational interpretative center are located. Named after John Muir because his historic walk during the 1870s from Indiana to Florida passed through this area, the Muir Refuge was created after World War II primarily for waterfowl and resident shorebirds. Consequently, most of the sites are on barrier islands along the coast. In more recent decades, however, several other tracts have been added some miles inland. A large tract (about 12,000 acres) borders Camp Fraser, and a smaller tract (4000 acres) lies on the western edge of New Scotland.

Under recent U.S. Fish and Wildlife Service ecosystem management and planning initiatives, Muir Refuge is developing plans to expand its programming to embrace biodiversity and more diverse outdoor recreation activities compatible with the refuge's goals. The refuge manager, Jolene Chan, is working closely with other federal and state agency personnel to make sure that the refuge is part of any future plans for wider-scale management. Jolene has been appointed by the Director of the U.S. Fish and Wildlife Service to a 3-year task force concerning the implementation of community-based management; consequently, the Muir Refuge will be prominent in the publicity on this subject nationally.

THE FOX SWAMP WILDLIFE MANAGEMENT AREA. The Fox Swamp WMA is a 20,000-acre wildlife area, acquired by the state in 1980. It is mostly forested, with a mixture of bottomland hardwoods, longleaf pine stands, and some remnant loblolly pine plantations. The state manages mostly for upland game, primarily wild turkeys that were reintroduced in the 1980s from West Virginia populations. The WMA also contains Fox Swamp, the major source of the Queen River. The state has named Fox Swamp, which occupies about 25% of the WMA, as a Natural History Preserve, which means that hunting, logging, and other activities are prohibited, except as needed to control pest species. Outside the Fox Swamp Preserve, the WMA is available for multiple uses, including logging, hunting, trapping, and fishing.

A similar swamp, owned by a paper company, forms the headwaters of the Dee River. Several smaller swamps are located along the Paumaussee River corridor, draining into the river.

HIGH TYMES. Sonny Tymes owns a 20,000-acre tract of land, which he calls High Tymes. It adjoins Camp Fraser along the Paumaussee River and extends south to the banks of the Queen River. The land was formerly a working plantation, and Sonny has restored the plantation house to its original state (along with modern conveniences, of course). The tract has a mixture of land types and uses, including active pine plantations, active farms (which Sonny leases to local farmers), and native longleaf and hardwood stands.

Sonny has donated the land in a revocable trust to the Low Country Land Conservancy (LC^2), as the cornerstone of LC^2's attempts to retain the historical patterns of land use and natural habitats in rapidly developing coastal communities. LC^2 is led by one of Sonny's college fraternity brothers, Andy Crawford. Sonny's trust is revocable because it has a clause that allows him to assess the progress, both strategic and implemented, of cultural and land conservation in the region in 10 years and withdraw his land if he is not satisfied. If he fails to withdraw the land during an established time interval, then the trust donates the land directly to LC^2 and the trust terminates. Sonny's intention, obviously, is for his gift to stimulate others to do the same, through any of several mechanisms, from land donations to conservation easements. Sonny has also retained ownership of a small amount of the land and the plantation home.

GOLF COURSE AND CONDOMINIUM DEVELOPMENTS. The PDQ region has been developing for recreation at a fast pace for the past 20 years. Several barrier islands are completely developed, with combinations of individual beach houses, low-rise motels, and high-rise condominiums. Several golf courses have been in the area for a long time, but five new courses have been built in the last 15 years along the I-7 corridor. In comparison to the old courses, the new ones are mammoth, typically including condominium development along the fairways, hotel and conference facilities, and associated amenities. The offices of real estate agents, banks, and builders are filled with charts showing planned developments, including three new golf course complexes. The one most likely to begin construction in the near future is scheduled for land just outside New Scotland, adjacent to the Muir Refuge and with five holes running along the Dee River. The golf complex will include three courses (designed by three of the biggest names in golf), and one will be a signature course with the promise of becoming "The Pebble Beach of the East."

The golf course developers (HighMark, Inc.) are interested in having their courses certified as ecologically sustainable by a national conservation group. Although they believe this is the right thing to do, they also know it would be a great marketing tool and would help convince the local community that they are caring partners in the landscape. Their project manager, Carol Slope, is perfect for implementing this idea; she has an undergraduate degree in wildlife management and a graduate degree in tourism management, both from the state university.

NATURAL RESOURCES AND ISSUES

The PDQ region has attracted attention because it is rich in renewable natural resources. The topics that follow represent some special aspects of the region, but other components are just as important.

Figure 1.22. *The PDQ region is famous for its hunting and fishing resources, including fishing for largemouth bass. (Photo courtesy of Dr. Richard Noble.)*

These include abundant fish (Figure 1.22) and game of various kinds (such as large populations of white-tailed deer and a spreading population of reintroduced wild turkey), major spring and fall flights of migratory waterfowl, and diverse birdlife that attracts ornithologists from around the country. In fact, many people refer to PDQ as meaning "ponds, ducks—and quiet!"

LONGLEAF PINE SAVANNAS. The dominant upland ecosystem is the longleaf pine savanna. Longleaf pine formerly occupied more than 60 million acres across the southeastern U.S. Today, only about 3 million acres remain. Within PDQ, scattered tracts cover about 100,000 acres, most of which are in Camp Fraser (about 50,000 acres), with other large stands in the Muir Refuge, in High Tymes, and on private forestlands scheduled for development. Major floristic components are an overstory of longleaf pine and an understory of wiregrass. Intermediate vegetation is sparse. The ecosystem is a product of fire; frequent cool fires suppress hardwood trees and shrubby vegetation, opening the middle canopy and promoting the growth of diverse annual and perennial plants.

The longleaf pine savanna is recognized as one of the most diverse floristic ecosystem types in the temperate world. Studies have shown that more than 100 vertebrates are associated with the longleaf pine habitat type, and many are specifically tied to the floral community of ground plants that occur in conjunction with longleaf pine overstory. Longleaf pine is also a highly desirable species for the wood products industry, producing high-quality and beautiful wood for finish work and furniture.

Two major management issues affect the viability of longleaf pine savannas. First, the control of wildfires has caused the succession of pine savannas to other stages, including dense understories of oaks and shrubs. New strategies for controlled burning on a regular basis are now being tried as a way to mimic natural disturbance patterns. Second, genetic diversity of longleaf pines in the PDQ region is among the lowest throughout the species' range (east Texas to the Atlantic). With the isolation of patches of longleaf pines in recent years, scientists fear that individual stands of longleaf pine will become increasingly monomorphic. With anticipated changes in global climate, considerable concern has been expressed that longleaf may become more susceptible to insect and disease attacks. Recent proposals have suggested mixing seed sources from throughout the range for planting in areas scheduled for longleaf pine restoration.

Camp Fraser, used for military training, is actually an inadvertent contributor to the continuation of longleaf pine ecosystems. Military training often ignites small, cool fires during artillery trials; these fires mimic natural fires that sustain longleaf pine trees. Natural resource managers at Camp Fraser are eager to see how they can work with the military trainers to build the uncertainty of training fires into a positive aspect of their ecological management plans.

POCOSINS AND PINE SALAMANDERS. Throughout the PDQ area, isolated habitats of upland wetlands occur. These habitats are called pocosins, a Native American word meaning "swamp on a hill." Pocosins are typically poorly drained and therefore tend to accumulate water that becomes acidic, developing peatlike soils. The soils also contain many underground channels and chambers, making them ideal habitats for reptiles and amphibians. They contain unusual assemblages of moisture-loving, acid-adapted plants, characterized by hollies and bayberries. Most pocosins are small in extent, usually no larger than 100 acres; they occur throughout the region, but are most abundant along the Dee and Queen Rivers. Most occur on private lands, and, through time, many have been cleared and drained because local people consider them the sources of snakes, salamanders, biting insects, and other "undesirable" organisms.

One organism of growing interest is the pine salamander (Figure 1.23). It is improperly named "pine" because it depends on the environmental conditions in pocosins. The name probably arose from the willingness of the pine salamander to make migratory journeys away from natal pocosins during cool wet weather in the fall, through

Figure 1.23. *The pine salamander is inappropriately named, because it spends most of its time in moist lowlands. (Photo by C. Kenneth Dodd, Jr.)*

ground litter and along moist depressions in the surrounding pine forest; this is where and when pine salamanders are most obvious.

The pine salamander is a relatively large amphibian, reaching about 8 inches at full maturity. Reproduction is in the spring, when eggs are laid in burrows; there is no truly aquatic phase. Pine salamanders mature at the end of their first summer and move away from natal sites in the fall. Evidence indicates that reproduction may occur in the fall, at dispersed sites throughout their habitat. Pine salamander populations have been declining from habitat loss and the increasing isolation of smaller and more distant pocosins. The pine salamander is an example of a species that is not legally protected yet (other than by regular state laws) and might need protection in another decade or so, but whose need for protection could be avoided by good management now.

THE AMPHIBIAN-REPTILE COMMUNITY. The pine salamander is only one example of a species that is tied to the native vegetative habitats of the PDQ region. A broader relationship ties a large number of amphibians and reptiles to the longleaf pine savanna. Extensive surveys of the herpetofauna have shown that among 290 species of native amphibians and reptiles found in the southeastern U.S., more than half (about 170) are found within the longleaf pine ecosystem. Many of these species are found nowhere else; amphibians that require temporary ponds for reproduction are particularly tied to longleaf pine habitats.

A large proportion of species are of special concern (endangered, threatened, rare, or declining), including the well-known gopher tortoise, whose burrows are home to more than 200 invertebrate and 65 vertebrate species. Studies of the status of these communities show that the diversity and individual species are generally stable where land conversions are not made and where forestry practices are conducted so as to protect the ground vegetation, surface organic layers, and structure of the near-surface soil layers. However, loss of habitat, fragmentation of habitat, and extensive disturbance of the surface all impair community diversity.

Extended droughts also have an impact on the amphibian-reptile community, especially when human activities exacerbate the loss of soil and surface moisture or compact the soil. Some studies also show circumstantial relationships of amphibian–reptile losses in areas where imported red fire ants are prevalent.

Figure 1.24. *The red-cockaded woodpecker has been the subject of intense research and management attention over the past two decades. (Photo by Phillip Doerr.)*

THE RED-COCKADED WOODPECKER. The symbol of the pine savanna is the red-cockaded woodpecker, which lives in large, living longleaf pines (Figure 1.24). Camp Fraser holds about 50% of all known red-cockaded woodpecker clusters on public lands in the state. The complex biology and history of red-cockaded woodpeckers are widely known throughout the profession and PDQ citizenry. The RCW (the bird has become such a commonly known phenomenon that it has its own acronym!) excavates cavity nests in living longleaf and related pines, generally older than 60 years. The breeding pair and some of the male offspring from the previous years remain at the nest site, forming a family unit called a group. Birds often build cavities near each other if suitable habitat is available, with up to 20 cavities together composing a cluster. The average cluster occurs on about 10 acres. The birds also need sufficient foraging habitat around the cluster, consisting of pines larger than 10 inches in diameter with an open understory; about 80 acres typically compose a foraging habitat. Of course, territory size depends on the number of birds in the cluster and the quality of the habitat.

Much attention has been focused on RCW conservation, as the bird is a federally endangered species. Because the species does not avoid humans, recovery efforts can proceed within human-dominated landscapes. Recently, state and federal agencies, forest-owning paper companies, other private landowners, and conservation groups have been working together to find solutions. Within the PDQ area, all groups have been discussing various ways to increase RCW habitat on protected lands, while making private landowners able to use their lands for traditional purposes (i.e., farming and forestry) and for development. The groups are considering large-scale, multiownership habitat conservation plans, individual conservation easements, and mitigation banking (where improvements on one land area can be used to offset losses on other areas). Through the PDQ Revival, the improvement of RCW conservation seems a strong possibility.

A nagging problem, however, is how to frame the conservation strategy. One approach is to focus on existing RCW clusters, protecting and improving the individual habitats so these populations can achieve increased birth rates and decreased death rates. A second approach is to focus on the landscape, attempting to increase the amount and quality of RCW habitat, including the creation of corridors among current clusters and encouraging old-growth management practices on various lands. A third approach is to focus on ecological processes that enhance RCW habitat; this approach favors the development of fire management practices that mimic natural fire patterns with the expectation that natural fire regimes will ensure the presence of ample RCW habitat. Conservation plans have been hampered by the failure of various participants to choose among these approaches.

The Coastal Fox. The coastal fox is a subspecies of the red fox that prefers to live and forage in saline and freshwater wetlands. Coastal foxes were common throughout the region until the 1920s, when farmers began killing them because of suspected predation on farm poultry and their high likelihood of carrying rabies (locals called them swamp dogs). A known population, growing slowly in numbers, lives in Fox Swamp, and occasional sightings of individual animals (sometimes pairs) are reported from other wetlands. The coastal fox is more secretive than most other wild canines, generally moving and foraging at night and avoiding open and populated areas.

The Gopher Tortoise. The gopher tortoise is a long-lived, slow-maturing terrapin that lives throughout the upland pine savanna region (it also lives in pine plantations very nicely). The tortoise lives in burrows that it builds in sandy, well-drained soils; the burrows may be inhabited for many years and are often 15–30 feet long. Adult tortoises tend to remain in a limited area but freely move among two to four burrows located close to one another, mostly during the warmer months. Young tortoises and males are the primary means of longer-distance population dispersal.

Studies of gopher tortoise burrows have shown that a remarkable number of invertebrates—more than 200 species—are found inside the burrows; some of these species are found nowhere else. Gopher tortoises require large areas of intact savannahs and have home ranges of about 4 acres. Young tortoises are heavily preyed on by domestic and feral dogs and cats and coastal foxes, and they are often run over by vehicles.

The Wild Turkey. Wild turkeys have been reintroduced into the PDQ ecosystem to support traditional hunting. Although they were native to the area, as in most southeastern states, wild turkeys almost became extinct in the early 1900s. Introductions began soon after World War II, and they have been highly successful. Both fall and spring hunting seasons now exist throughout the region. In recent years, however, reproductive success has been declining for unknown reasons. A series of warmer-than-average winters and summers is blamed, because of the increase in a whole series of pathogens and parasites.

A recent study of wild turkey genetics throughout the Southeast has shown that the DNA of the turkeys in the PDQ region has little heterozygosity and is closely related to the DNA of turkeys from West Virginia. (Careful analysis of stocking records confirms that the origin of most wild turkeys in this region was West Virginia.) A further study of DNA in the skin and feathers of museum specimens has shown that local birds had very different DNA than that of the West Virginia introductions. Furthermore, the local DNA is quite similar to a small population of turkeys residing in the Great Smoky Mountain National Park.

Wild turkeys are the first priority for a major sporting group in the region, the Back-Forty Forever (BFF, pronounced "beef"). BFF has always been a force in the PDQ region, supporting shooting sports and the rural lifestyle. They have great political muscle as well, through their long-time husband-wife team of lobbyists, Jim and P.J. Oakes. The Oakes are fifth-generation PDQ residents; they farm a small holding near High Tymes, and every major political figure for the past 20 years has stayed on their farm—to hunt, hike, or just relax.

The Shortnose Sturgeon. The shortnose sturgeon is the smallest of three sturgeon species that inhabit Atlantic coastal rivers. The Paumaussee River is an important historical habitat for shortnose sturgeons; catches were always higher in the river and Paumaussee Sound during the early 1900s than in any other river system in the southeastern U.S. Since the 1950s, however, catches and population levels have declined. Now there is a moratorium on catch, as the shortnose is a federally listed endangered species.

In the Paumaussee, the shortnose spawns throughout the river, well past Camp Fraser. Young shortnose presumably live in the river for several years, but they increasingly migrate to the Paumaussee Sound and adjacent coastal areas as they grow. They mature at 4–5 years, at a size of 20–24 inches. Full adult size is generally 36 inches or less, but all current national and state records for shortnose sturgeon come from the PDQ area (the current record is 46 inches and 42 pounds).

The reasons for the declining population are complex. Certainly they include overexploitation for commercial catch in earlier decades and the increasing pollution of the river and sound. Two characteristics of decline have been identified: Fewer adults have returned to spawn in the river through time, and average fecundity per pound of spawning female has been dropping since the 1970s.

PAUMAUSSEE RIVER STONE FISH TRAPS. The Paumaussee River is named for the Paumaussee tribe of Native Americans, who lived in dispersed groups throughout the region. Tribal descendants do not live together, and only a few of the elderly Paumaussees can still tell the stories of the tribal ways. A small group of Paumaussees living in New Scotland has formed to reinvigorate their heritage. One of their interests is the traditional practice of building circular stone fish traps in the lower parts of the Paumaussee River. The traps worked because juvenile shortnose sturgeon moved into them at high tides and were trapped there when the water receded. Remains of the stone traps are scattered throughout the shoreline areas of the Paumaussee River and Sound, but most are not evident to the untrained observer. The stone traps have become of intense interest recently to archeologists as well, either as the origin of similar traps and weirs used throughout the eastern U.S. or as a local adaptation of ideas imported from other regions.

TOXIC WASTES. Camp Fraser's current dump is located in the Dee River floodplain near the camp's eastern border. In earlier decades, this dump received an unknown but undoubtedly "rich" mixture of chemical wastes. Although the dumping of toxic chemicals has been significantly reduced, waste oil and solvents from the camp's motor pool are still discarded at the dump, along with organic wastes. The U.S. Department of Defense has proposed closing this site and opening a new one on the little-used southern area of the base near the Paumaussee River. Until then, the old site will remain in use. The final choice of a site will be the responsibility of General Aberdeen, but he is likely to agree with the Pentagon unless evidence convinces him that there is a better alternative.

The dump is likely to be a serious issue in the future. New Scotland's mayor is worried because the city's drinking water comes from a series of wells west of the town that need to be as far up the watershed as possible to protect from saltwater intrusion. A recent study by a graduate student at a nearby university has identified problems in nesting colonies of the black-billed tern located on Camp Fraser and downstream along the Dee River. The study found low nesting success, a high incidence of crossed bills and other deformities symptomatic of heavy metal contamination, and evidence of neurological and physiological symptoms characteristic of endocrine disruptors in the food chain.

PFISTERIA. Outbreaks of pfisteria-related disease and contamination have raised concerns about the role of intensive animal agriculture and the safety of the Queen and other rivers. Samples of fishes and invertebrates from occasional fish kills in the Queen River have revealed the presence of pfisteria colonies, as well as the lesions that are commonly reported to be associated with pfisteria infection. Widespread flooding in nearby drainages caused by hurricanes and the resulting pfisteria contamination have alerted New Scotland residents to what might be happening in their watershed.

Farmers and representatives of farming interests continue to refute cause-and-effect relationships between intensive animal agriculture and pfisteria, and they assure residents that their control mechanisms are sufficient to protect against storms or other hazards.

AGRICULTURAL HERITAGE PARK. The governor has developed a new concept known as state heritage parks (Figure 1.25). Rather than typical state parks in which the government owns a particular tract of land designed as a state park, the heritage park concept covers a region in which various sites will show the heritage of a specific industry. Local tourism businesses have proposed the PDQ region as an Agricultural Heritage Park. The idea is for various public and private sites (with subsidy) to preserve or develop examples of the agricultural and forestry industries and plantation life that were present in PDQ from colonial times to the present. Opportunities would exist for museums, living history demonstration areas, farms, and fields to highlight rare and native plants and animals associated with farming, as well as educational programming for schools in the PDQ region and statewide.

Figure 1.25. *State heritage parks preserve the rural landscape and culture, an interesting addition for the PDQ region. (Photo by Larry A. Nielsen.)*

AN EXPANDED TRAINING MISSION FOR CAMP FRASER. The state's federal representatives and senators view the defense budget as a great opportunity for the PDQ region. Therefore, they have proposed the expansion of training on Camp Fraser, with the goal of doubling training activities and offering training to foreign nationals who are allies of the U.S. The expanded training would require a tripling of the lands actively used for training, including more firing ranges, more armored vehicle courses, and increased use of advanced weapons systems. Early indications are that the best training grounds would primarily encompass longleaf pine savannas and pocosins, especially within the southern half of Camp Fraser. Until now, the southern part of the camp has been considered too distant from the central facilities to be very useful, and selling the lower portion has been discussed at various times in the past when federal budgets were tight.

Another possibility is the expansion of Camp Fraser's role in the development of training technology. This idea is being pursued by local development interests in New Scotland, who believe that new training technology would attract high-technology businesses to the PDQ region. Working with a planning grant from the governor's Partners for Progress initiative, which attempts to build public-private collaborations, Harold Smith is leading a group to craft a strategy for this development.

Getting a Grip on Ecosystem Management

TO SUCCESSFULLY WORK WITHIN COMPLEX SYSTEMS such as those outlined in the landscape scenarios—and, more important, in thousands of such scenarios in the real world—we must chart a different course into the future than has been done to date. But to effectively map a course into that future, we must clearly understand where we have been, so let's start with a look at the past.

The Evolution of Natural Resource Management Toward Ecosystem Management

Natural resource management in the United States changed a great deal throughout the twentieth century. Resource management was born of fears that valuable natural commodities—fishes, forests, water, game—were being overexploited. Losses of seemingly inexhaustible species, such as the bison and passenger pigeon in the nineteenth century, taught tough lessons about the blatant waste of resources and helped create a new profession that focused on the sustainable uses of these resources. The birth of the Progressive political era at the turn of the twentieth century brought with it a new concern for conservation and a public disgust of waste and overuse.

Aided by President Theodore Roosevelt, who elevated conservation to a high priority in his administration, Gifford Pinchot, the first Chief of the U.S. Forest Service, helped develop a new approach toward natural resources. His goal for management was based on the utilitarian ethic, to produce "the greatest good of the greatest number for the longest time." This so-called **resource conservation ethic** guided natural resource management for much of the first half of the twentieth century. Conservation, to Pinchot, was equivalent to the wise and prudent use of resources. He once stated that "the first great fact about conservation is that it stands for development" and that "the first duty of the human race on the material side is to control the use of the earth and all that therein is." His push for the efficient and fair use of natural resources for both present and future generations eventually led to the multiple-use concept adopted by the U.S. Forest Service and Bureau of Land

Management. Sustainable use and control were at the core of Pinchot's views.

Preceding this perspective, and in stark contrast, was another school of thought called the **romantic-transcendental conservation ethic**, which developed from the writings of Ralph Waldo Emerson, Henry David Thoreau, and John Muir in the late nineteenth century. Their collective view was that nature has uses other than human economic gain and that it in fact has inherent value independent of human use. They saw nature in a quasi-religious sense and believed that the protection of large areas of pristine habitat with no human extraction was justifiable as a celebration of God's creation. These two schools of thought served as the endpoints of a philosophical gradient from pure wilderness protection to a resource-based, utilitarian approach to nature.

A third perspective, known as the **evolutionary-ecological land ethic**, was introduced by Aldo Leopold in the 1930s and 1940s (Figure 2.1). Concomitant with the development of ecology and evolution as scholarly disciplines, this ethic proposed that nature was not simply a collection of parts, some to be used and others discarded based on their usefulness to humanity; nor was it a temple to be worshipped and left untouched. Rather, nature was seen by Leopold as a complicated, interconnected, functional system that is the result of long-term evolutionary change. Something like a fine piece of machinery that is not fully understood, the land mechanism could and should be used by humans, but its essential structure should not be fundamentally altered. This view is made clear in one of his most famous declarations: "To keep every cog and wheel is the first precaution of intelligent tinkering."

Figure 2.1. *Aldo Leopold, who provided some of the first conceptual underpinnings of the evolutionary-ecological land ethic and ecosystem management. (Photo courtesy of the Aldo Leopold Institute.)*

World War II and the post-war development boom saw the industrialization of natural resources and their intensive use. The baby boom, a robust economy, and a building frenzy as Americans moved from farms and cities to the suburbs witnessed heavy use of private and (especially) public lands, extraction of minerals, and acceleration of a recreational push into the great outdoors. The multiple-use paradigm of Pinchot was in its heyday, and it led to many of the resource issues that we deal with even today.

During the 1960s and 1970s, resource management changed substantially, based on new attention to other human interests in the environment besides extraction: recreation, endangered species, cumulative effects, and aesthetics. Stimulated by the writings of Rachel Carson, Jacques-Yves Cousteau, Paul Ehrlich, and others and events such as the degradation of Lake Erie and many instances of egregious toxic pollution, a new awareness of the magnitude of the human footprint spawned an environmental awakening and a concern regarding air and water pollution, habitat loss, and species extinctions.

In the 1980s, the new discipline of conservation biology added scientific rigor and new philosophical dimensions to resource management. New perspectives in ecological science, such as the recognition of the importance of periodic natural disturbances in ecosystems, also influenced natural resource management. Along with changing social and economic conditions, these various currents coalesced during the 1990s into changing perspectives in management agencies, resulting in the subject at hand—ecosystem management.

As this brief discussion suggests, ecosystem management is not a sudden *revolution*, thrust upon us by a signal event, but a slow *evolution*, one that has built upon decades of experience of thousands of individuals in natural resource management, on increasing sophistication and understanding in the ecological sciences, as well as on changing societal priorities. This evolution involves a continual process of learning and improvement so that new ideas and approaches are constantly being tested, evaluated, and rejected or implemented. Indeed, what we teach and learn and understand today should be outdated a decade from now as our collective base of knowledge and experience builds.

A COMPARISON OF TRADITIONAL MANAGEMENT AND ECOSYSTEM MANAGEMENT

This evolution during the last century has produced several contrasts between traditional (i.e., earlier) approaches to natural resource management and the ecosystem approach embraced here (Table 2.1). These are only generalities, but they do illustrate some major trends, both in our changing ecological understanding and in the application of that understanding to management situations.

First, traditional management tended to facilitate natural resource extraction (e.g., timber production, fishery and hunting resources, minerals, agriculture) that is typically the first concern of most societies. Ecosystem management expands these interests to include amenities (e.g., camping, birding, clear skies, clean water, nature appreciation), ecological processes, and biodiversity. The ecosystem approach also emphasizes that ecosystems that are intact and functional are necessary for the production of commodities and amenities over the long term, as well as the outdoor amenities that we seek. This idea—don't kill the goose that lays the golden eggs—has long been recognized in natural resource management, but the ecosystem approach encourages a better understanding of all the factors that contribute to keeping the goose alive, healthy, and productive.

Second, for much of the past century (until the late 1970s), academic ecologists and resource managers primarily subscribed to an equilibrium perspective of the natural world, believing that ecological succession led to climax communities that would remain stable for long periods of time. Disturbances (e.g., fire, floods) were viewed as events that reset the clock, pushing succession back to earlier stages, something to be avoided through proper management. This view has been replaced by one that recognizes the fundamentally dynamic, nonequilibrium nature of the world and that acknowledges natural disturbances as being essential parts of resilient ecosystems. Rather than trying to maintain stable, climax communities, we expect shifting mosaics of communities across the landscape, resilient to disturbances and changing over time. The former equilibrium or "balance of nature" view has been replaced by a "flux of nature" perspective, in which populations vary through

Table 2.1. *General Contrasts Between Traditional Natural Resource Management and Ecosystem Management*

Traditional Management	Ecosystem Management
Emphasis on commodities and natural resource extraction	Emphasis on balance between commodities, amenities, and ecological integrity
Equilibrium perspective; stability; climax communities	Nonequilibrium perspective; dynamics and resiliency; shifting mosaics
Reductionism; site specificity	Holism; contextual view
Predictability and control	Uncertainty and flexibility
Solutions developed by resource management agencies	Solutions developed through discussions among all stakeholders
Confrontation, single-issue polarization; public as adversary	Consensus building; multiple issues, partnerships

time and local extinctions and recolonizations are natural processes.

Third, traditional resource management tended to be more reductionistic and site-specific, solving immediate, local problems. Consequently, often it focused on species (e.g., how do we restore vanishing wood ducks?) within a geographic area that could be readily managed (e.g., a national wildlife refuge or a state forest preserve). This focus was reasonable at the time because resource management grew in response to obvious and first-order issues, such as the Dust Bowl or uncontrolled harvests. Ecosystem management, in contrast, tends to be more holistic and incorporates the larger spatial context. It tries not to focus exclusively on individual species or a given management area, but addresses multiple species and entire ecosystems in its vision and looks beyond political boundaries to consider entire, natural landscapes.

Fourth, traditional management tended to rely on prescriptions and tight control in its approach to natural resources. This approach arose partly from Western society's collective confidence in science and technology's abilities to solve problems and control situations. The result was great faith in our capacity to control ecosystems or species through tried-and-tested techniques taught in fishery, forestry, game, and range management programs. Ecosystem management, on the other hand, recognizes a growing understanding that natural ecosystems come with a huge degree of uncertainty (see Chapter 3) and that human control of systems is not only difficult but illusory (discussed in the next section). Instead, flexibility and an adaptive approach to management is what guides the ecosystem approach (Chapter 4). Doing so successfully requires the involvement of stakeholders (see Chapter 10), as well as various government agencies because of the large scale at which ecosystem management typically occurs.

Fifth, traditional management saw problem solving and decision making as the province of the resource management agencies themselves, disconnected from society at large. Throughout the twentieth century, the growth of professional specialties and expertise in various disciplines created an operating style in the United States in which professionals, from doctors to engineers to resource managers, were given authority to act on our collective behalf. That approach is no longer popular in any field. Consequently, ecosystem management emphasizes reaching solutions and making decisions through broad stakeholder involvement (a feature to be repeatedly discussed in subsequent chapters).

Finally, as a result of the above patterns, traditional management often resulted in confrontation and polarization based on single issues, with various publics viewed by authoritative agencies as adversaries. Ecosystem management, by contrast, relies on consensus building to alleviate confrontation, and ideally it seeks solutions to multiple issues through partnerships and broad stakeholder involvement. Rather than as an adversary to be dealt with, the public is welcomed as a diverse resource and effective problem solver; in short, citizens are partners in the effort.

In the remainder of this chapter we will develop these themes, which will in turn become more evident throughout the book. Solutions to situations you will wrestle with in the following pages will rely on the more holistic, inclusive, flexible ecosystem approach.

EXERCISE 2.1

Talk About It!

Select a high-profile local or national natural resource management issue with which you are familiar (e.g., logging in national forests, an endangered species conflict with land use, or a pollution issue). Are there aspects of management actions that fit the traditional management model? The ecosystem model? Which prove to be more effective? Which ones are easier to accomplish? Which are more accepted by society?

FROM COMMAND AND CONTROL TO ADAPTIVE ECOSYSTEM MANAGEMENT

Early in the exploitation of a natural resource, that resource usually is perceived as superabundant. Indeed, resources often are abundant, such as early in the human settlement of North America five centuries ago. When human population densities are low relative to the available resource base, there is little incentive to use them carefully or to be concerned with their long-term sustainability. People quickly learn the lessons of overexploitation, however, when their efforts to get more from a resource through ingenuity and technological innovation are met by declining yields. For example, as recently as 1954, a book by Daniel and Minot called *The Inexhaustible Sea* expounded on the limitless resources for humans to be found and exploited in the world's oceans. Now, only five decades later, and after continuing commercial fisheries collapses, losses of coral reefs, and expanding areas of biological "dead zones" developing in our oceans, we know how tragic that perspective was.

Although initial exploitation of natural resources is generally easy and seems to be accompanied by a lack of self-restraint, limitations eventually do present themselves. At that point, we must then begin to use the resources sustainably, and active management comes into the picture. This was the case for fisheries in the latter part of the nineteenth century, as nation after nation undertook studies and established agencies to look into the declines in fish yields. Along with these investigations, however, humans also used their collective ingenuity to enlarge the resource in the short term by fishing with steam power in bigger boats, farther from shore, with bigger nets. Through technology, they ended up shortening the time available before they had to become careful stewards of their resources.

EXERCISE 2.2

Think About It!

Can you think of examples of uncontrolled or careless resource consumption in your area and nationally? Think in terms of both aquatic and terrestrial systems, and plants and animals. How could a different management approach have improved the situation and used the resource more sustainably?

The overuse of resources partly results from using technological ingenuity to manipulate nature toward a specific goal (usually enhanced resource extraction), rather than understanding natural limits and using careful stewardship. This approach to problem solving is often called **command and control**, in which the precise control of events and outcomes is desirable and possible (Figure 2.2). *Command* means to have authority or jurisdiction over or to direct resources (including people) in a manner that achieves a desired outcome. *Control* means to manipulate, govern, manage, or regulate, in the sense of observing a given situation, assessing the extent to which it differs from the desired state, and taking actions that will drive the present state toward the desired outcome. Command is only possible if control exists.

Successful command and control rely on a good knowledge of the system (e.g., the physical laws of mechanics and electricity) and a high probability of regulating its behavior (e.g., the ability to control electricity's flow through wires and connections). When a command-and-control approach is applied to problems such as wiring a house, guiding an aircraft to a safe landing, or building a bridge across a river, it is quite effective. In fact,

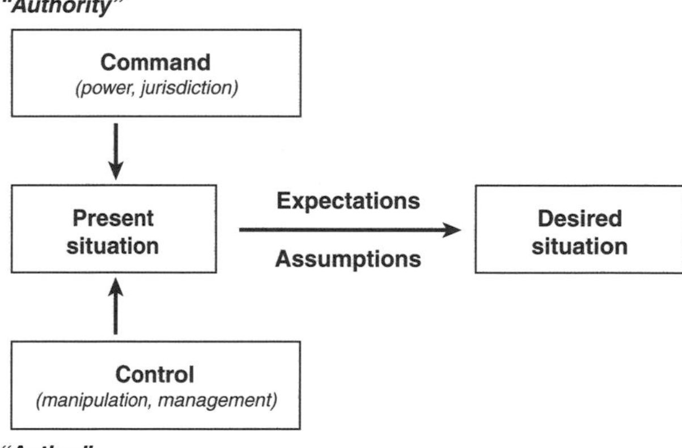

Figure 2.2. *A conceptual model of command and control. A* command *is an authoritative action involving power and jurisdiction.* Control *is an action that involves manipulation and management. Both are used to move from a present situation to a new, desired situation. Several expectations and implicit assumptions are necessary for this movement to be successful.*

we very much need command and control in such circumstances; imagine the mess if electricity decided to jump around outside wires, circuits, and appliances or if the work of electricians was not strictly reliable and regulated! However, when a command-and-control approach is applied to natural resource management—in which the objective is to constrain, control, or change nature—it often fails. Why?

Underlying successful command and control is a certain set of *expectations*: The solution is appropriate, feasible, and will work over relevant spatial and temporal scales. Furthermore, there is a set of implicit *assumptions*: The problem is well-bounded, clearly defined, relatively simple, often responds linearly to manipulation, and is without unforeseen consequences and externalities. We act as though these expectations and assumptions are accurate in the world of physics and engineering, and we're usually correct (correct enough, at least, for most situations; but remember that bridges do fall down, electrical fires do occur, and buildings do collapse occasionally). However, these conditions are never true in natural ecological systems because of their inherent complexities and accompanying uncertainties (discussed in Chapter 3). Here are several examples of ineffective command and control applied to natural systems, to help illustrate the dilemma:

- *Deer populations and predator removal.* On the Kaibab Plateau of northern Arizona, a management program was instituted early in the twentieth century to increase the deer herd and improve hunting by removing natural predators to reduce predation on mule deer. The predators were removed and the deer herd did indeed increase. Then they increased some more. They gradually exploded in number until they destroyed their food base, and the population eventually collapsed through starvation. By failing to recognize that deer would not stop reproducing just because the numbers rose to a level desired by hunters (i.e., by assuming a linear, negative relationship between population size and reproduction and by assuming that hunting would compensate for natural predation), managers failed to appreciate their lack of control over animal behavior.

- *Pacific salmon and hatcheries.* Throughout the Pacific Northwest of North America, various species and stocks of salmon are declining and going extinct for a variety of reasons: degradation of their rivers and streams through clear-cutting and the subsequent siltation of streams; the removal of downed trees and other coarse woody debris (sources of cover) from the streams; overfishing in the marine systems; and the building of huge dams that block spawning migrations of adults and increase the mortality rate of young salmon returning to the ocean. A popular solution to this problem has been a command-and-control, engineering-type approach: Build hatcheries to grow fish and put them back into the rivers. Although this action can be expected to increase fish numbers, it can never restore the system as a whole to its former state, with hundreds of unique stocks returning to their natal streams. In fact, the approach is generally acknowledged to have failed terribly because it further damages declining natural stocks and has hastened their extinctions—and yet it continues. By simplifying river systems to achieve certain benefits and ignoring the underlying causes for decline, managers and policy makers neglected the complexity that underlies salmonid communities, and they "commanded" many salmon stocks to extinction.

- *Controlling agricultural pests.* For the latter half of the twentieth century, farmers controlled crop pests with toxic pesticides. This method works to the extent that most insect pests are killed and crop harvests increase (ignoring for now the possible harmful effects of pesticides in the ecosystems at large or in the human body through consumption). But not all of the pests are killed; some survive the spraying, perhaps through genetic variations that enable them to effectively metabolize the toxins at the doses used. The result is that the next generation of insects is somewhat resistant to the toxin, and heavier doses or a new pesticide must be used. But a few individuals are genetically resistant and persist, and soon there is an "evolutionary arms race." The result is our inadvertent encouragement of and selection for pesticide-resistant genetic strains—evolution in action! By failing to recognize that the reaction of nature to an imposed limit is to evolve beyond it, managers create only short-term solutions. We may well understand this and in fact keep working on newer insecticides; however, unless the underlying strategy changes (such as toward integrated pest management), our attempts to command nature may continue to work in the short term but fail over the long term.

Why do such failures occur? The answer is because the natural systems in which they are practiced violate the implicit assumptions necessary for command and control to work: Natural ecosystems are not well-bounded or clearly defined; they are certainly not simple; responses to manipulation often are not linear, but involve thresholds; and there are many unforeseen consequences and externalities. Such complex systems are not conducive to a command-and-control approach. Thus, the expectations involved are not met, and management is perceived to have failed.

EXERCISE 2.3

Think About It!

Think of more examples of command and control, applied to natural systems, that have had negative consequences. How could management have been approached differently? Why would the alternative be more successful?

THE PATHOLOGY OF NATURAL RESOURCE MANAGEMENT

This command-and-control approach leads to a phenomenon that has been called the *pathology of natural resource management*. The pathology arises from our attempted control of variation in natural systems; such control often initially succeeds but then eventually fails, frequently with undesirable ecological and human consequences. The pathology may be formally stated as follows: *When the range of natural variation in an ecosystem is controlled, the system loses resilience when faced with new stressors.* In other words, when humans attempt to control the general behavior, degree of fluctuations, or range of extreme conditions of natural systems, those systems tend to become less resilient in the face of further perturbations, of either a natural or a human-induced nature (Figure 2.3). By "less resilient" we mean a system is less likely to retain its basic character after a perturbation and may change to a fundamentally new state. For example, a forest may transform to a shrubland or a grassland to a desert if it is not resilient when subjected to natural stresses such as fire or climate change.

Human civilization undeniably has prospered by reducing the variability of nature throughout our history. The trick lies in deciding when the reduction is wise and when it is not and in recognizing long-term costs along with the short-term gain. For example, we build smooth roads to reduce the variability in terrain so we can travel more quickly and more comfortably. We plant

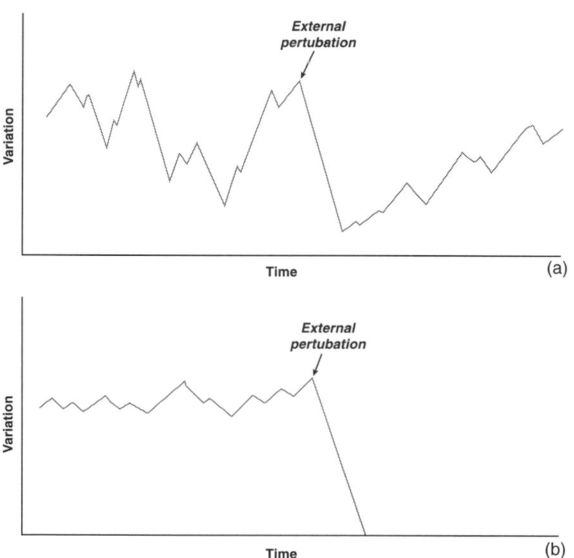

Figure 2.3. *The loss of resilience in an ecosystem due to the control of natural variation. (a) The behavior of an uncontrolled ecosystem fluctuates through a wide range of conditions. A perturbation to the system has an immediate effect, but it is "absorbed," and the system continues on as before. (b) In an ecosystem controlled by humans, behavior is tightly constrained within narrow limits. When a perturbation strikes, the system may change fundamentally to a new state and not be able to return to its former condition.*

crops of domesticated plants so that our harvest of food is more reliable than simply gathering what grows wild. We dam rivers to predictably control flooding, our water supply, and electrical generation. But this does not mean we should build roads everywhere, plant crops wherever we feel the urge, or dam every free-flowing river. Controlling variation in nature can have great negative consequences as well as benefits, and it is the consequences that often are ignored.

As an example of the pathology of natural resource management, consider that many ecosystems have normally and naturally burned on some periodic basis over millennia. Longleaf pine ecosystems in the southeastern United States have burned every few summers as the result of lightning strikes; in fact, their extraordinarily high diversity of herbaceous plants is largely dependent on these fires. Chaparral of the American Southwest has also burned periodically. However, in our desire to control, fire suppression in fire-prone ecosystems has been a common policy in the U.S. and elsewhere over the last century. Fires were put out when started because they were seen as destructive rather than rejuvenating events. We even had a national public relations campaign to implement the policy (the U.S. Forest Service's "Smokey Bear"). Fire suppression, of course, is an attempt to control ecosystems by reducing their variation, that is, by not allowing them to burn when and where they naturally would.

It is well known that fire suppression in fire-prone ecosystems results in a buildup of fuel loads and heavy undergrowth. These fuels, which normally would burn periodically in "cool" ground fires (say, every 5 years or so), now accumulate in the forests over time. Eventually the system escapes human control and fire breaks out. This results in a much hotter and more catastrophic fire than would normally occur, a hot crown fire that kills the trees and changes the character of that ecosystem. The forest was in fact *made less resilient* by the attempted control of fire by humans. In seeking stability through control, we sow the seeds of catastrophe in the process.

The pathology actually is more complex and has three distinct components:

1. A human-imposed external control.
2. Institutional changes to focus on the control.
3. Increased economic dependence on control and overcapitalization.

The pathology is initiated by some external control of the ecological system by humans, leading to a reduction in natural variation, as we have discussed. This initial control often is quite successful; in the example here, fire is perhaps controlled for decades. Control is followed by changes in the relevant institutions so that they focus on efficient and effective control. As the responsible institutions focus on control (in this case, the control of fire), society increasingly expects the institution to be effective in that control. Consequently, we cease to understand, and become distanced from, normal behavior of the natural system. Thus, fires are effectively controlled to the exclusion of understanding the role of fire in the natural ecosystem, and the effectiveness of the institution is measured by its success at control.

The third component of the pathology is increased economic dependence on the controlled system and overcapitalization within the area. People come to depend on fire control, for example, and build homes and businesses in the controlled region. This capitalization within the ecosystem places even heavier demands on and expectations for the controlling agent to be effective, thus feeding back toward even greater control. Once human communities are built in a fire-prone region, then fire control is absolutely critical. Of course, when a fire eventually breaks out, it can result in catastrophic ecological and economic losses because the combined ecological-human system is now completely unresilient to fire, made so by the original attempt at control. A bad year for fires can result in the replacement of an agency head or even changes in next year's appropriation!

The composite results of this pathology are less resilient and more brittle ecosystems, institutions more intently focused on control, and increasingly dependent economic interests attempting to maintain short-term and mid-term success (i.e., control). Continued success in this command-and-control approach depends on the ability to effectively con-

EXERCISE 2.4

Talk About It!

As another example, discuss the pathology relative to controlling river flows through damming, levy building, and dredging. How does attempted control result in a pathology? What are the three components of this particular example?

trol naturally variable systems over the long term. However, at some point natural variation is likely to exceed human abilities for control, and surprises will occur. The result is that, by pursuing command and control of natural resources, we may experience short-term success, but *we build long-term failure into the system!* Another perspective of the pathology—by C.S. Holling, who developed this idea—is presented in Box 2.1.

THE NEED FOR RESILIENCE

Is there an alternative to continuing on this command-and-control path? Would it be wiser in the long term (though admittedly more difficult in the short term) to avoid widespread attempts at controlling natural systems and instead understand their variation and behavior and utilize them accordingly? For example, rather than building communities in floodplains (with billions of dollars of commercial and residential development subject to destruction), might it be better to understand that these are highly fertile and productive systems *because of* their flooding and that perhaps they would be more appropriately used as wildlife refuges and for agriculture? If appropriate farming was conducted in these floodplains (rather than building cities), then less chemical fertilizer would be needed, and any periodic crop destruction by

BOX 2.1

A Perspective on the Pathology, from C.S. Holling (1995)

So this is the puzzle: The very success in managing a target variable for sustained production of food or fiber apparently leads inevitably to an ultimate pathology of less resilient and more vulnerable ecosystems, more rigid and unresponsive management agencies, and more dependent societies. This seems to define the conditions for gridlock and irretrievable resource collapse. . . . Moreover, those pathologies occur not only in examples of renewable resource management but also in examples of regulation of toxic materials or in examples of narrow implementation of protection for endangered species.

Crisis, conflict, and gridlock emerge whenever the problem and the response have the following characteristics:

- A single target and piecemeal policy.
- A single scale of focus, typically on the short term and the local.
- No realization that all policies are experimental.
- Rigid management with no priority to design interventions as ways to test hypotheses underlying policies.

The pathology continues and deepens when the reaction to conflict is to demand more data or more precision in data (e.g., for defense of lawsuits) and more certainty and more control of information and individuals.

The pathology is broken when the issue is seen as a strategic one of adaptive policy management, of science at the appropriate scales, and of understanding human behavior, not a procedural one of institutional control. This requires:

- Integrated policies, not piecemeal ones.
- Flexible, adaptive policies, not rigid, locked-in ones.
- Management and planning for learning, not simply for economic or social product.
- Monitoring designed as a part of active interventions to achieve understanding and to identify remedial response, not monitoring for monitoring's sake.
- Investments in eclectic science, not just in controlled science.
- Citizen involvement and partnerships to build "civic science" (Lee, 1993), not public information programs to inform passively.

floods might be more than compensated for by higher yields and lower costs at other times. (This is, after all, how the great civilization of Egypt developed along the Nile River.) Likewise, not building in fire-prone forests, and allowing them to burn at regular intervals, would benefit biodiversity while avoiding the astronomical costs of fire prevention, followed by the destruction of residential and commercial properties when control eventually fails.

As a result of the above assessment, what might be called a "golden rule" for natural resource management seems appropriate: *Natural resource management should strive to identify and retain critical types and ranges of natural variation in ecosystems, while satisfying the combined needs of the ecological, socioeconomic, and institutional systems.* Maintaining such variation would result in higher levels of resilience in ecosystems, permitting them to better absorb stresses. A major challenge here is determining and defining the types and ranges of natural variation in various ecosystems. It is far from an easy task and often requires historical data from or knowledge about a system, as well as long-term monitoring; this is an area of active research. It also begs the question of the proper historical period to emulate. Is it immediately post-Pleistocene, when humans had little influence on many ecosystems? Is it (in the U.S.) pre-Columbian, when Europeans had not yet appeared? Is it preindustrial, when environmental alterations were confined to human and animal power? These questions have not been answered, and the answers often come down to pragmatism.

C.S. Holling (in Lee, 1993) observed that "environmental quality is not achieved by eliminating change. The goal, instead, is resilience in the face of surprise." Creativity and innovation will be needed to develop solutions to our resource problems that permit such resilience. And this is partly what ecosystem management is all about: finding creative new ways to solve problems over the long haul and reducing our dependence on short- and medium-term command-and-control approaches that might do more harm than good. Maintaining high ecosystem resilience, rather than continued manipulations in the ignorance of ecosystem function, needs to be a central goal of natural resource management.

EXERCISE 2.5

Talk About It!

Try applying the golden rule to a management activity with which you are familiar (e.g., the use of pesticides to control mosquitoes, insecticides to control tree disease outbreaks in forest management, or the means and rates by which riverine flows are regulated below major dams). How might the management actions change as a result? What would be the ecological, economic, social, and institutional repercussions to such a change? Is the change realistic? What might it take to actually implement such a change?

Is there an alternative to command and control? How can we implement better, more responsive and flexible management actions that do not attempt to impose tight, predictive control on highly variable systems? We suggest that the answer is adaptive management, a topic to be discussed at length in Chapter 4 and employed throughout this book. For now, we propose that effective natural resource management that promotes long-term system viability should seek to *understand* and *accommodate* natural ranges of variation in ecosystems. It must work within constraints of that variation to the extent practicable for long-term success and sustainability, rather than pursue short-term gain through command and control. The ubiquitous presence of complexity, uncertainty, and surprise in nature encourages us to take an adaptive management approach.

A Model of Ecosystem Management

Most contemporary approaches to ecosystem management call for integrating scientifically based ecological understanding and socioeconomic perspectives and values. The Keystone National Policy Dialogue on Ecosystem Management (1996) and Yaffee et al. (1996) reviewed more than 100 examples of ecosystem management in which diverse stakeholders achieved at least some level of agreement about mutual goals, meeting both societal and ecological needs. As a first step in reconciling human interests and needs and the desire to protect and restore natural resources, we propose two overarching objectives of ecosystem management:

1. Retain, restore, and sustain ecosystem integrity.
2. Make the places we live, work, and play noticeably better today and in the future.

The first objective requires science to analyze and understand ecological issues, along with the consequences of proposed actions as diverse as economic development, commodity extraction, ecological restoration, or the creation of an ecological preserve. Objectivity requires recognizing uncertainty: We need to recognize the limits of our current science, the uncertainty inherent in ecological systems, and their nonequilibrium dynamics, potential resilience, and natural limits.

The second objective acknowledges the multiple interests and values of society. It also requires that we define what we mean by "better." We contend that "better" must encompass the long-term consequences and reciprocal relationships of decisions affecting both natural and human communities.

Failure to uphold the first objective in our decisions and actions invites disaster within those very ecological systems that are needed to sustain all life, including humans. Failure to consider the second objective will induce legal and political gridlock, and social backlash. Without the active support of society, the implementation of otherwise good ecosystem decisions is impossible, and we will fail to protect that which we need to sustain life.

Conflict between these two objectives is common. Resolution requires understanding the perspectives of ecologists, diverse interest groups, and government agencies. These perspectives represent three very different contexts from which to approach ecosystem management: ecological, socioeconomic, and institutional (Figure 2.4). Within

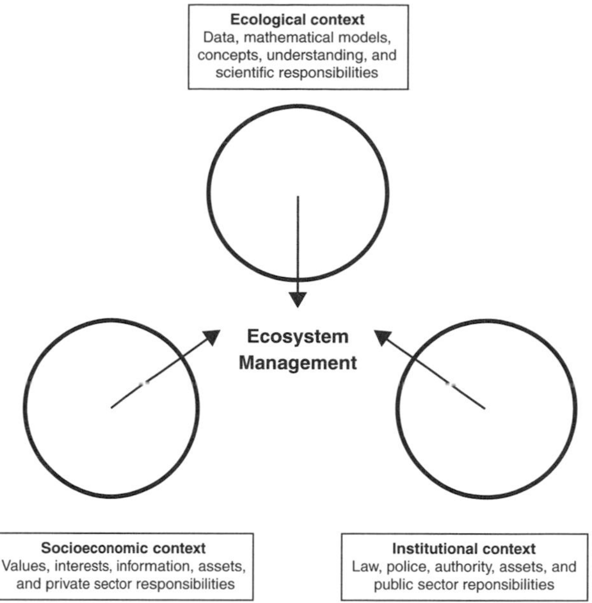

Figure 2.4. *The three contexts of ecosystem management. Interests from the ecological, socioeconomic, and institutional circles all have perspectives relevant to ecosystem management, and each has a limited understanding of the entire process.*

each context, people have different viewpoints, limited possession (and sometimes narrow understanding) of facts, and different capacities to pursue or prohibit actions that affect an ecosystem. The single best decision or course of action as seen from any one context may prove unworkable, politically impossible, or ecologically not sustainable. A blending of perspectives, knowledge bases, and realities from these three contexts represents the fundamental challenge of ecosystem management and is new territory for many natural resource managers.

A collaborative approach, which we call the **three-context model of ecosystem management**, considers ecological, socioeconomic, and institutional perspectives in the search for solutions acceptable to all (Figure 2.5). These three contexts form the core of the ecosystem approach and will be the focus of much of this book.

When attempts are made to draw the three contexts together for decision making, four areas of contextual overlap emerge (Figure 2.5, areas A–D,

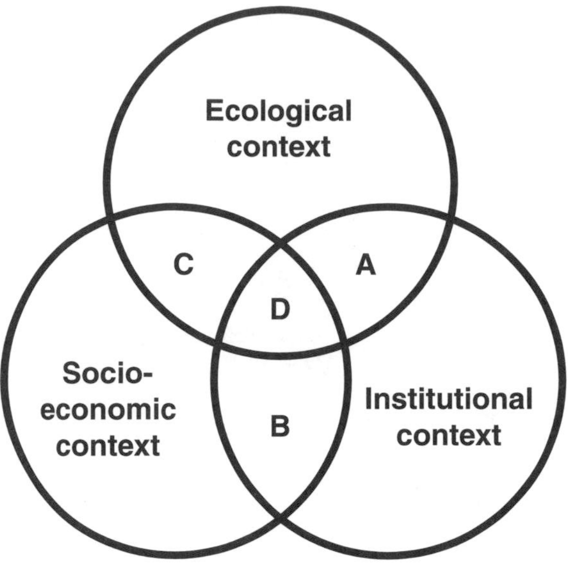

Figure 2.5. *The three-context model, showing partial overlap of the three contexts. A: zone of regulatory or management authority; B: zone of societal obligations; C: zone of influence; D: zone of win-win-win partnerships.*

described below). These provide key insights that help explain the relative roles of three important tools for ecosystem management—legal regulations, interpersonal influence, and partnerships. These tools will become central to everything that follows.

A: Zone of regulatory or management authority. This is the area where enforcement and regulation primarily dictate decision making. Government agencies have been granted legal authority by society to make decisions that may be counter to the goals or short-term interests of some individuals, but in the higher interest of society at large. For example, the National Environmental Policy Act, Endangered Species Act, Clean Water Act, National Forest Management Act, and other laws direct agencies to stop or substantially modify land use and development proposals, prohibit the discharge of pollutants or contaminants, and directly manage natural resources on public lands. Agencies exercise their enforcement authority through formal and defined processes, and they retain that authority unless it is redefined by a court or modified through legislation. In this zone, agencies—backed up by the law—call most of the shots.

B: Zone of societal obligations. Many areas of public policy do not directly affect the environment, but they serve the broader interests of society and must be adhered to by natural resource agencies. For example, by implementing civil service laws, ensuring equal opportunity in employment or accessibility for the disabled, or protecting cultural resources (such as Native American ceremonial sites, historic buildings, or archeological sites), agencies implement legitimate public policies, even though they require an expenditure of funding and time that would otherwise go toward environmental clean-up or the management of threatened resources. This zone does not directly influence natural resource management but is still an important obligation.

C: Zone of influence. This zone represents an area of decision making where interpersonal relationships and informal processes, rather than legal requirements, prevail. It is an area of opportunity that is often used by grass-roots organizers or politically connected developers, but is traditionally overlooked in much of natural resource management. It provides opportunities that are not available through either regulatory authority or formal partnerships. This approach relies on informal processes and the development of interpersonal relationships among interested stakeholders. For example, a local Trout Unlimited chapter may work with a farmer to clean up and improve a degraded trout stream that runs through his farm, or The Nature Conservancy may purchase land from a property owner who will not work with government officials. Such actions are done largely through interpersonal relationships and the development of trust. This zone prevails with respect to voluntary land health and stewardship efforts on private lands.

D: Zone of win-win-win partnerships. This is the main focal area for successful ecosystem management, the place where the three zones of influence overlap and the three contexts are best served. It represents partnerships among the three contexts of ecosystem management. Partnerships are especially useful when the issue at hand has a high degree of ecological uncertainty and socioeconomic interest. They flourish when the major stakeholders develop and share one or more mutual goals and seek actions that are compatible with prevailing ecological science, satisfy the major socioeconomic stakeholders, and are supported by a legal framework of law and agency policy. Many of the case studies used throughout this book are examples of such partnerships.

The degree of overlap of contexts varies with the complexity of the issue under consideration, the extent of our knowledge about the ecosystem, the concerns or polarity of the stakeholders, and the authority of the agency. For example, policy decisions within the boundaries of an agency-controlled wildlife refuge—where stakeholders perceive there will be little or no negative impact on their interests, ecological data are available, and the agency has clear authority—may require little involvement by anyone outside the agency. The decision can be made by the agency with minimal risk of outside challenge because it rests within the agency's regulatory and managerial authority. Similarly, some decisions on private land lie beyond the jurisdiction of an agency, do not require a partnership with multiple stakeholders, and are made solely by the landowner. In these cases, personal influence, peer pressure, or community goodwill may prove useful in influencing the decision.

Where ecological knowledge is lacking and uncertainty is high, or where people perceive they have a greater stake, more extensive dialog among stakeholders, scientists, and government is necessary to pull all three contexts together into a community-based partnership. Any actions or decisions made in these circumstances may be viewed as experiments and are best pursued as adaptive management, where ideas from people in all three contexts can be used to define the bounds of the experiment, its goals, and evaluation criteria. The participation and consent of the stakeholders and government are essential when the ecosystem issue is at a large ecological, social, or economic scale, such as salmon recovery efforts in the Pacific Northwest, wolf reintroduction in the Greater Yellowstone ecosystem, restoration of the Florida Everglades, or improving New York City's water supply by restoring ecological function to the watersheds in the Catskill Mountains.

A Closer Look at Ecosystem Management

With this background, we now move into a more formal examination of ecosystem management. To fully understand the concepts, we begin with definitions of two important and relevant terms that are germane to understanding ecosystem management: biodiversity and ecosystem. **Biodiversity** has been defined in several related ways, of which we present two:

> The variety of living organisms considered at all levels of organization, from genetics through species, to higher taxonomic levels, and including the variety of habitats and ecosystems, as well as the processes occurring therein (Meffe and Carroll, 1997).

> The variety of life and its processes; it includes the variety of living organisms, the genetic differences among them, the communities and ecosystems in which they occur, and the ecological and evolutionary processes that keep them functioning, yet ever changing and adapting (Noss and Cooperrider, 1994).

Actually, "the variety of life and its processes" is fine as a simple and clear definition. Regardless of the definition used, one point is critical: Biodiversity is much more than the number of species in a given area. In fact, species richness *is not* by itself biodiversity. It is *one component* of biodiversity, but the concept encompasses much more than species counts; it involves the entirety of life and the various processes it undergoes or participates in, such as nutrient cycling, predator-prey interactions, migrations, trophic dynamics, and evolutionary change.

An excellent way to understand biodiversity is to think about composition, structure, and function from the level of genes to the level of landscapes (Figure 2.6). Composition is "what is there," structure is "how it is distributed in space and time," and function is "what it does." Thus, one could ask what types of genes occur in a given area, how they are

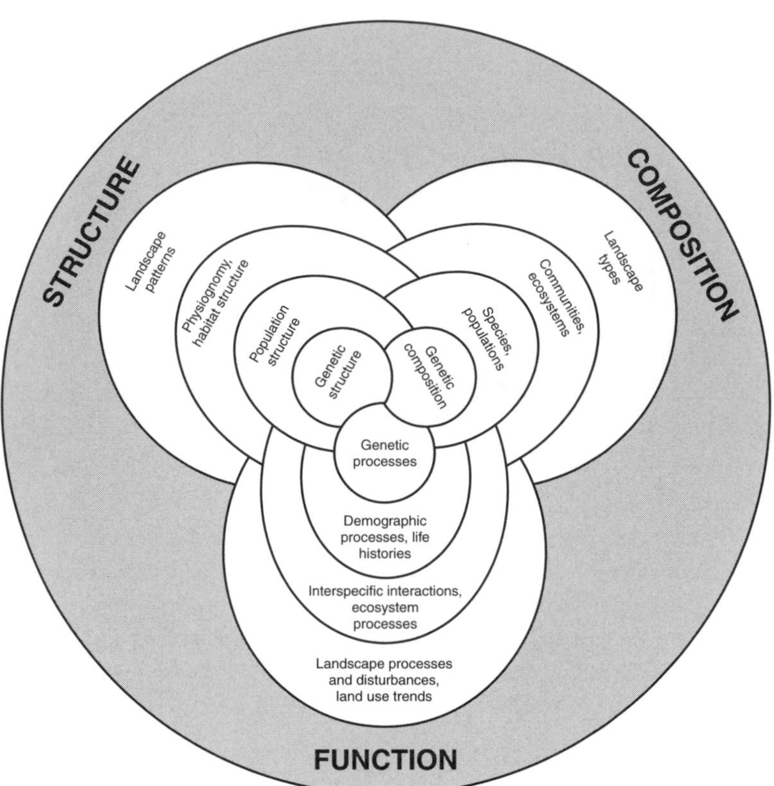

Figure 2.6. *A conception of biodiversity, illustrating the composition, structure, and function that exist from the genetic level to the landscape level of biological organization. (From Noss, 1990.)*

distributed within and among individuals across that area (or through time), and what the specific genes do. The same questions could be addressed at the population, species, community, ecosystem, and landscape levels. In total, this is biodiversity.

Ecosystem is defined here as "a *dynamic* complex of plant, animal, fungal, and microorganism communities and their associated nonliving environment interacting as an ecological unit" (italics added) (Noss and Cooperrider, 1994). Note the emphasis on "dynamic": An ecosystem is a fluid, changing entity that undergoes various processes, moves energy and materials, and changes over time. Note also that this definition does not designate a spatial scale, because an ecosystem is a *functional*, not a *spatial,* concept. The scale is whatever one chooses, from a rotting log in the forest, to a lake and its shoreline, to a watershed, to a continent. The important point is that it encompasses biotic and abiotic components of an area as a dynamic ecological unit. In addition, an ecosystem certainly includes humans as part of the system if they are present at the particular place and time.

Finally, we come to **ecosystem management**. Many definitions have been offered in the historical development of this concept, and they vary in their focus and completeness (Box 2.2). But several common threads run through them: large-scale, systemwide perspectives; emphasis on the composition and processes of ecological systems in all their complexities; integration across various spatial and temporal scales and human concerns (ecological, economic, and cultural); and participation by many stakeholders in consensus decision making, rather than command and control by a few. We prefer a broadly encompassing definition that incorporates many of the ideas present in these other definitions. We define ecosystem management as:

> an approach to maintaining or restoring the composition, structure, and function of natural and modified ecosystems for the goal of long-term sustainability. It is based on a collaboratively developed vision of desired future conditions that integrates ecological, socioeconomic, and institutional perspectives, applied within a geographic framework defined primarily by natural ecological boundaries (Meffe and Carroll, 1997).

BOX 2.2

Selected Definitions of Ecosystem Management

- Management of natural resources using system-wide concepts to ensure that all plants and animals in ecosystems are maintained at viable levels in native habitats and that basic ecosystem processes are perpetuated indefinitely (Clark and Zaunbrecher, 1987).
- The careful and skillful use of ecological, economic, social, and managerial principles in managing ecosystems to produce, restore, or sustain ecosystem integrity and desired conditions, uses, products, values, and services over the long term (Overbay, 1992).
- A strategy or plan to manage ecosystems to provide for all associated organisms, as opposed to a strategy or plan for managing individual species (Forest Ecosystem Management Assessment Team, 1993).
- The strategy by which, in aggregate, the full array of forest values and functions is maintained at the landscape level. Coordinate management at the landscape level, including across ownerships, is an essential component (Society of American Foresters, 1993).
- Ecosystem management integrates scientific knowledge of ecological relationships within a complex sociopolitical and values framework toward the general goal of protecting native ecosystem integrity over the long term (Grumbine, 1994).
- Any land-management system that seeks to protect viable populations of all native species, perpetuate natural disturbance regimes on the regional scale, adopt a planning time line of centuries, and allow human use at levels that do not result in long-term ecological degradation (Noss and Cooperrider, 1994).
- To restore and maintain the health, sustainability, and biological diversity of ecosystems while supporting sustainable economies and communities (U.S. Environmental Protection Agency, 1994).
- Ecosystem management is the integration of ecological, economic, and social principles to manage biological systems in a manner that safeguards long-term ecological sustainability. The primary goal of ecosystem management is to develop management strategies that maintain and restore the ecological integrity, productivity, and biological diversity of public lands (U.S. Department of the Interior, Bureau of Land Management, 1994).
- Protecting or restoring the function, structure, and species composition of an ecosystem, recognizing that all components are interrelated (U.S. Fish and Wildlife Service, 1994).
- Integration of ecologic, economic, and social principles to manage biological and physical systems in a manner that safeguards the ecological sustainability, natural diversity, and productivity of the landscape (Wood, 1994).
- The ecosystem approach is a method for sustaining or restoring natural systems and their functions and values. It is goal-driven, and it is based on a collaboratively developed vision of desired future conditions that integrates ecological, economic, and social factors. It is applied within a geographic framework defined primarily by ecological boundaries (Interagency Ecosystem Management Task Force, 1995).
- Ecosystem management is management driven by explicit goals, executed by policies, protocols, and practices, and made adaptable by monitoring and research based on our best understanding of the ecological interactions and processes necessary to sustain ecosystem composition, structure, and function. . . . Sustainability must be the primary objective, and levels of commodity and amenity provisions adjusted to meet that goal (Christensen et al., 1996).
- A collaborative process that strives to reconcile the promotion of economic opportunities and livable communities with the conservation of ecological integrity and biodiversity (Keystone National Policy Dialogue on Ecosystem Management, 1996).
- The application of ecological and social information, options, and constraints to achieve desired social benefits within a defined geographic area and over a specified period (Lackey, 1998).

Let's analyze this definition. First, ecosystem management is an *approach*, a way of getting something done—specifically, maintaining or restoring the composition, structure, and function (i.e., biodiversity) of natural and modified ecosystems (note we are not just dealing with pristine areas). It has a *goal:* long-term sustainability (or, what Aldo Leopold called "learning to live on a piece of land without spoiling it"). And that goal is based on a *vision* of desired future conditions that is developed collaboratively by interested parties (or stakeholders, discussed in Chapter 10). That vision should integrate considerations of the three circles of the basic conceptual model—ecological, socioeconomic, and institutional concerns—and is applied to a geographic place that is defined by ecological (rather than political) boundaries.

The three-context model presented above is a good way to conceptualize ecosystem management (see Figure 2.4). It embodies the major areas of concern, influence, and opportunities, and it will be referred to throughout this book. The intersection of the three circles is a "target" for ecosystem management (see Figure 2.5).

Another conceptual model distinguishes ecosystem management from more traditional forms of natural resource management: the three-axes model (Figure 2.7). The axes represent *space* (the spatial extent of concern), *time* (the time frames in which management actions are considered), and *inclusion* (who is involved in making decisions and what is considered). Traditional management tended to work close to the origin of these axes: within a relatively small spatial scale (e.g., within the boundaries of a national wildlife refuge or national forest), using a short time frame (e.g., the next budget cycle or perhaps a 5-year planning range), with few individuals included in decision

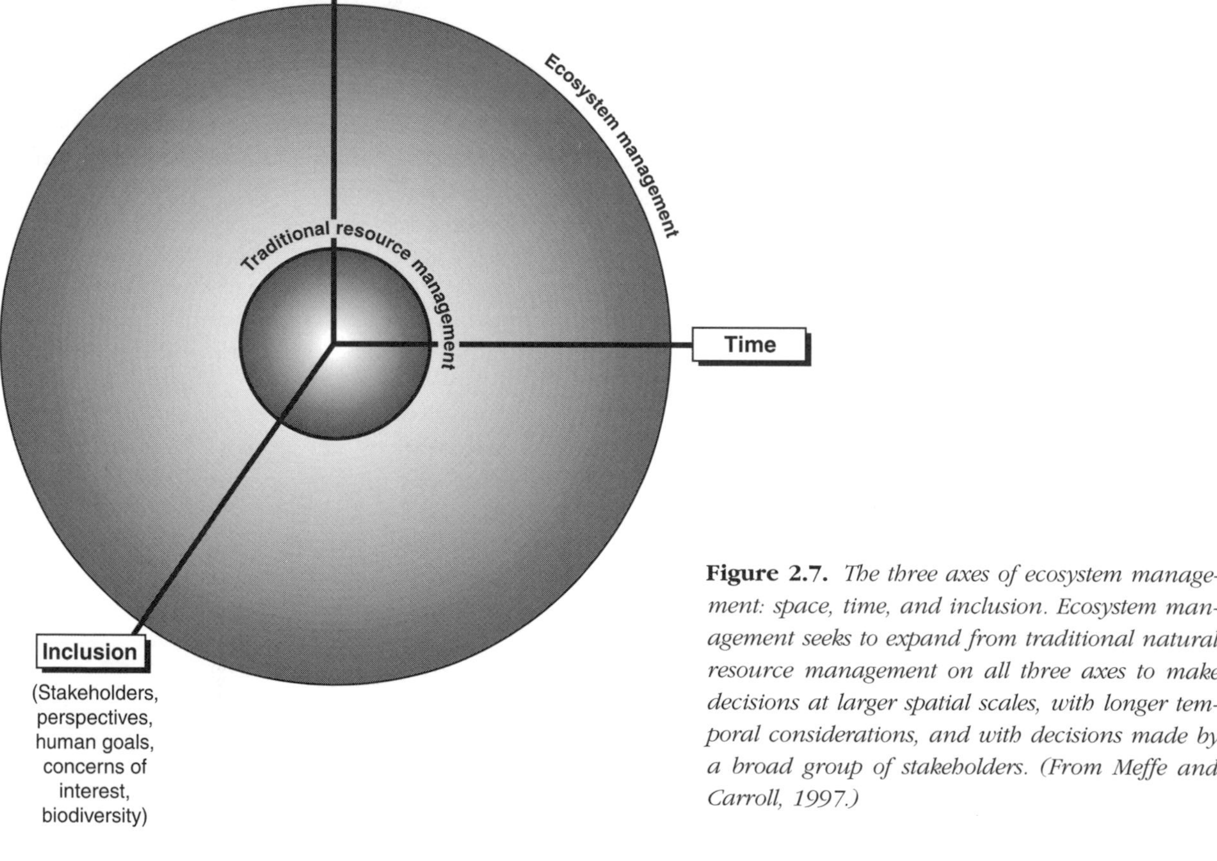

Figure 2.7. *The three axes of ecosystem management: space, time, and inclusion. Ecosystem management seeks to expand from traditional natural resource management on all three axes to make decisions at larger spatial scales, with longer temporal considerations, and with decisions made by a broad group of stakeholders. (From Meffe and Carroll, 1997.)*

making (e.g., the refuge manager or forest supervisor), and a few traditional users.

In contrast, ecosystem management expands all three axes to larger dimensions. It typically covers a larger spatial scale because it acknowledges that administrative lines drawn on a map do not capture true ecological complexity. Thus, to look beyond the boundaries of a politically designated refuge, for example, is to acknowledge that the refuge affects and is affected by surrounding lands. Ecosystem management also considers much longer time frames and asks what an area might look like decades into the future as a result of decisions made now. An ecological time scale, which includes short- to long-term natural fluctuations and cycles, is a more sensible and complete way to consider ecosystem conditions.

Finally, ecosystem management requires a broad sharing of decision making and the inclusion of many interests and stakeholders. This requires a willingness to give up some control and share "power"—not an easy task for individuals or institutions with decades of experience in top-down decision making or power politics within a community. But this change to shared decisions makes sense for at least two reasons. First, if individuals are involved in reaching decisions, they then have ownership in them and are more likely to abide by and support those decisions. They may even voluntarily act beyond what the minimum of the law requires with respect to land stewardship. Second, more brains working on a problem mean greater potential for innovative solutions. Many stakeholders working on an issue increases the likelihood of creative solutions being considered.

EXERCISE 2.6

Collaborate on It!

In the scenario you are using, what are some important spatial, temporal, and inclusion boundaries that might need to be expanded? How might you begin this expansion process? How would you know if you are being effective and working at the right scales?

COMMON MISCONCEPTIONS ABOUT ECOSYSTEM MANAGEMENT

Two common misunderstandings accompany ecosystem management at the political level, and they can be invoked to avoid using this approach altogether. First, some critics ask how anyone could possibly do ecosystem management when we usually cannot unambiguously delineate ecosystems. They argue that "ecosystem" itself is a fuzzy notion, there are no strict geographic boundaries, and thus ecosystem management cannot possibly be a viable concept. We respond that ecosystem boundaries, at least strict boundaries, are in fact irrelevant to this approach because *ecosystem management ultimately is not a geographic place but an approach to problem solving*. The geography of ecosystem management is whatever makes sense for the particular situation. Often, the watershed is a good and sensible unit in which to conduct ecosystem management, but that may vary from place to place. Ecosystem management could include a very large region, such as the Greater Yellowstone ecosystem, which encompasses many watersheds and political boundaries, or a much smaller area, such as a local township and its surrounding farms, forests, and other lands. Many successful efforts begin at even smaller scales. The boundaries are much less important than the approach used.

Second, some people have the impression that ecosystem management involves large-scale manipulations, as though major land areas are to be managed or controlled. That is not necessarily the case. In fact, much of ecosystem management involves stakeholders deciding how to manage human behavior—the limits to using a given place in the context of the larger physical setting, as well as the larger vision for the area. In fact, "ecosystem management" perhaps is a bit of a misnomer. A more accurate description might be "adaptive management at the landscape scale," for that is really what this is all about.

EXERCISE 2.7

Collaborate on It!

The class should split into five groups and be assigned one of the following statements to complete. Each person should independently think about the statement and write down several responses, then discuss these within the group (many will probably overlap). After organizing these thoughts, select the best four or five. Each group will then report back to the class as a whole.

1. The best reasons to use an ecosystem approach might be . . .
2. The best reasons to NOT use an ecosystem approach might be . . .
3. Managers have probably avoided using an ecosystem approach because . . .
4. We could recognize the ecosystem approach being used if we saw . . .
5. We will know the ecosystem approach works when . . .

After completing the lists, discuss what your collective results mean. Are there both pros and cons to using an ecosystem management approach? Are there reasons to be skeptical about it? Hopeful? How could you measure success or failure of such an approach?

BOX 2.3

Organizational Behavior and Ecosystem Management

Cortner and Moote (1999) summarized the situation with agencies in this way:

> Agency cultures are a substantial barrier to ecosystem management . . . Agencies become wedded to routine and deeply resistant to any alteration that doesn't agree with their own professional view of what should be done. Issues become framed as "them versus us," and . . . divisions between the professional expert and the public are sharpened. Incentives and rewards systems in resource management agencies traditionally have been heavily weighted toward commodity production; efforts toward improving ecological conditions have not been rewarded. Management incentives also exist to control information. When faced with conflict, conformity rather than dissent and innovation is rewarded. Agency cultures have yet to foster a spirit of cooperation and a willingness to give up resources and hence power to other agencies and entities. Agencies have been reluctant to shift from linear step-by-step approaches to public participation to those that are flexible, open, and encourage a rich public discourse. Innovation and new forms of leadership have been impeded by hierarchical decision-making structures, the risk aversion found in upper levels of decision making, and standards for organizational promotion. However, efforts to diversify the workforce by discipline, gender, ethnicity, and philosophy have brought new attitudes and perceptions that are providing support for new approaches. Moreover, employee loyalty is increasingly not to the organization but to issues such as protection of resources or to the employee's own sense of personal ethics.

Information, Organizational Behavior, and Command and Control

The way that organizations and institutions behave, learn, evolve, treat information, and conduct their daily business has great relevance to ecosystem management because the ecosystem approach ultimately will be implemented by institutions of one sort or another. If you do any natural resource management you will be part of some organization, so it is never too early to consider how organizations should work in order to maximize their effectiveness. To manage adaptively, and to avoid the pitfalls of the command-and-control approach, an organization must promote certain behaviors and discourage others. However, many institutions have such a long history of deeply bureaucratic structure, and a heavy, top-down form of control, that they will need substantial changes to become more effective (Box 2.3). Government institutions, large private organizations, and universities seem to be deeply entrenched in bureaucracy, which discourages many of the behaviors necessary for successful ecosystem management.

A comparison of three organizational types—

Table 2.2. *Stereotypical Organizational Types and Behaviors*

Pathological Organization	Bureaucratic Organization	Adaptive Organization
Don't want to know	May not find out	Actively seek information
Messengers are "shot"	Messengers are listened to—if they arrive	Messengers are trained and rewarded
Responsibility is shirked	Responsibility is compartmentalized	Responsibility is shared
Bridging is discouraged	Bridging is allowed but neglected	Bridging is rewarded
Failure is punished or covered up	Organization is just and merciful	Failure results in learning and redirection
New ideas are crushed	New ideas present problems	New ideas are sought and welcomed

Source: Modified from R. Westrum, 1994. An organizational perspective. Designing recovery teams from the inside out. Pp. 327–349 in T.W. Clark, R.P. Reading, and A.L. Clarke (eds.). *Endangered Species Recovery. Finding the Lessons, Improving the Process.* Island Press, Washington, D.C.

pathological, bureaucratic, and adaptive—clarifies the differences between the status quo and what is needed (Table 2.2). In a pathological type of organization (one that is driven by fear, reductionism, myopia, excessive external and internal accountability, and top-down control), the basic message is that information is trouble (because it can upset the status quo) and should be shunned. Thus, in the pathological organization, messengers (those who bring conflict and problems to light) are figuratively "shot" (punished for their efforts, perhaps through transfers to undesirable places); responsibility is shirked; communication bridges within or outside the organization are discouraged; failure by risk takers is punished; and new ideas are seen as dangerous and thus crushed immediately.

A bureaucratic type of organization is not as overtly in fear of information and may even talk a good game, but it is so complex and compartmentalized that little actually can be accomplished—sometimes by design. Consequently, important information may never come to light, messengers can be lost in a maze of complexity, and responsibility is passed along to others and compartmentalized. Although communication bridges are allowed and even desired, they are often ineffective. Failure is dealt with in an understanding and merciful manner (but not learned from), and new ideas present problems and implementation challenges because of the bureaucratic complexity.

An adaptive type of organization, on the other hand, thrives on information and actively seeks it at every level. Consequently, its people are *trained* to be messengers, and responsibility is broadly shared. Communication bridges within and outside of the organization are sought and rewarded. Perhaps most importantly, an adaptive organization realizes that failures are important learning opportunities and are openly shared. In fact, the only real "failure" is when an unwanted outcome is hidden from view; in that case,

EXERCISE 2.8

Talk About It!

Consider a federal government agency such as the U.S. Fish and Wildlife Service. Discuss how it might react to the following situations if operating as each of the three organizational types discussed:

1. An endangered salamander is discovered in an area that is contiguous with a national wildlife refuge but that is slated for a housing development.
2. A captive breeding program for a severely declining small mammal species fails because of a disease epidemic.
3. An internal report suggests that an aquifer in an agricultural area will likely be drawn down severely within a decade. Such a decline would endanger local springs containing several endemic snail species.

EXERCISE 2.9

Collaborate on It!

Split into four groups. Each group will be assigned one of four tasks: to address one of two aspects of their scenario using either a traditional or an ecosystem management approach.

SnowPACT

- Management of the Kachina Arch Resource Management Area
- Effects of possible bison reintroduction

ROLE Model

- Management of the Bingham Wildlife Refuge
- Effects of cranberry farming

PDQ Revival

- Management of the John Muir Wildlife Refuge
- Effects of toxic wastes from Camp Fraser

Each management issue will be dealt with using either a traditional (one group) or an ecosystem (second group) approach. In your group, discuss some of the major elements of how you would manage for that particular situation using the approach assigned to you. Thus, for the traditional approach, pretend that you never heard of a holistic, ecosystem approach, and develop management actions accordingly. Alternatively for the ecosystem approach, go ahead and use a more holistic perspective, as we've discussed it here.

What are the main issues and major concerns driving management from the particular perspective you are assigned? How might management proceed based on your approach? What are some of the first things you would do? Develop these ideas, and then present them to the class.

How do the two approaches differ for each management issue? Are there any similarities? Does one approach produce more tangible products more quickly? If so, what does this say about the slower approach?

mistakes are likely to be made repeatedly, and institutional learning never occurs. Rather, an adaptive organization understands that all outcomes and activities must be open to view so that the maximum learning potential can be achieved. And to further that cause, new ideas and risk taking are sought and rewarded.

Many of the professionals we have dealt with in ecosystem management believe their organizations are solidly in the bureaucratic camp, with definite hints of the pathological behavior and some glimmers of the adaptive side. And they almost unanimously agree that they would like to work for an adaptive organization and that all should be pulling toward the right side of Table 2.2. This is happening in many institutions, albeit slowly. One thing that will help create more adaptive, learning organizations is the influx of fresh new ideas from younger people who are willing to challenge long-held assumptions and the status quo in productive ways. We hope that such fresh ideas will develop in the ensuing chapters.

References and Suggested Readings

Agee, J.K., and D.R. Johnson (eds.). 1988. *Ecosystem Management for Parks and Wilderness*. University of Washington Press, Seattle.

Callicott, J.B. 1990. Whither conservation ethics? Conservation Biology 4:15–20.

Carson, R. 1962. *Silent Spring*. Houghton Mifflin, Boston, MA.

Chen, L., and S.L. Yaffee. 1999. Three faces of ecosystem management. Conservation Biology 13:713–725.

Christensen, N.L., A. Bartuska, J. Brown, S. Carpenter, C. D'Antonio, R. Francis, J. Franklin, J. MacMahon, R. Noss, D. Parsons, C. Peterson, M. Turner, and R. Woodmansee. 1996. The report of the Ecological Society of America Committee on the scientific basis for ecosystem management. Ecological Applications 6:665–691.

Clark, T.W., and D. Zaunbrecher. 1987. The Greater Yellowstone Ecosystem: The ecosystem concept in natural policy and management. Renewable Resources Journal, Summer: 8–16.

Cortner, H.J., and M.A. Moote. 1999. *The Politics of Ecosystem Management.* Island Press, Washington, D.C.

Daniel, H., and F. Minot. 1954. *The Inexhaustable Sea.* Collier Books, New York.

Ehrlich, P.R. 1968. *The Population Bomb.* Ballantine, New York.

Forest Ecosystem Management Assessment Team (FEMAT). 1993. Forest ecosystem management: An ecological, economic, and social assessment. Report of the Forest Ecosystem Management Assessment Team, Multi-agency Report.

Grumbine, R.E. 1994. What is ecosystem management? Conservation Biology 8:27–38.

Gunderson, L.H., C.S. Holling, and S.S. Light. 1995. *Barriers and Bridges to the Renewal of Ecosystems and Institutions.* Columbia University Press, New York.

Holling, C.S. 1995. What barriers? What bridges? Pp. 3–34 in L.H. Gunderson, C.S. Holling, and S.S. Light (eds.). *Barriers and Bridges to the Renewal of Ecosystems and Institutions.* Columbia University Press, New York.

Holling, C.S., and G.K. Meffe. 1996. Command and control and the pathology of natural resource management. Conservation Biology 10:328–337.

Interagency Ecosystem Management Task Force. 1995. *The Ecosystem Approach: Healthy Ecosystems and Sustainable Economies. Vol. I: Overview; Vol. II: Implementation Issues; Vol. III: Case Studies.* National Technical Information Service, U.S. Department of Commerce, Springfield, VA.

Johnson, N.C., A. J. Malk, R.C. Szaro, and W.T. Sexton (eds.). 1999. *Ecological Stweardship. A Common Reference for Ecosystem Management. Volume I.* Elsevier Science, Ltd., Oxford, U.K.

Keystone National Policy Dialogue on Ecosystem Management. 1996. Final Report. The Keystone Center, Keystone, CO.

Knight, R.L., and S.F. Bates (eds.). 1995. *A New Century for Natural Resources Management.* Island Press, Washington, D.C.

Knight, R.L., and P.B. Landres (eds.). 1998. *Stewardship Across Boundaries.* Island Press, Washington, D.C.

Knight, R.L., and G.K. Meffe. 1997. Ecosystem management: Agency liberation from command and control. Wildlife Society Bulletin 25(3):676–678.

Lackey, R.T. 1998. Seven pillars of ecosystem management. Landscape and Urban Planning 40:21–30.

Lichatowich, J. 1999. *Salmon Without Rivers: A History of the Pacific Salmon Crisis.* Island Press, Washington, D.C.

Lee, K.N. 1993. *Compass and Gyroscope. Integrating Science and Politics for the Environment.* Island Press, Washington, D.C.

Leopold, A. 1938. Engineering and conservation. Lecture delivered to the University of Wisconsin College of Engineering, reprinted in S.L. Flader and J.B. Callicott (eds.), 1991. *The River of the Mother of God and Other Essays by Aldo Leopold.* University of Wisconsin Press, Madison.

Leopold, A. 1949. *A Sand County Almanac and Sketches Here and There.* Oxford University Press, New York.

MacKenzie, S.H. 1996. *Integrated Resource Planning and Management: The Ecosystem Approach to the Great Lakes Basin.* Island Press, Washington, D.C.

Meffe, G.K. 1992. Techno-arrogance and halfway technologies: Salmon hatcheries on the Pacific coast of North America. Conservation Biology 6:350–354.

Meffe, G.K., C.R. Carroll, and contributors. 1997. *Principles of Conservation Biology,* 2nd ed. Sinauer Associates, Sunderland, MA.

Nehlsen, W., J.E. Williams, and J.A. Lichatowich. 1991. Pacific salmon at the crossroads: Stocks at risk from California, Oregon, Idaho, and Washington. Fisheries 16:4–21.

Noss, R.F. 1990. Indicators for monitoring biodiversity: A hierarchical approach. Conservation Biology 4: 355–364.

Noss, R.F., and A. Cooperrider. 1994. *Saving Nature's Legacy: Protecting and Restoring Biodiversity.* Defenders of Wildlife and Island Press, Washington, D.C.

Overbay, J.C. 1992. Ecosystem management. Pp. 3–15 in Proceedings in the National Workshop: Taking an Ecological Approach to Management. USDA Forest Service Publication WO-WSA-3, Washington, D.C.

Pickett, S.T.A., and P.S. White (eds.). 1985. *The Ecology of Natural Disturbance and Patch Dynamics.* Academic Press, Orlando, FL.

Pickett, S.T.A., V.T. Parker, and P.L. Fiedler. 1992. The new paradigm in ecology: implications for conservation biology above the species level. Pp. 65–88 in P.L. Fiedler and S.K. Jain (eds.). *Conservation Biology: The Theory and Practice of Nature Conservation Preservation and Management.* Chapman and Hall, New York.

Pinchot, G. 1947. *Breaking New Ground.* Harcourt, Brace, New York.

Salafsky, N., R. Margoluis, and D.E. Blockstein. 1999. Integrated science for ecosystem management: An achievable imperative. Conservation Biology 13: 682–685.

Samson, F.B., and F.L. Knopf (eds.). 1996. *Ecosystem Management: Selected Readings.* Springer, New York.

Sexton, W.T., A. J. Malk, R.C. Szaro, and N.C. Johnson (eds.). 1999. *Ecological Stwardship. A Common Reference for Ecosystem Management. Volume III.* Elsevier Science, Ltd., Oxford, U.K.

Society of American Foresters. 1993. *Sustaining Long-Term Forest Health and Productivity.* Society of American Foresters, Bethesda, MD.

Szaro, R.C., N.C. Johnson, W.T. Sexton, and A.J. Malk (eds.). 1999. *Ecological Stewardship. A Common Reference for Ecosystem Management. Volume II.* Elsevier Science, Ltd., Oxford, U.K.

Tenner, E. 1996. *Why Things Bite Back: Technology and the Revenge of Unintended Consequences.* Knopf, New York.

U.S. Department of the Interior, Bureau of Land Management. 1994. Ecosystem management in the BLM: From concept to commitment. Bureau of Land Management, Washington, D.C.

U.S. Environmental Protection Agency. 1994. Integrated Ecosystem Protection Research Program: A Conceptual Plan. Working Draft, 89 pp., U.S. Environmental Protection Agency, Washington, D.C.

U.S. Fish and Wildlife Service. 1994. An ecosystem approach to fish and wildlife conservation. Internal working draft, December 1994, U.S. Fish and Wildlife Service, Washington, D.C.

Wondolleck, J.M., and S.L. Yaffee. 2000. *Making Collaboration Work. Lessons from Innovation in Natural Resource Management.* Island Press, Washington, D.C.

Wood, C.A. 1994. Ecosystem management: Achieving the new land ethic. Renewable Resources Journal 12(1):6–11.

Yaffee, S.L., A.F. Phillips, I.C. Frentz, P.W. Hardy, S.M. Maleki, and B.E. Thorpe. 1996. *Ecosystem Management in the United States: An Assessment of Current Experience.* Island Press, Washington, D.C.

CHAPTER 3

Incorporating Uncertainty and Complexity into Management

HUMANS INHERENTLY SEEM TO SEEK CERTAINTY IN life. For example, we are trained in school from an early age that there are right and wrong answers on tests. We memorize established facts and relay them back when asked, and if we do that well enough, we progress through the educational system. We are examined through true-false or multiple-choice tests, which imply great certainty—a response is either correct or incorrect—and allow for very little "fuzziness" in our world. This conditioning teaches us early on that, if we can only find the *one* right answer, we will succeed. That message is reinforced throughout life, and we continue to seek certainty in many things we do. Thus, we expect physicians to make exact and correct diagnoses, we want our political leaders to legislate based on a good grasp of future conditions, and we even expect our weather forecasters to be right for the next 5 days. And we are angry when these diagnoses and prognostications are not borne out; the rained-out picnic can result in some choice words for the weather forecaster! We become confused when there seems to be more than one right answer, or no right answer.

This search for certainty is led by a strong confidence in science and technology. Because some aspects of our civilization, based on physical science, have been so successful, we have come to expect that all parts of life should be equally certain. However, deep down we also seem to understand that very little in our world is really very certain. In fact, the complex world in which we live offers a great deal of uncertainty, complexity, vagueness, and unknowns—despite humanity's best efforts to paint a very different picture. This is the true world of the natural resource manager.

Complexity and uncertainty are deeply inherent aspects of ecology in general and natural resource management in particular. The ecologist Frank Egler is credited with saying that "ecosystems are not only more complex than we think, but more complex than we *can* think." The pathology we discussed in Chapter 2 illustrates some of the reasons why, and here we will go into further detail. One of the messages we wish to convey is that we can try to minimize uncertainty and plan for it to some degree, but we cannot avoid it; it will always be part of the management equation. However, we actually can make it work *for* us if we try to understand it, work with it, and include uncertainty as a "buffer" in management decisions.

EXERCISE 3.1

Think About It!

Picture a local ecosystem—natural or modified—with which you are familiar, and take a few minutes to think about some of the complexities and uncertainties in that area. Consider the levels of biodiversity ranging from genes through landscapes and the composition, structure, and function at each level. Think about the unknown species (e.g., the number of microbial or nematode species in that area), the biochemical reactions occurring, the interspecific interactions of soil organisms, the flow of nutrients through the system, the hydrological patterns, and so forth. Then discuss your thought experiments as a class. Which of these could be made more certain with observation and study? Which will always be unknown and unknowable? With all the unknowns and complexities at every turn, it's not surprising that we have an easier time safely sending people into space than understanding how ecosystems work!

Sources of Complexity and Uncertainty in Natural Resource Management

The many and seemingly confusing reasons for high complexity and uncertainty in natural resource management are better understood if they are organized into groups. We briefly discuss four general categories of phenomena or events that result in complexity and lead to uncertainty in natural resource management (Table 3.1). These are not exclusive categories, and there may be other sources of complexity and uncertainty that you can think of, but they should get you started in considering how such factors might affect natural resource management.

Table 3.1. *Categories of Phenomena or Events that Result in Complexity and Contribute to Uncertainty in Natural Resource Management.*

Category	Phenomena
Environmental variation	Environmental uncertainties Natural catastrophes
Biological variation in small populations	Genetic uncertainties Demographic uncertainties
Non-independence of events and interactions	Indirect or cascading effects Synergistic effects Cumulative effects
Uncertainties in the human realm	Human-caused catastrophes Insufficient knowledge Noise Basic human behavior

CATEGORY 1: ENVIRONMENTAL VARIATION

Basic environmental variation consists of two major categories—environmental uncertainties and natural catastrophes.

ENVIRONMENTAL UNCERTAINTIES. These are the many uncertainties occurring in nature—ranging from the level of a species' microhabitat to the ecosystem level—that result from various unpredictable events, including weather, food supply, changing populations of competitors, predators, and disease-causing organisms. Specific characteristics and effects of these uncertainties depend on the environmental change being measured, the taxa involved, the ecosystem, and the scale of measurements made. Examples are the uprooting of trees in a forest, thereby opening up light gaps and changes local vegetation structure; the flooding of a river, which can change community composition; the patchiness of a fire that spreads its effects in an unpredictable way through a forest; or the unpredictability of disease outbreaks or colonization by a new predator, which can alter species composition in an ecosystem (Figure 3.1).

(a)

(b)

(c)

Figure 3.1. *Examples of environmental uncertainties. (a) A tree blow down opens up a light gap, changing the vegetation dynamics in that local area for years to come. (Photo by G.K. Meffe.) (b) A flooded river affects the flora and fauna, as well as the river's geomorphology. (Photo by William J. Matthews.) (c) A fire greatly influences vegetation dynamics and thus animal communities. (Photo by Nancy Deyrup.)*

> EXERCISE 3.2
>
> **Talk About It!**
>
> As a class, list some other common examples of environmental uncertainties pertinent to an ecosystem near you. You should be able to come up with 25 in a few minutes.

NATURAL CATASTROPHES. These are extreme cases of environmental uncertainty, such as hurricanes, very large fires, and volcanic eruptions (Figure 3.2). They usually are rare, short-lived, intense events that are geographically broad in their impact. A good example is Hurricane Hugo, which swept through South Carolina in 1989. It came through Francis Marion National Forest, which held one of the largest populations of the endangered red-cockaded woodpecker. Literally overnight, because its nesting habitat was destroyed, this seemingly secure population was significantly reduced by this natural catastrophic event, thereby changing the face of management for this species.

(a)

(b)

Figure 3.2. *Examples of natural catastrophes. (a) The volcanic eruption of Mt. St. Helens in 1988 vastly changed the landscape for thousands of square miles. (Photo by Charlie Crisafulli.) (b) The effects of Hurricane Hugo in Francis Marion National Forest, South Carolina, in 1989. (Photo by Rebecca R. Sharitz.)*

CATEGORY 2: BIOLOGICAL VARIATION IN SMALL POPULATIONS

Uncertainties in biological variation in small populations arise from both genetic and demographic effects.

Genetic Uncertainties. These are uncertainties at the level of population genetic structure resulting from founder effects, population bottlenecks, genetic drift, inbreeding, or migration (to be discussed in Chapter 5). Genetic problems leading to population declines can arise from a loss of genetic diversity or unique alleles, inbreeding depression, or a loss of local adaptations through outbreeding. Such problems typically occur only in very small populations, on the order of tens of individuals or less.

Demographic Uncertainties. These are random events in very small populations that can negatively affect population persistence or the reproductive abilities of individuals. For example, very small populations may have highly skewed sex ratios or age structures, resulting in single-sex populations or only older, postreproductive individuals. The final factor causing the extinction of the dusky seaside sparrow in south Florida, for example, was a demographic event: The last six remaining birds were all males, not a situation with much of a future. Of course, other factors—primarily habitat destruction—led to this precarious point, and an unusual demographic event was merely the final blow in a series of problems that led to extinction.

CATEGORY 3: NONINDEPENDENCE OF EVENTS AND INTERACTIONS

Many events and biological interactions act in nonindependent ways. These include indirect or cascading effects, synergistic effects, and cumulative effects.

Indirect or Cascading Effects. These are effects on species or ecosystems that are not a direct result of a particular stress or event, but rather are less predictable, indirect effects, often ones that cascade through trophic levels. For example, the removal of a predator from an ecosystem can affect its prey species, which might increase in numbers as a result. That increase, in turn, could affect plant species composition when herbivory increases. This change in plant species also could change the insect composition of the area, which in turn can affect the transmission of insect-borne diseases. Such complexities can seem endless.

Synergistic Effects. These arise when there are multiplicative effects resulting from two or more events or stresses that occur together. For example, two toxic chemicals might independently have little effect on a species, but together they may be lethal. Multiple stresses, such as drought and disease outbreak, are known to be much more devastating to forests than if they occur separately (Figure 3.3). There are well-known positive synergistic effect as well. For example, in agriculture, adding fertilizer and lime to soils results in much better productivity than adding either of the two separately.

Figure 3.3. *A synergistic effect: pine bark beetle destruction of loblolly pine trees in north Florida in 2001. Several years of drought stressed the trees, making them vulnerable to beetle invasions (note the dozens of small bore holes [top]), which killed many thousands of trees that are typically resistant to the beetle during well-watered periods. (Photos by G.K. Meffe.)*

EXERCISE 3.3

Collaborate on It!

In small groups, discuss some possible cascading and synergistic effects for your scenario. Select one or two of each to present to the entire class. Would any of these have serious enough repercussions that they should receive special attention from managers and planners?

Cumulative Effects. This is a large class of effects that have additive results over time or space. These can range from the accumulation and concentration of materials through the food chain, to accumulated effects on streams by the runoff of fertilizers and pesticides from many small farms, to repeated environmental perturbations such as storms or droughts that affect species or ecosystem structure. A given event, by itself, may have little to no measurable effect on a system, but over time or space, the cumulative effects can become large. This is sometimes referred to as "the tyranny of many small events."

A special and very important class in this category is *cumulative spatial effects.* These occur when many small decisions are made or events occur in a region that add up to major changes across the landscape. Therefore, management units should not be treated spatially independently of other management units or private lands because they affect one another in some way. These effects are a very strong argument for working across local ownerships or political boundaries in an ecosystem-wide fashion. For example, in fragmented forests the survival of species in patches is increasingly dependent on immigration among patches in a metapopulation fashion (discussed in Chapter 7). Perhaps the ultimate cumulative spatial effects occur with respect to neotropical migratory birds. Protection of summer nesting habitat in Connecticut will not protect a species if its wintering habitat in Mexico is destroyed (or vice versa). Similarly, protection of a Florida coral reef is futile unless recruitment zones elsewhere in the Gulf Stream, where many larval organisms are produced, are also considered. These spatial effects also require that decisions about land use are not made piecemeal but rather in a regional context. The cumulative effects of the loss of 5- and 10-acre parcels across a landscape must be considered, for example, and not just the small, and seemingly trivial, local effects at each site.

CATEGORY 4: UNCERTAINTIES IN THE HUMAN REALM

The human realm presents many challenging sources of uncertainties arising from human-caused catastrophes, insufficient knowledge, "noise" in our measurements, and basic aspects of individual and group human behavior.

Human-Caused Catastrophes. These are the many damaging events of human origin that can alter natural systems, such as oil spills, other toxic dumping, large-scale water diversions such as in the Caspian and Aral Seas, and the destructive effects of war. Such effects can be very local or broadly distributed across the landscape.

Insufficient Knowledge. Nature is more complex than we can even imagine, and we are always learning something new that revises our perspective. Current knowledge certainly will be replaced by newer perspectives, and it is tempting to do nothing to avoid doing the wrong thing because we lack sufficient understanding. But decisions typically need to be made now, and natural resource managers need to make the best decisions they can based on existing information, with the knowledge that actions need to be revisited and plans will need to reflect new information as it becomes available.

The problem of insufficient knowledge is reflected in a bias in the way statistical analyses are conducted. A Type I statistical error occurs when it is claimed that an effect of a "treatment" is real or significant, when in reality it is not. Thus, we might claim that increased phosphorus in a lake will cause eutrophication and possible species loss, when in fact it does not. The probability of a Type I error is what is reported as the familiar "alpha level" for a statistical test, and usually it is set at 5% or less. Thus, we wish to have less than a 5% chance of a Type I error—making a claim for an effect when there actually is none. This is the way scientists usually think; they are conservative about interpreting their knowledge.

At the same time, we often ignore a Type II error—claiming a treatment has no effect, when in fact there is an effect. This is much more relevant to the world of decision making, where the concern is failure to act when action is needed. For example, we may make a claim of no likelihood of a species' extinction or no significant effects from global warming when in fact there are such effects.

A Type II error can have much broader repercussions on management and humanity than a Type I error, which merely results in management actions that might not be necessary.

"NOISE." Often it is difficult to separate "signals" of real and meaningful patterns from multiple sources of random variation. For example, one of the current global environmental debates concerns the real meaning of global temperature increases. Were the 0.5–1.0°C increases in temperature in the twentieth century a signal of true global warming, or were they the result of perhaps long-term, random fluctuations in temperature? Separating reality from the artifacts of natural variation or our limited measurement techniques is a continual challenge. Such difficulties are often used politically as reasons for postponing decision making. The global warming example is a case in point: The federal government of the United States has been slow to respond to this potential problem, citing uncertainty in the meaning of temperature data and potential "noise" as reasons not to embark on expensive solutions. Many other nations have accepted the current state of our knowledge as a real signal and are trying to take action.

BASIC HUMAN BEHAVIOR. As if all the biophysical sources of complexity and uncertainty were not challenge enough, added on to them are many aspects of social, institutional, and economic changes, fluctuations, and uncertainties. Public values change over time (Box 3.1), economies go through boom-and-bust cycles, institutions come and go and their missions change through time—all while the human population and its demands continually grow. Thus, grass-roots groups ranging from those with a strong protectionist focus to property rights advocates will emerge; unemployment rates will go up or down and affect monetary flows, which in turn influence tax dollars available for the environmental sector: Federal, state, and local government administrations can change every 2–4 years, bringing with them new priorities and directions. Industries can arrive to or depart from a region, changing employment rates, cash flow, and environmental demands and influences. And social movements and values help determine the general tenor of society and its willingness to deal with various issues. These and countless other events in the human realm will influence, perhaps more than any other source, the various outcomes of ecosystem management efforts.

BOX 3.1

An Example of Changing Human Values

Early in our history, Americans valued rivers for where they led: Rivers were transportation systems that took us to and from our destinations, and they carried our products. Later, we valued rivers more for what they carried away: pollution. Our solution to pollution was dilution in our major rivers. We also sought to tame the rivers for power and irrigation in the name of the public interest, and we spent vast amounts to "tame" and "conquer" rivers and the power they offered.

More recently, our society has begun to look at rivers differently. We have designated some as "wild and scenic," and we are beginning to remove dams to restore others and recover endangered salmon stocks. Our values change with time—and with that, public policy changes. As with the many other factors that affect ecosystems, when and how values change is uncertain, but the playing field indisputably is a dynamic one.

Dealing with Complexity and Uncertainty

Complexity and uncertainty are inherent parts of all living systems, and together with the human aspects of uncertainty, they must be taken into account in ecosystem management from the outset. Consequently, it is best to expect uncertainty and plan for it, rather than be surprised and crippled by it. But if systems are so complex and uncertain, how can we plan for and deal with the unexpected? There are several ways:

1. *Many people holistically thinking about the system, as well as the events that are likely to occur there.* Some uncertainties can be anticipated and

planned for. For example, along the southeastern coast of the United States, hurricanes are frequent events. One cannot say with much certainty that a severe hurricane will strike in a given year at any particular place. But long-term records could indicate that there is, for example, a 3% chance of a hurricane hitting because three have occurred there in the last 100 years. Combined with consistent indications that we are in a period of increased hurricane activity, this threat seems real. With this knowledge that a hurricane is a distinct possibility *sometime* in the coming decades, it would be prudent to take that into account in planning for natural and human communities in that region. Are there vulnerable populations of endangered species, for example, that would be adversely affected by such a storm? If so, is there anything that can be done to ensure their safety? Perhaps new populations should be started elsewhere for redundancy. What about the water absorption and drainage capabilities of the region? Should construction permits be denied in low-lying regions to protect those areas (and human developments) and reserve them instead for wetlands and absorption functions? Would there be sewage and toxic overflow problems if 20 inches of rain fell in 24 hours?

As another example, consider public and political support for a local, state-owned natural area that is part of your larger ecosystem management plan. What happens if the present "nature-friendly" governor is replaced in 3 years by one who does not appreciate the natural, cultural, and recreational values of the area but instead might want to log it for the timber jobs, or convert the land to housing and commercial development, and generate quick revenues? Is there anything that could be done now to head off such a prospect and maintain ecological and community stability in that area? Perhaps bringing state government agencies into local planning processes and having them discuss with local community members the ecological and social values of the intact forest would ensure against short-term changes in political fortunes that would upset long-term plans. Or perhaps a sustainable forestry program, with a slow, steady schedule for selection logging, would keep jobs, sustain the ecosystem, and build broad support that would not be politically acceptable to oppose. Or maybe there is a need to develop programs that retrain workers so that they have other opportunities and need not depend solely on timber jobs. These and other such "thought experiments" can be conducted for any of the types of uncertainties likely to be relevant to a given area. Planning can then incorporate such possibilities, thereby reducing the effects of unpredictable events.

Careful consideration of major biophysical, socioeconomic, and institutional components of a system can reveal potential issues, problems, changes, and challenges. Some of these can be prepared and planned for, although of course not all "surprises" can be anticipated. But some advanced thinking and planning can help alleviate larger problems later.

2. *Mathematical and computer modeling and statistical analyses that account for random events.* Models of how the world works, even though artificial, can be valuable heuristic and predictive tools if the input data are of good quality. Using models, we can easily conduct many "what if" experiments with little cost or effort. What if global temperatures increased by 3°C? How would that affect sea levels and rainfall patterns? What if grizzly bears were introduced to a wilderness area? What is their likely population growth rate and probability of encounters with humans? What if an annual economic expansion of 3% occurs for 10 years? How would that affect the demands for water, waste disposal, building materials, and outdoor recreation in our community?

By manipulating variables in models, we can simulate scenarios and develop likely outcomes with given probabilities of occurrence. Statistical techniques such as error analysis (wherein random errors are introduced into models to simulate random events) and sensitivity analysis (assigning different values to model parameters to identify which variables in the system are most sensitive to manipulation) enable us to explore the behavior of a modeled system and reduce some of the likely surprises.

3. *Being prudent and making decisions with "buffers" that account for uncertainty.* Lee (1993) put this very simply: "What does it take to be prudent when there is uncertainty? First, recognize the possibility of surprise. Second, plan and act to detect and to correct avoidable error." When we have a lot of data about a situation—especially long-term data—and have rigorously studied its many parameters, we may feel especially confident about the system's likely behavior and our management decisions pertaining to it, and we can make decisions with little maneuvering room. However, in many cases the data are minimal or of poor quality, or short term, and we do not understand the system very well. As a general rule, the fewer the data or the greater the uncertainty we have, the more conservative should be decisions or management plans, and the more flexible we need to be in changing directions in response to new information.

When uncertainty is high, buffers should be included in our decisions. For example, in marine fisheries, population models of harvested species are used to set harvest quotas. Of course, there is always some level of uncertainty associated with these models, and variances are provided. When variances are especially high, and if there is concern for the viability of the stock, it would be prudent to set lower quotas unless more information allowed for larger harvests. A buffer against uncertainty would protect the stock against biologically unjustified takes.

4. *Employing adaptive management, which allows for and in fact seeks changes in direction as new information becomes available.* Adaptive management very simply means that natural resource management and policy are approached as experiments that should teach lessons. As Lee (1993) has stated:

> Because human understanding of nature is imperfect, human interactions with nature should be experimental. Adaptive management applies the concept of experimentation to the design and implementation of natural resource and environmental policies. An adaptive policy is one that is designed from the outset to test clearly formulated hypotheses about the behavior of an ecosystem being changed by human use. . . . If the policy succeeds, the hypothesis is affirmed. But if the policy fails, adaptive design still permits learning, so that future decisions can proceed from a better base of understanding.

Adaptive management is the topic of the next chapter and will be discussed in greater detail there.

Because planning and management are not deterministic or prescriptive processes but probabilistic ones, likely sources of uncertainty can be included in management plans. If management is flexible and adaptive, then it can respond more effectively to uncertainties and surprises and be less buffeted by the many complexities that nature and the human condition have to offer.

Although complexity and uncertainty can be intimidating and seem to make life difficult for natural resource managers, they actually may be viewed as the "energy sources" that drive management and make it interesting and challenging. If nature were predictable and there never were surprises, then professions in natural resource management would be dull and uneventful. The many challenges associated with the unknown and changing conditions are what make these fields so fascinating.

EXERCISE 3.4

Collaborate on It!

Split into small groups to focus on your scenario. Your group should consider one of three foci: (1) broad, ecosystem-level complexities and uncertainties; (2) complexities and uncertainties specifically relevant to small populations of concern; or (3) complexities and uncertainties associated with the human dimensions (social, economic, institutional). Develop a list of four or five possible problems and issues that could be faced in the coming years as a result of the complexities in and uncertainties of your scenario, specifically at your assigned level of focus. Can advanced consideration and planning minimize or eliminate the problems? How? Share your insights with the rest of the class.

References and Suggested Readings

Baydack, R.K., H. Campa III, and J.B. Haufler (eds.). 1999. *Practical Approaches to the Conservation of Biological Diversity.* Island Press, Washington, D.C.

Better Policy and Management Decisions Through Explicit Analysis of Uncertainty: New Approaches from Marine Conservation. Eight papers in Conservation Biology 14 (Oct. 2000).

Cortner, H.J., and M.A. Moote. 1999. *The Politics of Ecosystem Management.* Island Press, Washington, D.C.

Funtowicz, S.O., and J.R. Ravetz. 1990. *Uncertainty and Quality in Science for Policy.* Kluwer, Dordrecht, Netherlands.

Gunderson, L.H., C.S. Holling, and S.S. Light (eds.). 1995. *Barriers and Bridges to the Renewal of Ecosystems and Institutions.* Columbia University Press, New York.

Lande, R. 1988. Genetics and demography in biological conservation. Science 241:1455–1460.

Lee, K.N. 1993. *Compass and Gyroscope. Integrating Science and Politics for the Environment.* Island Press, Washington, D.C.

Ludwig, D., R. Hilborn, and C. Winters. 1993. Uncertainty, resource exploitation, and conservation: Lessons from history. Science 260:17, 36.

Orians, G.H., et al. 1986. *Ecological Knowledge and Environmental Problem Solving. Concepts and Case Studies.* National Academy Press, Washington, D.C.

Pickett, S.T.A., V.T. Parker, and P.L. Fiedler. 1992. The new paradign in ecology: Implications for conservation above the species level. Pp. 65–88 in P.L. Fiedler and S.K. Jain (eds.). *Conservation Biology: The Theory and Practice of Nature Conservation, Preservation, and Management.* Chapman and Hall, New York.

Experiences in Ecosystem Management:
Ecosystem Management in Policy and Practice

Steven L. Yaffee

THROUGHOUT THE EARLY AND MID-1990S, MANY scientists, managers, and policy makers embraced the idea of shifting public resource management policies to an ecosystem-based approach, or ecosystem management (EM). The underlying principles of this approach included: considering larger spaces and longer time frames; incorporating greater understanding of the complexity of ecological and social systems; integrating ecological, socioeconomic, and institutional aspects of a situation through a goal-setting process that elevated the importance of ecological integrity; coping with uncertainty and change through adaptive management; and working collaboratively across geographic and organizational boundaries (Grumbine, 1994; Christensen et al., 1996; Kohm & Franklin, 1997; Yaffee, 1999).

Between 1992 and 1997, eighteen federal agencies adopted many of these core elements of an ecosystem-based approach (Congressional Research Service, 1994). Agencies in several states worked to shift their natural resource agencies to adopt an EM perspective. Ecosystem management had boosters in high places, most notably Secretary of the Interior Bruce Babbitt and USDA Forest Service Chief Jack Ward Thomas. In their view, ecosystem management was a "way out" of single-species controversies, such as had occurred over the management of old-growth forests for the northern spotted owl (Yaffee, 1994). By following a more integrative and larger-scale management approach, we could balance competing interests and avoid future "train wrecks." Ecosystem management was a way "to have it all."

Has ecosystem management worked? How have federal and state agencies incorporated it into their policies and management approaches? How has it fared in practice? What successes have occurred? What accounts for them? What obstacles have stood in the way? What lessons can we take from experience to improve the practice of EM in the future?

Policy-Level Response

At a policy level, ecosystem management encountered significant resistance. Indeed, by 1999, the term "ecosystem management" was largely abandoned in federal agency policy statements. This occurred for several reasons. First, the concept of an ecosystem was hard for policy makers to understand. Second, political symbols that became attached to the term were problematic. For development interests, ecosystem management was seen as an early Clinton/Babbitt/Thomas policy that led to restrictions on the use of public lands. For community interests, ecosystem management seemed to threaten private property rights. For environmentalists, EM was viewed as management dressed up to appear environmentally sensitive. In a legal and political system that rewards extreme positions, a policy that emphasizes science-based balancing and integration of interests satisfied no one. "Sticker shock" also undercut the evolution of an ecosystem management approach as formal government policy. To truly move federal agencies toward EM would require major change, costly both in dollar terms and in the resistance of agency employees, and threatening to agencies that guard their fragmented jurisdictions carefully.

As state agency policy, ecosystem management also was problematic. States such as Missouri, Florida, Minnesota, and Michigan adopted various forms of an ecosystem-based management approach, and some achieved important internal organizational changes and enhanced planning in subregions. But there were major barriers to these changes, and conflicts with the structure and cultures of state agencies led to resistance. Significant public opposition, fanned by property rights fears, limited agency accomplishments. The difficulty of measuring EM-based performance also was problematic for states with budgeting tied to performance. When faced with these political and performance problems, changes in state-level political leadership resulted in the elimination of these innovative programs. For example, newly elected Florida Governor Jeb Bush appointed a secretary of environmental protection who eliminated the Office of Ecosystem Management and reorganized along more media-based lines. In Missouri, a new department of natural resources director similarly eliminated the Coordinated Resource Management program.

While the term "ecosystem management" has seen decreasing use and its wholesale imposition on federal and state agencies was problematic, the good news is that many components of an EM approach are still very much alive in both federal and state agency policies. In particular, managing across boundaries using collaborative partnerships among many landowners and interests is still an explicit objective of most of the federal resource management agencies. Rather than ecosystem management, the new resource management paradigm more often goes under the label of a "watershed approach," "place-based management," or "collaborative stewardship." Their core principles are often the same, but they have shed the political burdens associated with the ecosystem management label. Indeed, practicing "stealth" ecosystem management as the fundamentals of an ecosystem-based approach evolve may be wiser than creating a visible target and political symbol that can be attacked.

Ecosystem Management in Practice

While the debate raged in the policy arena, many people were experimenting with EM approaches on the ground. The fact that there was not a policy consensus on the dimensions or desirability of an EM approach allowed innovation that might have been suppressed by a "one size fits all" approach, had one been adopted (Haeuber, 1996). Changes in political and administrative realities made it clear that resource management as usual would not succeed (Yaffee, 1997). As a result, managers tried various elements of EM to deal with the very real problems that faced them on the ground.

In 1995, we began studying 105 of these EM efforts, including projects distributed widely across the United States (Yaffee et al., 1996). Four years later, 75% of these efforts were still under way. Only 10% ended because of a lack of resources or changes in political direction (Brush et al., 2000). Some of these experiments in ecosystem management were large-scale, policy-driven approaches, such as the Chesapeake Bay and Everglades restoration efforts. Others—the habitat conservation plan for the Karner Blue Butterfly in Wisconsin, for example—responded to the incentives created by such policies as the Endangered Species Act. Many others were driven by community-based partnerships such as the Applegate Partnership in southern Oregon, where decades of conflict produced a desire to try something else and a place-based approach organized on a watershed scale made sense (see Box 10.6). Others were promoted by nongovernmental organizations such as The Nature Conservancy, which adopted an ecoregional planning process to deal with areas surrounding core preserves.

Almost all of these projects produced changes on the ground. Most showed improvements in the ways that people and organizations interact, an important outcome given the high levels of conflict and resistance evident at many of these sites in the early 1990s. Eighty-six percent of the projects reported better communication and cooperation, and 75% evidenced increased awareness of the ecosystem and the involvement of new stakeholder groups in management activities. Although some of these process outcomes were evident in the initial survey in 1995, significant increases occurred as the projects matured. Levels of trust and respect, public education, and awareness all increased dramatically in our 1999 survey.

If all EM accomplished were better social dynamics in critical ecoregions, I would be reluctant to view it as a success. But many projects reported specific ecological outcomes in 1999, a dramatic change from our first survey in 1995. Expanded scientific understanding, ecological restoration results, increased native species populations, and improvements in "overall ecosystem integrity" were reported by a majority of sites.

What enabled these projects to achieve these outcomes? In 1995, project participants ranked the use of a different, more collaborative process as the most important factor. They also pointed to the need for agency and public support and available resources. While all these items were seen as important 4 years later, the presence of dedicated, energetic individuals jumped to first place. The character of the management plan and political support also increased dramatically as key factors in facilitating progress (Figure A).

Interestingly, the obstacles facing projects dropped considerably. While organized opposition and a lack of resources were reported by half the projects in 1995, these levels dropped by half in 1999. Most other project-related obstacles, such as problems with the project's process and insufficient scientific information, also fell. Obstacles that increased were related to the underlying ecological problems: the magnitude of the ecosystem stresses and the amount of development pressure (Figure B).

Overall, the experience to date is quite hopeful. EM projects are evidencing more support and less opposition over time. They are maturing and changing, and are increasingly showing ecological results. But the evidence suggests that improvements in

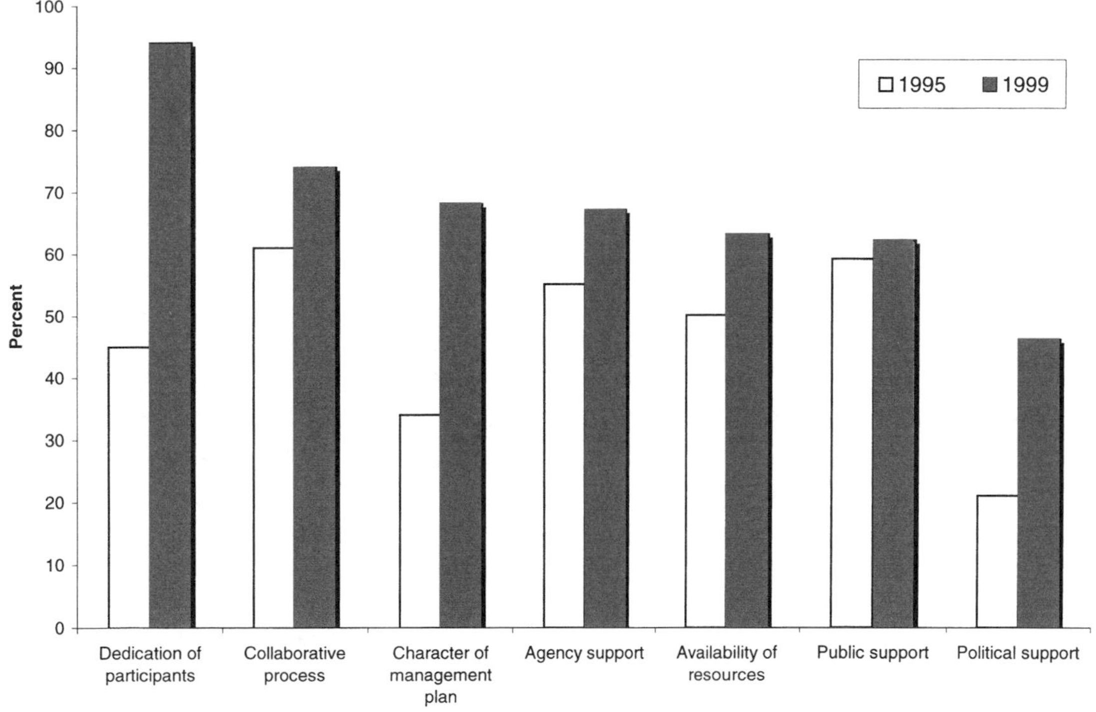

Figure A. *Factors facilitating the progress of ecosystem management projects across the United States. (From Brush et al., 2000.)*

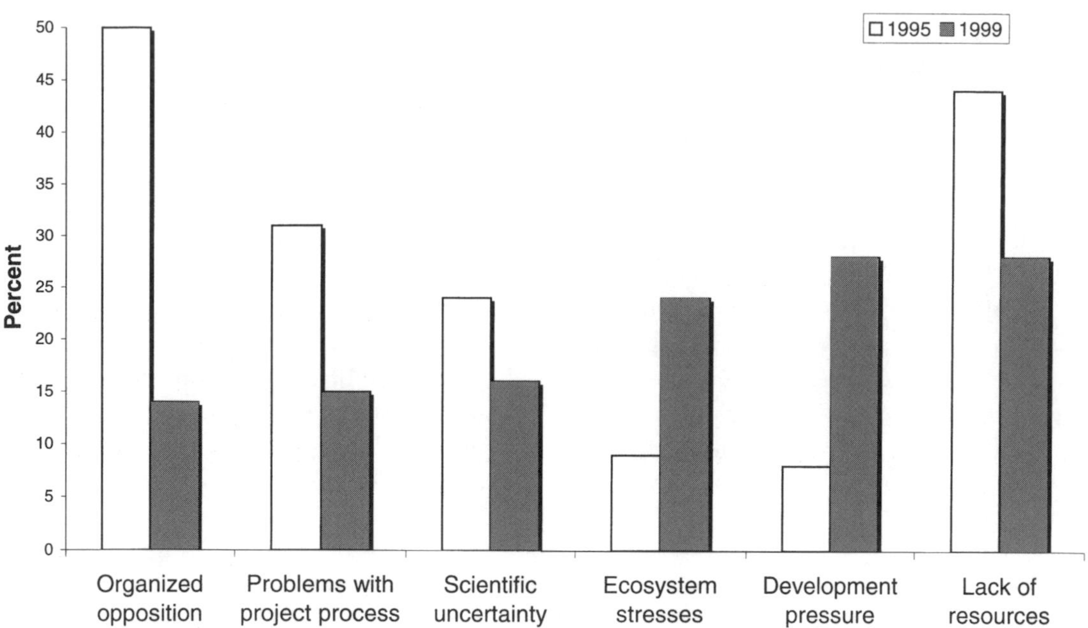

Figure B. *Obstacles facing ecosystem management projects across the United States. (From Brush et al., 2000.)*

social processes and ecological outcomes are linked; that is, it is hard to improve the ecological situation on the ground without improving the ways people and agencies interact. By improving the social climate, a foundation is laid that enables managers to achieve a more sustainable style of management.

Lessons for Success

Managers can use the lessons from this decade-long experience with many EM projects to inform their strategies and organizational approaches. It is clear that projects need to move through a stepwise evolutionary process—a life cycle—starting with early activities to initiate a project, continuing with planning and preliminary implementation activities, to more complete implementation. Ultimately, this stepwise process generates ecological effects.

Through this process, strategies and outcomes change over time. Many project managers start with outreach to agencies and other affected groups, move into a plan development phase, and then use the plan as a key organizing feature. Many highlight the use of pilot activities—small successes—in the early to middle stages of a project. Most managers noted the importance of mobilizing dedicated people to sustain the effort as the projects matured. In a similar way, process outcomes change as the projects evolve: from improvements in interagency and multiparty communication to broader involvement to increased trust and reduced opposition.

The shift in strategies is partly a response to changes in the major problem facing the on-the-ground efforts. First they have to overcome resistance and skepticism by dealing with opposition, building communication, and developing scientific understanding. They then have to work at sustaining the effort, by developing needed resources and a cadre of individuals committed to the project's success. Most recently, it appears that the obstacles project participants face relate to the magnitude of the underlying stresses. As understanding and communication develop, the difficulty of dealing with the underlying ecological stresses and development pressures has become more apparent.

This evolutionary process is exactly what we

hoped would happen. By dealing with the human and organizational issues that make effective decision making difficult, the process allows participants to focus on the underlying substantive problems confronting the resource. In doing so, it enables managers to improve the on-the-ground situation.

The underlying messages from a decade's worth of experience with ecosystem-based management approaches are simple but important:

1. EM works. It has the ability to improve social dynamics in a stressed area, which allow ecological improvements to occur.
2. Effective EM takes time. It requires patience, diligence, and annual commitments of expertise and funds—all of which are challenging to achieve.
3. Being effective at EM means managing the process well (Yaffee, 1996; Wondolleck and Yaffee, 2000). Project leaders need to be aware of project life cycles and strategies appropriate to different stages of the process.
4. Resource professionals need a broad set of skills in order to be effective at EM. If human process improvements are necessary to ultimately bring about ecological improvements, managers and scientists must learn how to communicate, and how to design, participate in, and lead collaborative processes. Without that understanding, it is less likely that EM approaches will result in our ultimate objective: mitigated stresses on natural systems, resulting in biodiversity protection and a better quality of life for humans.

Discussion Questions

1. Suppose you were the governor of a state who was interested in drafting a state ecosystem management policy. What would be its key elements? Which provisions would deal with public lands? Which ones would be included that affect the management of private lands? Who would you expect to support or oppose the draft policy, and why?
2. Discuss the politics of ecosystem management. Why has it been challenging to get EM adopted and implemented at the state and federal levels?
3. Suppose you were interested in starting an ecosystem management project.
   a. What would be your first steps to initiate and provide momentum to the effort?
   b. What would you expect to be the major challenges facing you at different stages during the life of the project? How would you deal with them?
   c. How would you evaluate the success of the project at different stages throughout its life cycle? How would you expect process outcomes to relate to ecological outcomes?

References

Brush, M., A. Hance, K. Judd, and L Rettenmaier. 2000. Recent trends in ecosystem management. Unpublished Master's Project report. School of Natural Resources & Environment, University of Michigan, Ann Arbor.

Christensen, N.L., A.M. Bartuska, J.H. Brown, S. Carpenter, C. D'Antonio, R. Francis, J.F. Franklin, J.A. MacMahon, R.F. Noss, D.J. Parsons, C.H. Peterson, M.G. Turner, and R.G. Woodmansee. 1996. The report of the Ecological Society of America committee on the scientific basis for ecosystem management. Ecological Applications 6:665–691.

Congressional Research Service. 1994. Ecosystem management: Federal agency activities. Report #94-339 ENR. U.S. Library of Congress, Washington, D.C.

Grumbine, R.E. 1994. What is ecosystem management? Conservation Biology 8:1–12.

Haeuber, R. 1996. Setting the environmental policy agenda: The case of ecosystem management. Natural Resources Journal 36:1–28.

Kohm, K.A., and J.F. Franklin. 1997. Introduction. Pp. 1–5 in K.A. Kohm and J.F. Franklin (eds.). *Creating a Forestry for the 21st Century: The Science of Ecosystem Management.* Island Press, Washington, D.C.

Wondolleck, J.M., and S.L. Yaffee. 2000. *Making Collaboration Work: Lessons from Innovation in Natural Resource Management.* Island Press, Washington, D.C.

Yaffee, S.L. 1994. *The Wisdom of the Spotted Owl: Policy Lessons for a New Century.* Island Press, Washington, D.C.

Yaffee, S.L. 1996. Ecosystem management in practice: The importance of human institutions. Ecological Applications 6:724–727.

Yaffee, S.L. 1997. Why environmental policy nightmares recur. Conservation Biology 11:328–337.

Yaffee, S.L. 1999. Three faces of ecosystem management. Conservation Biology 13:713–725.

Yaffee, S.L., A.F. Phillips, I.C. Frentz, P.W. Hardy, S.M. Maleki, and B.E. Thorpe. 1996. *Ecosystem Management in the United States: An Assessment of Current Experience.* Island Press, Washington, D.C.

Adaptive Management

BEING ADAPTIVE MEANS BEING SUCCESSFUL. AN ADAPtive person recognizes when a change is coming, diagnoses the meaning of the change, makes a plan, and puts the plan into action. Adaptive people keep their "antennae up," watching whether or not the new plan is working. By adapting, they are able to prosper in the new situation, while their nonadaptive counterparts fall farther and farther behind. In other words, an adaptive person *learns*.

The same is true for people working together on ecosystem management—we must be adaptive. An adaptive organization is one that actively engages its environment, seeking information (positive or negative), rewarding those who bring it, and learning from it. Following that idea, the successful ecosystem management team will discover and adopt new ways to perform, embracing new partners, changing goals or interests, and using new processes to achieve its objectives. A nonadaptive group, in contrast, would probably do the same things over and over, perhaps getting better and better at doing the wrong things.

Younger people often complain that "old timers" want to use old tools, whether shovels or Chevys or chi-square tests, because they know how to use them. To learn, however, we need to accept new knowledge when it comes along, not ignore it. We also need to be willing to try new approaches to

EXERCISE 4.1

Collaborate on It!

In a group, discuss how an adaptive person and a nonadaptive person might react to the following situations (one for each scenario):

- ROLE Model: As a resource manager for the Bingham Wildlife Refuge, being assigned to a team that will create an Internet presence for the ROLE Model.
- SnowPACT: As a member of the SnowPACT Community Circle, learning that Congress is considering a bill to cede KARMA to the Semak Nation.
- PDQ Revival: As a waterfowl analyst in the U.S. Fish and Wildlife Service's national office, being transferred to a position of manager of the John Muir Wildlife Refuge in the PDQ region.

Figure 4.1. *Both sign painters in this cartoon may have stopped learning. How might they work together to create a learning opportunity in this situation? (Reproduced with permission of Nick Hobart.)*

solve problems and take advantage of opportunities, even when we are not sure how they will turn out (Figure 4.1). Fortunately, ecosystem management has come along simultaneously with a new approach to learning that enables us to be adaptive while we work—adaptive management.

Adaptive Management: Another Way to Learn

Through time, humans have developed many ways to learn. From superstition to scientific experimentation, each is useful in some circumstances and not very good in others. We can categorize learning into three well-known types, along with the new type we call adaptive management.

The most basic way to learn is through tradition. **Tradition** includes the transfer of knowledge through myth, lessons of elders, parental guidance, taboos, formal ceremonies, apprenticeships, and classroom education. Tradition simplifies learning, generally to emphasize important lessons. In resource management, for example, traditional learning focuses on how animal populations respond under conditions that are stable, nonmigratory, well-bounded, and instantly responsive to harvest levels. Although we know those conditions do not exist in nature, we use this simplified model to teach the fundamental concept of a "renewable resource."

A second way to learn is through **trial and error**. On-the-job-training, expert opinion, and the "college of hard knocks" are forms of trial-and-error learning. Trial-and-error learning allows advancement beyond tradition, as we use our experience to test our previous knowledge. Because it is based on individual experience, however, it is site-specific and learner-specific. For example, this is often the type of learning in aquaculture. Beyond the few general rules for raising fish, most practicing aquaculturists know how to manage their facilities to raise fish with maximum efficiency. But they often have difficulty expressing their implicit wisdom, and they hesitate to do so because what works for them might not work for someone else.

Whereas trial-and-error learning depends on a person and place, a more elaborate way to learn is called the **scientific experiment**. This method is considered the best way to gain explicit knowledge in Western cultures because it is objective, explicit, replicable, and, therefore, presumably valid anywhere, anytime, for everyone (Box 4.1). However, scientific experiments have limited utility because they are reductionist. They only work in relatively simple situations—where one variable can be changed while everything else is held constant, where the response to the experiment is rapid and easily measured, and where the experimental subjects can be chosen at random, used, and discarded. For instance, experiments can be designed and readily conducted using rainbow trout fingerlings that are grown in hatcheries in large numbers for pennies each, but how would we feel about conducting experiments on bowhead whales, which are rare, essential to native culture, roam throughout the Bering Sea, and cannot be kept in experimental aquariums?

In learning to perform ecosystem management, therefore, we have a dilemma. Tradition cannot prepare us for ecosystem management, because we are working in new situations without traditions. Trial-and-error learning is good, because it allows individuals to make some mistakes and im-

BOX 4.1

The Scientific Experiment

The scientific experiment follows a formal process that has become a model for objective, disinterested decision making. Although descriptions vary, the process follows these fundamental steps:

1. Observe a phenomenon in nature long enough that you develop a question about how it works.
   I've never seen rooted plants grow in water that is very deep; in fact, in really muddy ponds, plants only grow right along the shore.
2. State a hypothesis that describes how one part of the phenomenon (a cause) influences another part (an effect).
   Plants will only sprout if the light level reaching the water bottom is above a certain level.
3. Design a process that varies the cause so that its effect also varies in response; be certain that only this one cause is changing.
   Plant seeds for water plants in different jars, all with the same amount of the same soil and water, but with shading cloth over the top of the jars that restricts light levels to 0.75, 0.5, 0.25, or 0.0 of the sunlight. Make five jars at each light level.
4. Set up rules for deciding that the cause is actually producing the effect.
   If at least one plant sprouts in one jar at a light level, then there is enough light.
5. Run the process several times, and measure the outcomes.
6. Based on how the outcomes match the rules, conclude that the cause-and-effect relationship is real or not.
7. Report all the results so that others can repeat the experiment if they wish.

prove their performance, but the learning is not readily transferable to other places and people. Scientific experimentation can give us precise and universal answers to small questions, but it cannot help us address complex questions on an ecosystem scale.

Two ecologists, C.S. Holling and Carl Walters, began to address this quandary in the 1960s, by combining the advantages of trial-and-error and scientific learning. They called their approach adaptive management. In the most direct terms, **adaptive management** is the process of treating management as an experiment. By doing this, the practicality and importance of trial and error are added to the rigor and explicitness of the scientific experiment, producing learning that is both relevant and valid.

EXERCISE 4.2

Talk About It!

In groups, discuss the relative qualities of the three types of learning in regard to the characteristics listed below (add others, if you wish), and fill in the table. We have completed the first two rows for you.

Characteristic	Type of learning: Tradition	Trial-and-Erro	Scientific Experiment
Speed of Learning	Slow	Faster	Fastest
Resistance to change	High	Lower	Lowest
Ease of teaching to others			
Suitable for stable situations			
Suitable for complex situations			
Benefits			
Costs			
Others			

Active Adaptive Management

Walters and Holling developed a comprehensive way to apply adaptive management, known as **active adaptive management** (Figure 4.2). Their process follows a series of steps that resembles the scientific method but is applied in management-type settings. We will illustrate active adaptive management using a hypothetical forest ecosystem example.

Imagine an ecosystem management goal of sustaining a healthy hardwood forest that allows timber harvest, outdoor recreation, and biodiversity conservation. A desired aspect of such an ecosystem might be large and widely dispersed

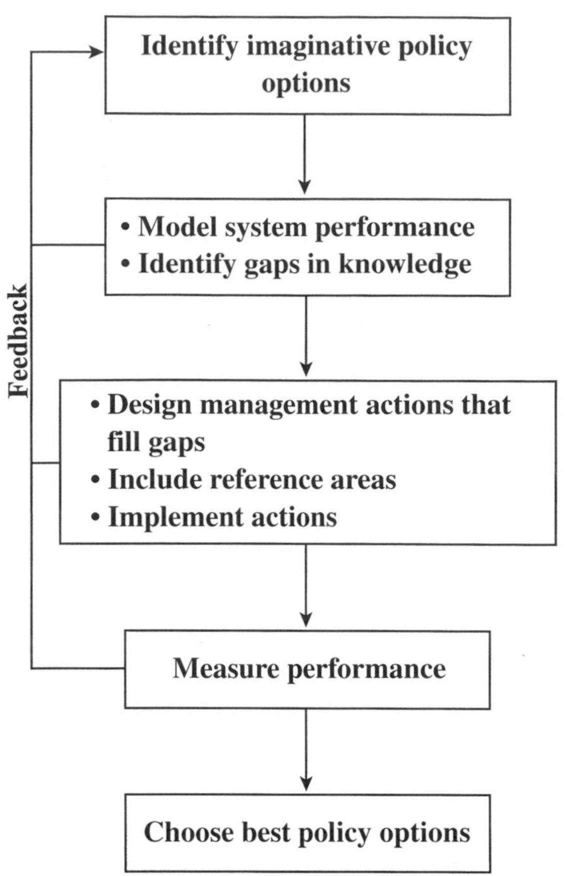

Figure 4.2. *The active adaptive management process, as developed by Walters and Holling. This model resembles a scientific experiment; the difference is that the adaptive management experiment is conducted at large scales, in actual working situations that lack strict control, and on socioeconomically important topics.*

late-successional stands of native hardwoods. Stakeholders may become concerned about the threat of gypsy moths, which are increasing in abundance in an adjacent state and moving toward this ecosystem. The gypsy moth is an introduced insect whose caterpillars eat tree leaves, defoliating large forest areas and sometimes causing massive death of trees (Figure 4.3). Various stakeholders, including public agencies, forestland owners, tree nursery owners, and park managers, might agree that a comprehensive, long-term approach is needed. However, serious questions might exist about what is an effective control of the moths, how to conduct the control process, and what unintended impacts those controls might have. Stakeholders might choose active adaptive management to investigate these questions.

The active adaptive management process begins with listing *imaginative policy options* that might be attempted (see Figure 4.2). In this example, the stakeholders might have several options for controlling gypsy moths: blanket insecticide spraying along the state's eastern border; spot treatments where moths are detected at some threshold density; building a "moth fence" by harvesting trees along the path moths are likely to travel; introducing a biological predator on moth caterpillars; helping pay for any treatment in the adjacent state before the problem crosses the border; or accepting their inability to control moths and salvaging what they can after the damage occurs. From these

Figure 4.3. *Insect invasions, like those of gypsy moths, occur on a large scale, in uncontrollable settings, with the potential for widespread forest damage. Deciding how to control them effectively is a likely topic for active adaptive management. (Photos by Larry A. Nielsen.)*

policy options, one that seems likely to succeed would be chosen. For this example we have chosen blanket insecticide spraying.

The next step is an in-depth and *explicit modeling* of how the system functions and how it might respond to the new policy. For example, imagining how insecticide spraying might affect a community of native species would require a computer model linking population trends and species interactions. As part of this process, *gaps in knowledge* are identified—things that are not known and that prevent the new approach from being implemented with confidence. For example, spring and summer spraying might have different effects on other insects and, therefore, on birds that eat the insects. This step is vitally important, because active adaptive management relies on using management actions to learn more about the system, so that better decisions can be made in the future.

Once the most critical gaps in knowledge have been determined, *management actions* are designed to gather information about the gaps; this step becomes the mechanism for learning. For example, if gaps exist in understanding how one aspect of the community of species (e.g., the food chain from insects to passerine birds to birds of prey) would respond to widespread spraying in spring versus summer, two schedules of spraying might be implemented—one in spring and one in summer. Spraying would be scheduled at sites where gypsy moth infestations are low and high and where bird populations and feeding habits have been studied previously (or at least where healthy bird populations are known to exist). The management experiment would include several sites to be sprayed, chosen randomly, along with several sites where no spraying would occur, *providing a reference* for judging changes. Rather than being conducted on a small patch of ground or in an aviary, the treatments would be conducted across large areas as part of the ongoing ecosystem management process.

This is an essential point: Active adaptive management differs from scientific experimentation because the actions are conducted at a management scale (e.g., thousands of acres) rather than at an experimental scale (e.g., a few square meters). Therefore, the treatments would be real, and the learning would be relevant to future management decisions. This scale of action also demonstrates why active adaptive management is often controversial. In some cases the treatments can be so large as to become public issues, affecting access to and use of public and private resources.

As the management actions proceed, data are collected at the treated and reference sites in order to *measure performance*. These data need to address all the questions that were raised during planning, that will be relevant to learning about the system, and that will detail the efficacy of the treatment. The experiment and data collection must continue until the planned ending date, even if it is several years ahead. For example, if the gypsy moth model suggested that changes in the food chain would not be evident until 3 years after spraying stopped, it would be necessary to collect data that long and to refrain from spraying those areas again.

Based on the outcome of the experiment, the model would be modified to include the new knowledge, and it, along with the direct experimental results, would help decide the best *policy option*. In this example, if the new understanding showed that summer spraying were better, then summer spraying would be continued and spring spraying would cease.

A cautionary note: Although we describe active adaptive management as something that should continue to its planned end, there may be times when special circumstances cause the experiment itself to be terminated. A devastating environmental event (a strong storm or an unanticipated outbreak of disease or parasite) may require cancellation of the experiment because it would no longer yield results useful for learning. Also, implementation of the treatments may yield intermediate outcomes (e.g., the death of many organisms or economic disruption) so extreme that the lesson is learned quickly or the experiment becomes too risky to continue. For example, imagine that a very valuable site, such as a large old-growth stand, had been sprayed once and was now untreated while collecting data for 3 years; if an extreme outbreak of gypsy moths (or another insect) occurred in the stand, new spraying might be conducted, regardless of the loss of the data.

THE GLEN CANYON DAM

The Glen Canyon Dam, completed in 1963, impounds the upper Colorado River in northern Arizona, forming the massive Lake Powell reservoir in southern Utah (Figure 4.4). Glen Canyon Dam also regulates the flow of water in the Colorado River and Grand Canyon National Park. For many years, stakeholders have been concerned that flow regulation has an impact on the environment in the Grand Canyon, causing the depletion of sand bars and similar structures, and changing the water temperatures, thereby affecting fishes and aquatic invertebrates.

In response, new federal laws passed in the 1990s required that the Glen Canyon Dam be operated as an adaptive management program. The Glen Canyon Dam Adaptive Management Work Group (AMWG) was created to organize and implement the adaptive management processes. The AMWG is the decision-making group, composed in the true spirit of ecosystem management with representatives from federal agencies, Native American tribes, state agencies, environmental groups, recreation interests, and contractors who buy power produced by the Glen Canyon Dam. The AMWG includes two other subgroups: a Technical Working Group that helps the AMWG frame the technical issues and opportunities for adaptive management; and the Grand Canyon Monitoring and Research Center, which implements the adap-

Figure 4.4. *The Glen Canyon Dam impounds Lake Powell and controls water flow into Grand Canyon National Park. The dam has substantial influence on water flow regimes, which affect the downstream conditions in the Colorado River ecosystem. (Photo by Larry A. Nielsen.)*

tive management projects and collects, interprets, and reports data regarding the outcome of experiments and baseline conditions in the Colorado River ecosystem.

The Glen Canyon program is essentially active adaptive management. The AMWG has developed conceptual models for the Colorado River ecosystem, fulfilling a fundamental requirement of active adaptive management. It also considers its actions within the context of identifying innovative policy options and gaps in knowledge. For example, the AMWG has been considering how water temperatures in the river downstream can be better managed by changing the water intake strategy at the dam. It has held numerous workshops and discussion groups regarding alternative designs for a "temperature control device," what data are needed to evaluate those designs, and how an experiment might be conducted. The Grand Canyon Monitoring and Research Center collects both baseline and experimental data, maintains the ecosystem models, and updates the models with the knowledge learned from experiments.

However, the program departs from strict active adaptive management in several revealing ways. First, of course, there is only one Glen Canyon Dam, so having reference dams and rivers is impossible. The best they can do is monitor before and after a management action. Second, randomness must be abandoned because there is only one location, and time is limited. For example, a decision to try an innovative spring water release schedule would probably be implemented as soon as possible, rather than randomly choosing one year out of the next five during which to perform the experiment. Third, many variables change along with the management action, including weather, flow regime, biological populations, and human use (all the uncertainties mentioned in Chapter 3 apply here). Therefore, questions of cause and effect will always occur.

A recent adaptive management experiment illustrates these compromises. The experiment responded to an idea that low steady summer water releases from the dam would benefit the endangered humpback chub that lives in the Grand Canyon region of the Colorado River. The experiment called for a continuous low flow of 8000 cubic feet per second from June through September. However, because the total annual release of water through the dam is determined by the anticipated runoff (based on the snowpack accumulated in the previous winter and the spring rainfall), the experiment could only be conducted in a year when the predicted runoff was low. After waiting for 5 years, these conditions finally occurred in the winter of 1999–2000, allowing the experiment to be planned for summer 2000. Conditions remained suitable throughout the spring and summer, and the planned steady low-flow releases occurred throughout the summer. However, the experiment would have been stopped immediately if flood control, power needs, or other primary operational factors reached emergency levels. With the completion of the low-flow experiment, biological data on the response of the humpback chub and other aquatic organisms can now inform the full Adaptive Management Working Group on the future decisions for dam operation.

IDAHO ELK MANAGEMENT

The Idaho Department of Fish and Game has implemented an ambitious project to use active adaptive management to manage elk populations better (Figure 4.5). Its project, designed in 1992, demon-

Figure 4.5. *Managing elk populations is important throughout the United States. This large herbivore, popular with hunters, forages extensively on natural lands and farm fields, and is a symbol of wild beauty. (Photo by Larry A. Nielsen.)*

strates the possibilities and difficulties of actually using the technique in its full expression. Hunting antlerless elk is a basic part of the department's elk management program, because hunters want to shoot elk and because the harvest reduces elk damage to farm crops. Biologists sought a better understanding of elk population dynamics so that the harvest could be regulated more accurately, resulting in optimal production and consequent benefits to hunters and landowners. They adopted active adaptive management, following the general process developed by Holling and Walters.

They began with a specific goal—to optimize the harvest of antlerless elk. They constructed a complex model of elk population dynamics, including the full range of influences on populations, from weather to breeding success (Figure 4.6). Their model analysis revealed a critical gap in knowledge about the relationship of antlerless elk harvest to total antlerless elk mortality. Hunting mortality could be additive (thereby increasing total mortality), compensatory (replacing natural mortality but not adding to it), or compensatory up to a certain level and then additive. These situations gave very different results for population sizes in their model and also different results at low and high harvest levels. Based on their model, the biologists determined that by implementing zero, low, or high harvest levels in different areas, they could learn the true relationship between hunting mortality and total mortality.

They then implemented an active adaptive management project by selecting 11 elk management units for treatment. The difference between management experiments and true scientific experiments is revealed in their design. The experimental units were large, covering more than 10% of the

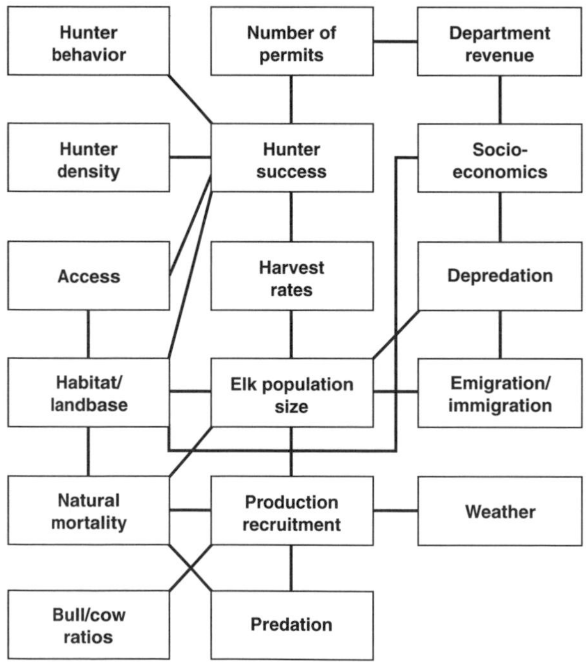

Figure 4.6. *A diagrammatic representation of the elk population model that formed the basis for the active adaptive management experiment conducted in Idaho.*

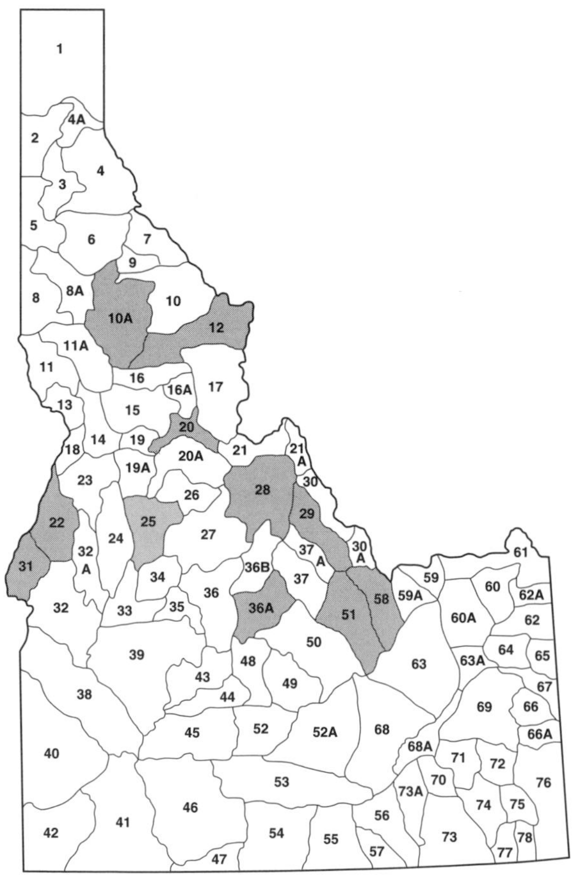

Figure 4.7. *The active adaptive management process for elk in Idaho used 11 different management units (hatched areas on the map), comprising 10% of the state's land area.*

state (Figure 4.7). Because the harvest is a function of the number of hunters afield and their success, harvest rates could not be controlled exactly. Therefore, biologists had to control the harvest as well as possible by increasing or decreasing the number of hunting permits available in each experimental area. The "zero" treatment actually had to include some hunting opportunity (for public relations purposes), set at 2–3%; the low and high harvest rates ranged from 6% to 10% and from 14% to 28%, respectively. Also, because other management goals were also being sought at the same time, a totally random selection of experimental areas was not possible.

After 6 years of implementing the active adaptive management experiment, data showed that higher harvest rates actually improved calf/cow ratios and improved overall recruitment to the populations, indicating that harvest mortality was compensatory. To ensure that the experimental results were valid statewide, however, additional studies were undertaken using GIS data to reveal widespread patterns in harvest rates and population characteristics, and further adaptive management experiments were continued in one management region. Based on the results of the active adaptive management experiment, the Idaho Department of Fish and Game set a policy that the target for antlerless elk harvest rates in all 78 management areas would be a minimum of 10% of the population.

Passive Adaptive Management

Although active adaptive management has been discussed extensively, it has been tried in full only a few times. Consequently, practitioners have developed a less demanding form called passive adaptive management. **Passive adaptive management** means that some aspects of an experiment are missing, but learning is still a major objective of the activity and the activity is conducted as much like an experiment as possible. The most frequent differences between active and passive management are in the earlier steps, usually skipping the construction of elaborate models and choosing sites for treatments nonrandomly. In these cases, the management action or sites may be selected based on more immediate needs, such as the expressed desires of stakeholders, legal requirements, court decisions, or economic requirements. Working within these constraints, however, reference areas still can be included and performance measured.

THE NORTHWEST FOREST MANAGEMENT PLAN

A massive and explicit commitment to passive adaptive management has occurred through the Northwest Forest Plan (NWFP), the coordinated management of 24 million acres of federal lands in Washington, Oregon, and northern California. The NWFP is the outcome of the set of issues and decisions surrounding (1) the protection of late-successional forests that are home to northern spotted owls, marbled murrelets, and many other species and (2) the sustainability of the logging industry upon which many Pacific Northwest communities depend (Figure 4.8). This situation became so controversial in the late 1980s and early 1990s that President Clinton held a "forest summit" in the region in April 1993; he created the Forest Ecosystem Management Assessment Team (FEMAT) to forge an interagency ecosystem management plan. The FEMAT's report created a true ecosystem-level approach (the NWFP) that has guided federal land management since it was signed into law as the "Record of Decision" on April 13, 1994.

The plan explicitly commits federal land management agencies to use adaptive management as a fundamental element of their programming. The FEMAT members realized that accomplishing the twin goals of environmental protection and economic sustainability required answers to many questions about how timber management (e.g., tree planting, thinning, and harvesting; road building) influenced endangered species, biological communities, and ecological functions. Decades might pass before traditional experiments could answer these questions, during which all commercial forestry activities would stop—an unacceptable situation for most stakeholders. Therefore, the NWFP created

Figure 4.8. *The public lands in the Pacific Northwest have been a focal point for ecosystem-level thinking, as goals for wildlife habitat and timber harvesting have come into conflict. An innovative ecosystem-scale project, the Northwest Forest Plan (NWFP), chose to enhance learning about the ecosystem by including adaptive management areas in land allocations. (Photo by Larry A. Nielsen.)*

the opportunity "to develop and test new management approaches to integrate and achieve ecological, economic, and other social and community objectives." In allocating lands across the region for specific purposes, the plan assigned 1.5 million acres in ten "adaptive management areas."

Since 1994, many passive adaptive management experiments have occurred on these lands. The Little River Adaptive Management Area, for example, the smallest of the ten areas, covers approximately 83,000 acres in southern Oregon. The emphasis for adaptive management in Little River is to learn how to integrate intensive timber production with the restoration and maintenance of high-quality riparian habitat. Box 4.2 describes several projects demonstrating the range and scope of the Little River program. These activities also illustrate why we classify the NWFP as passive adaptive management. Because the actions are restricted to formal adaptive management areas, they are not randomly placed and may not occur where learning is optimal. Also, actions within adaptive management areas must comply with a series of federal and state laws and plans that take precedence over experimental designs.

THE NORTH AMERICAN WATERFOWL PLAN

The U.S. Fish and Wildlife Service practices a form of passive adaptive management in its program to manage the waterfowl harvest (Figure 4.9). Since 1916, the waterfowl harvest has been managed via an international treaty among Canada, the U.S., and Mexico. Working within guidelines that establish the general length and timing of the waterfowl

BOX 4.2

Adaptive Management in the Northwest Forest Plan

The Little River Adaptive Management Area is one of ten designated land areas devoted specifically to "managing to learn and learning to manage" within the Northwest Forest Plan. The 83,000-acre area is 63% federal lands (BLM and U.S. Forest Service) and 37% private lands, mostly managed as industrial forests. The following projects illustrate the range and purpose of management in the Little River AMA (adapted from the Little River AMA Web site). As you read about these projects, consider how they match the characteristics of passive adaptive management.

The Fall Creek Riparian Restoration

The Fall Creek watershed burned in 1987, destroying the riparian conifers and also the upland seed sources. The goal of the adaptive management project is to use tree planting as a means to reestablish the original riparian condition, including snags and downed logs, on a 20-acre site.

The Glide School Partnership for Education and Ecosystem Management

The goal of this project is to provide local students with experiences in natural resources and ecosystem management that are practical and teach problem solving. The project also seeks to generate usable water quality information through student monitoring. In the project, students at the school will be given portable instruments and taught the interagency protocols for collecting data. They will measure water quality parameters and report them to the agencies.

Restoration of the Umpqua Mariposa Lily

The Umpqua mariposa lily is an endemic flower in Oregon and is classified as a state endangered species. The adaptive management project seeks to develop proactive methods for maintaining or increasing populations through prescribed burning, tree girdling, and thinning of competing vegetation.

Sampson Butte Commercial Thinning

This project seeks to learn how a proportional thinning approach to retaining trees across all diameter classes can improve the growth and vigor of individual trees, thereby contributing to greater structural diversity in the forest. Such diversity is desirable to enhance and prolong habitat conditions suitable for northern spotted owl foraging and dispersal and to increase understory plant diversity in riparian and upland areas.

The Withrow Timber Sale

This project explores how forest structure can be improved or restored to minimize fire risks and maintain habitat for late-successional species. A combination of thinning, group openings, prescribed understory fire, and snag creation will be used to mimic the natural fire regime on warm-dry slopes with the Little River watershed.

hunting season, four Flyway Councils assess the status of waterfowl populations and other data, and then suggest harvest guidelines for the respective governments. In the U.S., these guidelines are then adopted as hunting regulations by the states through which the waterfowl travel.

Harvest regulations are based on a model of waterfowl population dynamics that includes breeding population sizes, harvest levels, migration, and other characteristics; this is the model that adaptive management requires. Because of the continent-wide range of migratory waterfowl, the large number of harvest sites and conditions, and the array of species, the models that underlie management still contain great uncertainty. The USFWS waterfowl biologists, therefore, have utilized a statistical method (the Markov decision process) to learn how each of their harvest regulation decisions affects subsequent population characteristics. They continually adjust their models to incorporate the new knowledge as it develops.

One specific question under consideration by waterfowl managers is the right size and number of management units. Having many small management units increases the ability to tailor regulations for specific species, times, and conditions, but it

Figure 4.9. *Managing migratory waterfowl in North America involves several nations and many states, across the entire continent. Passive adaptive management is an appropriate approach, taking advantage of the opportunities to learn, while recognizing that the situation is too complex and politically sensitive to allow active adaptive management. (Photo by Larry A. Nielsen.)*

also makes monitoring, management, and regulation setting more costly and cumbersome. Having fewer large management units decreases costs, regulation variations, and hunter confusion, but it may reduce the ability to optimize overall population characteristics and hunting satisfaction. Biologists are using the data collected in each season to model how various objectives for waterfowl management turn out with fewer large or more small management units, thereby learning more each year about the optimal spatial scale for management.

Why is this process passive, rather than active? First, biologists collect data about harvest levels and population characteristics based on the regulations that are enacted, rather than choosing regulations specifically to generate the data they need. Second, they cannot set up true reference areas (places where changes in the regulations are not enacted); instead, they must rely on baseline data from previous years as a less powerful comparison for deducing cause and effect. Third, future policy choices are not specifically tied to the outcomes of the enhanced learning; decision makers pay attention to the improved models, but they are not obligated to modify harvest regulations accordingly.

Adaptive Management as Documented Trial and Error

Elaborate adaptive management like these examples is highly desirable, but it is seldom accomplished. Trial-and-error learning can also resemble adaptive management, as long as the learners collect data, analyze them objectively, and share their learning with others; this approach is called **documented trial and error**. We can use this type of learning with single, direct, short-term questions simply by being curious, rigorous, and explicit in our work (Figure 4.10). Virtually any task can be turned into a learning opportunity by asking a few questions, such as "What other ways to do this job could be better? How will I know which way is best? How can I test my ideas?"

Imagine a community-based ecosystem management plan that included establishing a series of gardens in a local park. The goal of the gardens might be to restore native wildflowers that would nurture a diverse insect community—including, of course, the butterflies that attract many visitors. This activity might ordinarily be guided by an Extension Service bulletin, with volunteers planting and caring for the gardens and being pleased or disappointed by the results. Although seemingly straightforward, many decisions need to be made, such as what plants to

Figure 4.10. *Documented trial and error can be used in the most simple cases, by individuals or groups, whether professional or volunteer. A curious mind and the willingness to be explicit about the outcomes are all that are necessary. (Photo by Larry A. Nielsen.)*

grow, what cultivars to use, when to plant, and how to thin and prune.

The principles of adaptive management could teach participants how to manage their gardens more successfully in the future. A number of garden plots might be planted and treated differently, even if the differences were slight. Backyard and schoolyard gardens could be used as well as public gardens, engaging more citizens, schoolchildren, and teachers. A plan for monitoring the garden's performance could be devised in which butterfly observers and other citizens measured use by butterflies, bees, and insects of various kinds. Community volunteers could collect data during routine visits to the plots, and scouting groups, 4-H clubs, and school classes could conduct projects to assemble and analyze the data. By sharing these with nearby communities, a new appreciation could be developed for what works, when, and where—and next year's gardens could be greatly improved.

Adaptive management is not restricted to biological work, but it can be just as useful in the socioeconomic and institutional realms of ecosystem management. Imagine trying to find the most effective way to reach stakeholders about proposed management actions. Based on discussions with planners and media experts, an ecosystem team might develop a stakeholder response model suggesting that people tending to agree with a proposal respond at higher rates to mailings, but that those tending to oppose a proposal respond at higher rates to newspaper stories. An adaptive management experiment might include using different notification styles in two or more communities and judging the rates of agreement and disagreement.

Remember also that a goal of adaptive management is learning, even at the cost of some additional time and expense. If this example had involved a regulatory change being proposed by a state or local government, laws would require that all stakeholders enjoy the same opportunity to comment. Consequently, a follow-up set of contacts would be needed in the communities, to ensure that everyone got the message. But consider how much would be learned if the two notification strategies produced equal results; individual mailings could be abandoned as a notification strategy in the future, saving unnecessary time and expense in the long run.

EXERCISE 4.3

Collaborate on It!

Describe how active adaptive management, passive adaptive management, and documented trial and error might be used in the following situations (one for each scenario):

- ROLE Model: Government agencies want to remove the old dams on Bent Creek in the Round Lake ecosystem, but there are concerns about changes in hydrologic conditions, sedimentation rates, and toxic chemical release, as well as the unimpaired migration of fishes and other organisms from Round Lake.
- SnowPACT: Native Americans of the Semak Nation want to introduce bison onto their lands in the Snow River watershed, but other residents are worried about the impact of disease, escape, and the destruction of other lands, as well as competition with other native animals.
- PDQ Revival: A project has been proposed to place artificial nesting boxes for red-cockaded woodpeckers in suitable trees at various locations in the PDQ region. In other places, nest boxes have been successful, but they have never been tried in environments quite like those in this watershed.

Conditions Necessary for Successful Adaptive Management

As the previous examples illustrate, adaptive management can be a tricky process. Although many active adaptive management projects have been planned and even partially implemented, most have withered on the vine. For one reason or another, adaptive management is often considered too costly—in time, money, or lost decision-making freedom—to have been widely adopted in its full form. Therefore, ecosystem teams should

BOX 4.3

Making Adaptive Management Work

Adaptive management, whether practiced formally or informally, works best under conditions that foster learning and changing. If all or most of these conditions are not present, it is better not to start!

Ecological Conditions

- Large differences in response between treatments and reference areas are likely.
- Data collection is relatively easy and inexpensive.
- Results will develop relatively quickly and clearly.

Socioeconomic Conditions

- Stakeholders agree on the desired outcomes of management, but disagree on the means to achieve them.
- Stakeholders are interested in and committed to the process.
- Most stakeholders agree on the facts and underlying models of performance.
- Nonscientific knowledge, from many sources, is included in the modeling and design.
- Sociological and economic knowledge, conditions, and concerns are included in the design.
- Communication is continuous.

Institutional Conditions

- The sponsoring organization is committed to learning.
- Eventual management decisions will be linked to the outcomes of the adaptive management experiment.
- Funding and leadership are stable, so later implementation is likely.

carefully assess the conditions necessary for success as they consider starting an adaptive management process (Box 4.3).

ECOLOGICAL CONDITIONS

Adaptive management works well when large differences in the performance of the system are predicted. Given our inability to control many variables at work at an ecosystem scale, the outcomes of adaptive management will have lots of "noise." Unless the expected results of adaptive management are quite large, the noise will probably confound the data, and the data will not be conclusive.

Data collection must be practical. A critical step in adaptive management is collecting data about the system's performance, at both treatment and reference sites. But decision makers will balk if the costs of data collection are too high. Common sense suggests that a monitoring program costing millions of dollars will seldom be fully funded or mandated. Therefore, an adaptive management program depending on costly monitoring will probably fail.

The last ecological condition is also related to data collection: An effective adaptive management project must result in observable changes that occur relatively quickly and clearly. Although we all like to ponder consequences that might occur decades from now, an elected official or community group is unlikely to wait that long before making other decisions. Similarly, an effect that is hopelessly tangled up with other phenomena, so that the outcome requires lots of explanation and hedging, has little chance of influencing decision makers.

SOCIOECONOMIC CONDITIONS

The socioeconomic conditions surrounding adaptive management are extraordinarily strict. After all, adaptive management involves the resources that people use and enjoy; unless stakeholders are convinced of the value of the work, they will resist. Kai Lee, in his 1993 book on managing natural resources in complex settings, notes that adaptive management can be used when people agree on the desired outcomes of management but disagree on how to get there. However, if people disagree on the desired outcomes—some want resource development, some want preservation—no amount of adaptive management will help the situation. In other words, adaptive management can help determine *facts* but it cannot reconcile *values*.

In order for adaptive management to succeed,

stakeholders must be interested and involved. Because they have the opportunity to disrupt an adaptive management project at any time, stakeholders hold the key to success or failure. Helping citizens, community leaders, and elected officials understand the process is vital before proceeding. A crisis—legal, economic, safety—often stimulates adaptive management because it helps people put aside their differences and work together on their common interest. If people all agree that something needs to be done, they may be ready to learn by doing.

All stakeholders need to accept the basic model on which an adaptive management process is based. Consequently, they need to be involved in the construction of the model. To avoid later arguments about the validity of the model, stakeholders need to see what goes into the model and add their own knowledge. Moreover, knowledge that comes from nonexperts needs to be given equal weight with knowledge offered by scientists and other specialists. Anglers, hunters, hikers, farmers, boaters, loggers, and others who spend much time on the land and water have experience that is relevant in adaptive management, especially in the initial stages of imagining policy options and constructing models. The knowledge of native peoples, learned across centuries and passed down through generations in stories and rituals, is equally legitimate and meaningful.

The policy options to be tried and the measures used to evaluate them must include sociological and economic elements. Indeed, much adaptive management is undertaken to allow economic or socially desirable activities to continue, or to alleviate strain on communities and businesses caused by environmentally oriented regulations. Therefore, most stakeholders (including the most influential ones) will be interested in what adaptive management means for the people who live, work, play, and worship in an ecosystem (Box 4.4).

Finally, information about an adaptive management project needs to be fully and widely shared while it is in progress. From a consideration of policy options to the final assessment of outcomes, stakeholders need to be kept informed. This is important, because people are likely to forget that an experiment is going on; if they like the treatment, they will want it to continue forever, and if they do not like the treatment, they will keep trying to stop it. Neither case is good for learning and improvement.

BOX 4.4

Adaptive Management Proposed—and Rejected

Lake Michigan's salmon sport fishery is an artificially maintained ecosystem where an exotic prey species, the alewife, is kept in check by stocking exotic predators, Pacific salmon. Over the past half-century, however, this ecosystem has become the basis of a thriving recreational salmon fishery. Consequently, effective management of the ecosystem is important to the economy and quality of life of many communities in Wisconsin, Illinois, Indiana, and Michigan

A question asked by fisheries scientists is: "At what level of predator stocking will the prey base collapse, and, if that happens, what trophic cascade occurs to the levels of zooplankton (a food of the alewife) and phytoplankton?" This is a great question, of substantial theoretical interest, and an adaptive management project to investigate it was actually designed in the early 1990s. Scientists proposed to continually raise salmon stocking rates until the alewife population crashed.

Citizens and business people dependent on the resource criticized the idea because it would put a sport fishery valued at several hundred million dollars per year at risk. Eventually the agency responsible for managing the fishery rejected the proposal. Although the well-intentioned scientists prepared a viable active adaptive management project, they could not get the necessary stakeholder agreement to proceed.

INSTITUTIONAL CONDITIONS

Once the adaptive management design is right and the stakeholders are all on board, the job is only partially done. The sponsoring institution, whether an agency, a nongovernmental organization (NGO), or a community group, must nurture the adaptive management project and grow with it.

The most important condition is that the

institution must be committed to learning as a goal of this action. A pathological or bureaucratic organization is not likely to embrace adaptive management. Even if it promotes the idea publicly, such an organization is probably using adaptive management as a buzzword, designed to make good public relations.

Groups also need to be ready to change their practices based on the outcomes of adaptive management. Even if an organization agrees to perform adaptive management projects, the payoff comes when the new knowledge is incorporated into actions. Organizations need to be open to change, including adopting new ideas as well as discarding old ones, when the proof arrives.

And, of course, organizations need stable leadership and financial commitment in order for adaptive management to succeed. Because adaptive management extends across budgetary years and ownership boundaries, it will overlap the terms and territories of mid-level group managers, organizational boards of directors, and elected officials. The champion of adaptive management must have the stability, prestige, financial control, passion, and persistence to ensure that the learning train stays on the track.

Adaptive management also reinforces perhaps the most adaptive of all traits—humility. Albert Einstein said that as a circle of light expands, the circumference of darkness also increases (Figure 4.11). That metaphor is perfectly appropriate for ecosystem management. As we try to learn, we are always humbled by how much more the learning reveals to us, in opportunities and uncertainties.

Figure 4.11. *Adaptive management, like all forms of learning, always raises more questions than it answers. Practitioners of ecosystem approaches to conservation will thrive if they embrace not only the circle of light, but also the circumference of uncertainty.*

References and Suggested Readings

Espy, M., and B. Babbitt. 1994. Record of decision for amendments to Forest Service and Bureau of Land Management planning documents within the range of the Northern Spotted Owl. U.S. Department of Agriculture and U.S. Department of the Interior, Washington, D.C.

Gratson, M.W. 1999. Managing cow elk harvest rates by experiment in Idaho: Preliminary results, practical considerations, and challenges. Pp. 21–22 in G.B. MacDonald, J. Fraser, and P. Gray (eds.). *Adaptive Management Forum: Linking Management and Science to Achieve Ecological Sustainability.* Science Development and Transfer Branch, Ontario Ministry of Natural Resources, Peterborough, Ontario.

Gratson, M.W., J.W. Unsworth, P. Zager, and L. Kuck. 1993. Initial experiences with adaptive resource management for determining appropriate antlerless elk harvest rates in Idaho. Transactions of the North American Wildlife and Natural Resources Conference 58:610–619.

Holling, C.S. 1978. *Adaptive Environmental Assessment and Management.* Wiley, New York.

Johnson, F., and K. Williams. 1999. Protocol and practice in the adaptive management of waterfowl harvests. Conservation Ecology 3(1):8 [online].

Lee, K.N. 1993. *Compass and Gyroscope—Integrating Science and Politics for the Environment.* Island Press, Washington, D.C.

Lee, K.N. 1999. Appraising adaptive management. Conservation Ecology 3(2):3 [online].

McLain, R.J., and R.G. Lee. 1996. Adaptive management: promises and pitfalls. Environmental Management 20:437–448.

Pinkerton, E. 1999. Factors in overcoming barriers to implementing co-management in British Columbia salmon fisheries. Conservation Ecology 3(2):2 [online].

Shindler, B., and K.A. Cheek. 1999. Integrating citizens in adaptive management: A prepositional analysis. Conservation Ecology 3(1):9 [online].

Walters, C.J. 1986. *Adaptive Management of Renewable Resources*. Macmillan, New York.

Walters, C.J. 1997. Challenges in adaptive management of riparian and coastal ecosystems. Conservation Ecology 1(2):1 [online].

Walters, C.J., and R. Green. 1997. Valuation of experimental management options for ecological systems. Journal of Wildlife Management 61:987–1006.

Williams, B.K., and F.A. Johnson. 1995. Adaptive management and the regulation of waterfowl harvests. Wildlife Society Bulletin 23:430–436.

PART II

The Biological and Ecological Background

Genetic Diversity in Ecosystem Management

ECOSYSTEM MANAGEMENT IMPLIES BIG THINKING—BIG in time, in space, and in inclusiveness of decision making. When approaching conservation issues at this scale, one might question whether genetics has a role to play. After all, it seems a major leap from genes to landscapes and human endeavors, so why should we be concerned with genetics when dealing at such a large scale? First, recall Figure 2.6, where biodiversity was illustrated as the composition, structure, and functions associated with levels of biological organization from genes to landscapes. Genes are the fundamental building blocks of all higher levels of biological organization, including ecosystems. If genetic composition, structure, or function is substantially altered, it could have a rippling effect throughout populations, species, and ultimately ecosystems.

Second, we know from the Fundamental Theorem of Natural Selection that the rate of evolutionary change in a population is proportional to the amount of genetic diversity available. No (or little) variation means no (or little) natural selection, which means no (or little) opportunity for species to adapt to changing conditions. Loss of genetic diversity can mean evolutionary stasis for a population or species, and premature extinction.

Third, a consensus among population geneticists holds that the individual fitness of organisms increases with genetic variation (or, perhaps more importantly, decreases with reduced genetic variation). Reduced fitness may mean lowered reproductive success and possible population declines.

Fourth, a tremendous local and global biotic resource is eroded as genetic variation disappears. Genes are, after all, the blueprints for life, and as they are lost, the basis for life on Earth becomes impoverished.

Finally, genetics has direct relevance to the problems of managing endangered species, which are themselves important components of the overall ecosystem approach. Almost by definition, populations of endangered species are small, and small populations tend to lose genetic diversity. This loss can lead to a higher probability of population extinction. Not only are endangered species populations small, but they typically occur as isolated populations across a fragmented landscape. This isolation can lead to a loss of genetic diversity within populations, as well as a loss of gene flow among populations, which alters geographic patterns of genetic diversity. In this way, genetics is linked directly with landscapes: The natural and modified landscape patterns in an ecosystem can have direct and important consequences on genetic variation and its spatial distribution, maintenance, or loss.

For these reasons, genes are as relevant to ecosystem management as any other level of biological organization. Ecosystem management is not focused at any particular level of biological hierarchy but encompasses them all. Consequently, we will begin our examination of the biological basis for ecosystem management with genetics and work our way through populations and species to landscapes.

What Is Genetic Diversity?

Genetic diversity exists at three fundamental levels: within individuals, among individuals within a population, and among populations. Let's explore each of these in turn.

Every individual organism that results from sexual reproduction carries genetic variation inherited from its mother and father. In humans, our 46 chromosomes are paired, with 23 coming from each parent. These "matching" chromosomes have (for the most part) identical *types* of information on them, but the specific *forms* of that information may differ. For example, there is a place on a particular chromosome where eye color is encoded, but the specific instructions (e.g., for blue, brown, green eyes) may differ between the two chromosomes. Likewise, genes for hair coarseness, skin tone, bone development patterns, biochemical pathways, and tens of thousands of other traits all are encoded and at least potentially carry different forms of expression on the two chromosomes of a pair (each derived from one parent). These different forms constitute within-individual genetic variation.

There is also variation among individuals within an interbreeding population. Every individual organism (except for identical twins or clonally produced individuals) differs genetically from every other individual of that species. This is easily detectable simply by looking at the people in a classroom or an office. Distinguishing between two or more individuals is not difficult, and this is partly a result of among-individual genetic variation. The sum total of that variation in an interbreeding population is considered to be the **gene pool** for that population.

Finally, different populations of organisms may differ genetically as a result of partial or complete genetic isolation from other populations of the same species. This can result in local adaptations. For example, a low-elevation and a high-elevation population of a lizard species may differ genetically, with these differences a reflection of their adaptations to different climates and environments at these different elevations (or, perhaps, simply due to random differences between the populations). The more isolated a population, the better the chance that it differs genetically from other populations of the species. Long-term isolation ultimately can lead to speciation.

Each of these forms of genetic diversity can be an important resource for the species and its role in ecosystem function. Each level of diversity deserves consideration, and possibly protection, in management programs.

Before proceeding further into the world of conservation genetics, you may wish to review definitions of some basic genetics terms (Box 5.1). These may be familiar to some but vague to others, so a quick review for proper understanding at this point is probably helpful. These terms and concepts will be used in further consideration of the role of genetics in ecosystem management.

A LOOK AT HETEROZYGOSITY

Heterozygosity (Box 5.2) represents much of the focus of conservation geneticists for two reasons: Heterozygosity often is positively correlated with fitness, and it typically declines in small popula-

BOX 5.1

Some Important Genetics Definitions

Locus: A physical location on a chromosome occupied by a specific gene. A given chromosome can contain thousands of gene loci, each responsible for encoding a particular trait.

Allele: One of a pair of genes at a particular gene locus. Alleles can come in one or more forms or variants (e.g., blue eye [*b*] vs. brown eye [*B*] alleles). There are possibly many alleles that can occur at a given locus (a_1, a_2, . . . a_n), although any individual would only carry two alleles of the many possible alleles available in that population.

Homozygous: Having two of the same alleles at a given locus (e.g., a_1a_1 or a_2a_2). The individual is genetically invariant at this locus because it can produce only one kind of gamete at that locus (e.g., only a_1 or only a_2). If all of the gene loci in a given individual were homozygous (an unlikely scenario), that individual would be said to have no within-individual genetic diversity.

Heterozygous: Having two different alleles at a given locus (e.g., a_1a_2 or a_3a_5). The individual is genetically variable at that locus because it produces two different kinds of gametes at that locus (e.g., a_1 *and* a_2). This is the basis for within-individual genetic diversity.

Fitness: The relative contribution of an individual's genotype to the next generation in context of the population gene pool. Loosely, fitness can be considered relative reproductive success. An individual is said to have high fitness if it produces more surviving offspring than other individuals in its population. An individual that does not leave any surviving offspring (i.e., its genes are not represented in the next generation) has zero fitness (ignoring, for the time being, shared genes from relatives passed on to the next generation).

Overdominance: The case in which a heterozygote at a given gene locus has higher fitness than either homozygote; that is, the condition of heterozygosity confers a fitness advantage over the homozygous condition. This is often what conservation geneticists are trying to protect.

Allelic diversity: The number and relative abundance of different alleles among individuals found at a given locus (or across a series of loci) in a population.

Heterozygosity: The proportion of measured gene loci that are heterozygous in an individual or averaged among individuals in a population (see Box 5.2). It is generally expressed as H, and % heterozygosity can range from 0 (all loci are homozyous) to 100 (all loci are heterozygous).

tions. Consequently, as populations decline in size as a result of human activities (e.g., habitat conversion, overharvesting, or the introduction of exotic competitors or predators), heterozygosity levels may decline, thereby possibly affecting the fitness of individuals in that population. Let's take a closer look at this issue.

Because heterozygosity is thought to correlate with fitness, higher individual or populational heterozygosity is considered to be advantageous to those individuals or populations. This is nearly impossible to validate clearly in nature because true fitness (relative reproductive success) is very difficult to measure. Instead, geneticists have measured parameters that seem to be correlated with fitness, so-called *fitness correlates.*

One example of the relationship of heterozygosity levels with fitness involves three populations of lions studied in the Serengeti (Tanzania, H = 3.1), the Ngorongoro Crater (Tanzania, H = 1.5), and the Gir Forest (India, H = 0.0). These three very different levels of heterozygosity corresponded to different reproductive measures in male lions (Table 5.1). Lions in the Serengeti (with the highest heterozygosities) had the highest sperm counts, the highest number of motile sperm per ejaculate, the highest testosterone levels, and the lowest percentage of sperm abnormalities. Lions of the Gir Forest (lowest heterozygosities) had the opposite pattern, indicating that male reproductive characteristics were, by comparison, poor. Although this a correlative study, it

BOX 5.2

Measuring Heterozygosity

Geneticists estimate heterozygosity in individuals or populations in a variety of ways. They all involve taking one or more tissue samples of the individuals being investigated. This method used to be lethal, because tissues such as muscle, liver, heart, and other organs were taken. More sensitive techniques have been developed that use blood samples, hair, skin slime from fishes, or even gut epithelial cells found in feces.

In one of the original and still popular techniques, the collected tissues are physically and chemically broken down and placed in a starch gel where they are exposed to an electrical field. The various enzymes and proteins in that tissue (which are gene products and indicative of the alleles encoding for them) migrate differentially in the electrical field based on their individual properties, including size and electrical charge. When the electrical field is stopped, they have migrated different distances. The gel is then biochemically stained for a particular gene product, a process that results in a color on the gel where the enzyme or protein stopped. The relative positions of these resultant "blotches" of color indicate the different alleles that resulted in the gene products. Thus, the allelic composition of an individual at a given gene locus can be determined (Figure A).

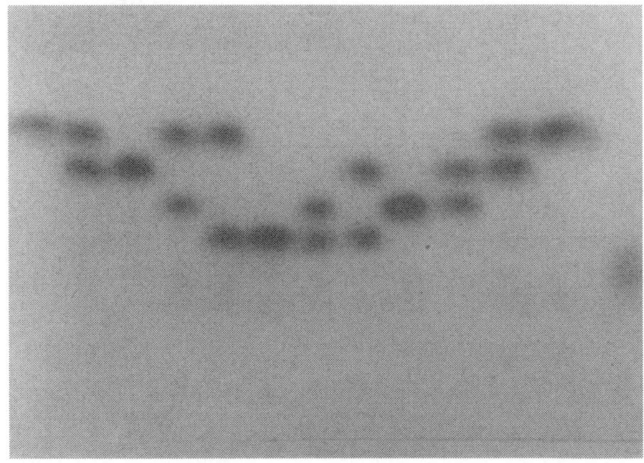

Figure A. *An example of genetic variation as seen on a starch gel. This is the gene locus aconitate hydratase (an enzyme) taken from the livers of 12 chinook salmon. There are four different alleles shown, with various heterozygotes (two bands) and homozygotes (one band) possible. Designating these as a, b, c, and d, on the basis of how far they traveled in the gel, the 12 individuals are as follows: a/a; a/b; b/b; a/c; a/d; d/d; c/d; b/d; c/c; b/c; a/b; a/a. (Photo by Paul Abersold.)*

If this process is conducted across a large number of loci (often several dozen or more), an overall estimate of the proportion of heterozygous loci for that individual can be made. If this procedure is repeated across many members of a population, an estimate of population-level heterozygosity can be made.

Obviously this approach is merely an estimate. A given species can have from tens to hundreds of thousands of gene loci, and we can measure only a relative few. Thus, an assumption is made that these loci are representative of the larger genome. However, many of the loci used are known to be important enzymes in cellular respiration, indicating that they have important functions in the organism and may be subject to strong selective pressures.

Table 5.1. *Correlation of Genetic Variation and Reproductive Parameters in Three Lion Populations*[1]

Parameter	Serengeti, Tanzania	Ngorongoro Crater, Tanzania	Gir Forest, India
Heterozygosity (%)	3.1	1.5	0.0
Reproductive measures			
Sperm count ($\times 10^{-6}$)	34.4 ± 12.8	25.8 ± 11.0	3.3 ± 2.8
% sperm abnormality	24.8 ± 4.0	50.5 ± 6.8	66.2 ± 3.6
No. motile sperm per ejaculate ($\times 10^{-6}$)	228.5 ± 65.5	236.0 ± 93.0	45.3 ± 9.9
Testosterone, ng/ml	1.3–1.7	0.5–0.6	0.1–0.3

[1]Data are the mean ± standard error of mean.

Source: Data from O'Brien et al. (1987b, 1987c), Wildt et al. (1987a), Yuhki and O'Brien (1990), and Gilbert et al. (1991).

does support the notion that levels of within-individual genetic variation can have fitness consequences.

This example illustrates an important aspect of heterozygosity: Relative, rather than absolute, levels can be most important. All three lion populations, as far as we know, are presently viable, having adapted to their local habitats. But a change in H could reduce their viability. If the low H seen in some populations is the result of long-term events and the population has done well, then it may be of minor concern relative to the population that experiences sudden declines in H and losses of genetic diversity.

Other examples of relationships between fitness and heterozygosity exist, as do some counterexamples. Overall, there is enough information to suggest that a relationship often exists between higher heterozygosity and higher fitness in many organisms, which means that a loss of heterozygosity could present conservation and management problems.

There is also an acknowledged relationship between population size and heterozygosity levels: Smaller populations tend to have lower heterozygosities than larger populations. This is demonstrated by a coniferous tree species from New Zealand (Figure 5.1a) and the red-cockaded woodpecker from the southeastern United States (Figure 5.1b). In both cases, population heterozygosity levels decrease significantly as population sizes become smaller, implying that heterozygosity (and thus fitness) is lost in small populations.

EXERCISE 5.1

Think About It!

The example presented here used male reproductive characters as fitness correlates. What are some other possible indicators of fitness that could be measured effectively in wild organisms?

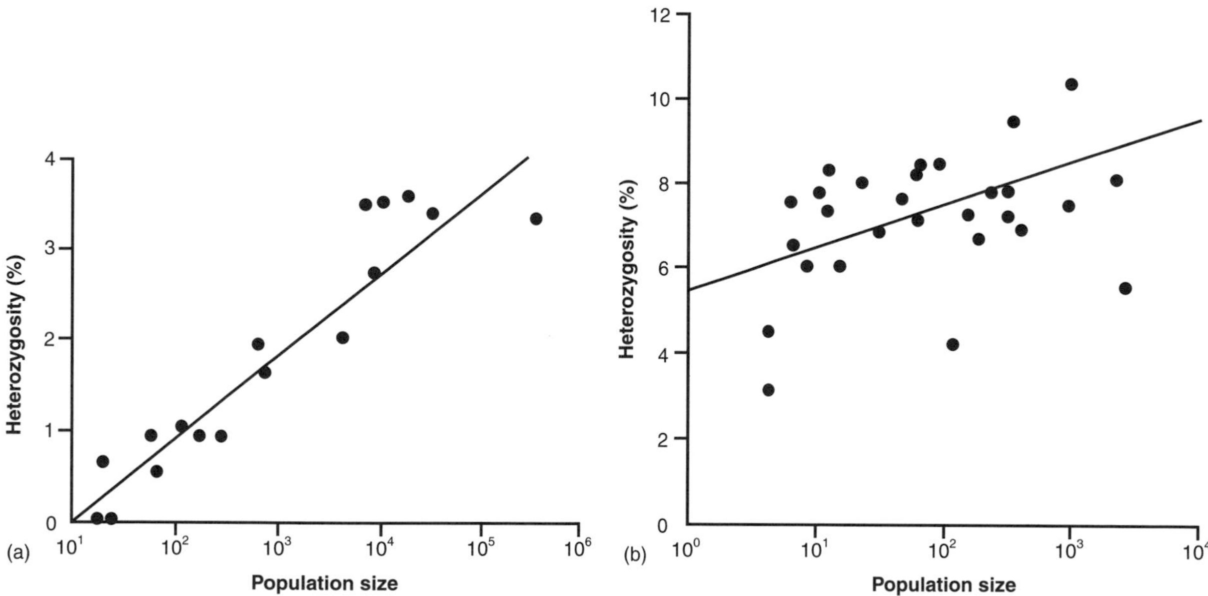

Figure 5.1. *Relationships between heterozygosity and population size in two species. (a)* Halocarpus bidwillii, *a coniferous tree from New Zealand (r = 0.94). (From Billington, 1991.) (b) The red-cockaded woodpecker* (Picoides borealis) *from the southeastern United States (r = 0.48). (From Stangel et al., 1992.)*

How Is Genetic Diversity Lost?

The two central problems in conservation genetics, and thus of concern in ecosystem management, are:

1. The loss of genetic diversity in small populations, which can result in reduced evolutionary flexibility and declines in fitness.
2. Changes in the natural distribution of genetic diversity among populations. These can occur either by artificially isolating formerly connected populations (possibly resulting in small, isolated populations and loss of within-population heterozygosity) or by artificially mixing formerly isolated populations (which can destroy local adaptations).

THE LOSS OF GENETIC DIVERSITY IN SMALL POPULATIONS

Population size is the critical factor in retaining genetic diversity; it is well established that small populations lose genetic diversity faster than large populations. Why? Before we can answer that, we must first introduce a concept called the **genetically effective population size (N_e)**, because it is not the actual, or **census population size (N_c)**, that is critical to genetic considerations, but the genetically effective size that counts. Here is why.

Population genetic models are based on "idealized" populations, which are defined as those with equal sex ratios, equal progeny production among females, random mating, and no selection occurring. Those features rarely, if ever, describe real populations; one sex may outnumber another, some females may be very productive while others are barren, in many species mating is anything but random, and of course natural selection is always operating. Furthermore, counts of individuals mean little where genetics is concerned, because *only those individuals actually reproducing matter at all.* Thus, prereproductive individuals do not count because they are not yet contributing genetic material to the next generation, and likewise any adult that does not reproduce is not considered either.

Such deviations from the idealized world of mathematically based genetic models must be accounted for, and this adjustment is through the use of N_e. To understand N_e, an analogy with the wind chill factor helps: N_e is to N_c as the wind chill temperature is to the actual temperature. For example, if you are outside in a 30°F temperature with no wind, it is an effective temperature of 30°. However, if the wind is blowing at 10 mph, then the effective temperature is lower, in this case, 16°F. It is *as though* the temperature were less than it really is. The genetically effective population size (N_e) is similar. Because of considerations such as unequal sex ratios or uneven progeny distribution, when we consider the genes being produced in the next generation, it is *as though* (in a genetic sense) there were fewer individuals reproducing than there really are. This is why N_e, rather than N_c, is used in conservation genetic analyses.

A common adjustment of N_c is for unequal breeding sex ratios, and it is calculated as follows:

$$N_e = 4(N_m \bullet N_f)/(N_m + N_f). \qquad [5.1]$$

This simple adjustment is an approximation that accounts for the actual successful matings going on in a population. For example, suppose a population of 200 individuals consisted of 100 males randomly mating with 100 females; the census size of the population (N_c) would be 200, and N_e would be calculated as

$$(4 \bullet 100 \bullet 100)/(100 + 100) = 40{,}000/200 = 200.$$

In this case, with equal sex ratios and random mating, N_e equals N_c. Now consider a population of 200 ungulates with a strong harem system, where 20 males mated with 180 females. In this case, N_c is also 200, but N_e would be calculated as

$$(4 \bullet 20 \bullet 180)/(20 + 180) = 14{,}400 / 200 = 72.$$

This population of 20 males and 180 females would be the genetic equivalent of 36 males and 36 females randomly mating! The genetically effective size is significantly smaller than the census size.

This should make intuitive sense. If you consider that a given gene pool—the total of all the genetic material of a population—comes from the previous generation, it makes sense that the amount of diversity of that gene pool would be a function of the number of parents whose genes are represented there. In the extreme case, if only one male fathered all of the offspring of the next generation, then each individual's genetic constitution would reflect only that one male, and the genetically *effective* population size would be smaller than if several males, with their own genetic diversities, were represented.

EXERCISE 5.2

Talk About It!

Consider a fish hatchery that breeds endangered salmon for release back into the wild. If you wanted to keep genetic diversity high in the next generation, how might you do this? What would be the effect of artificial insemination, in which, say, the sperm from one or two males is used to fertilize the eggs stripped from many females? How about the effect of one very fecund female whose eggs are fertilized by a large number of different males?

As you can see, the number of individuals actually involved in reproduction, and thus passing their genes to the next generation, rather than the total number of individuals in the population is what really counts genetically. In addition to an adjustment for uneven sex ratios, other adjustments can be made for the effects on N_e of unequal progeny distribution among females, or large changes in population size over time, but we will not cover these here. Suffice it to say that demographic, behavioral, and other events affect the successful passing of genes to the next generation, and this should be taken into account when considering population size. In nearly all known cases, N_e is smaller than N_c, often substantially smaller. In many cases where it has been rigorously estimated, N_e is on the order of 10–30% of N_c. For example, in two species of primates in Kibale National Park, Uganda, the red colobus and the red-tailed guenon, the estimated N_e/N_c ratios were 0.35 and 0.18, respectively. Consequently, from a genetic perspective, population size generally is quite a bit smaller than it may appear from an overall census.

Now that we have established the concept of genetically effective population size, we can discuss how genetic diversity is lost in populations. The loss of genetic diversity can result from four factors that are all a function of genetically effective population size and that all mathematically work basically in the same way: the founder effect, the demographic bottleneck, genetic drift, and inbreeding.

THE FOUNDER EFFECT. When dispersing individuals of a species begin a new population in a new area (such as colonizing birds or insects on a remote island), they are the "founders" of that population. All of the genetic material for all future generations (discounting future mutations or future immigrants) is contained in those founding individuals. The fewer the number of founding individuals, the less genetic diversity is available, a principle called the **founder effect**. This principle recognizes that the founders of a new population carry only a random subset of the genetic diversity represented in the larger, parental population from which they dispersed.

The strength of the founder effect is inversely proportional to the size of the founding population, and it is mathematically expressed as

$$1/2N_e. \qquad [5.2]$$

Thus, the proportion of genetic variation that is lost from the gene pool of the parental population when a founding event occurs is the inverse of twice the genetically effective population size. For example, if five males and five females begin a new population and randomly breed among themselves (N_e = 10), then the proportion of genetic diversity lost from the larger population is 1/20 (i.e., 1/[2*10]), or 5%. If only one male and one female start the new population, the proportion of variation lost is 1/4, or 25%. The other way to express this is to consider the proportion of variation remaining in (or the proportion carried by) the founders, which is simply

$$1 - [1/2N_e]. \qquad [5.3]$$

In these two examples, 95% and 75%, respectively, of the genetic variation present in the larger, parental population would be present in the first generation of the founder population.

EXERCISE 5.3

Talk About It!

Discuss the relevance of the founder effect to species reintroduction programs in which organisms are placed back into habitats from which they have disappeared. How would the number, sex ratio, and genetic history of the reintroduction stock influence the proportion of genetic variation that will be available to that new population? What can be done to maximize the amount of genetic variation contained in these reintroduced populations?

THE DEMOGRAPHIC BOTTLENECK. When a population experiences a significant, usually temporary reduction in size, either from natural causes (e.g., abiotic disturbance, disease, heavy predation) or human causes (e.g., overhunting, habitat destruction, oil spill), it is said to have undergone a **demographic bottleneck**. The result is effectively the same as the founder effect: all subsequent genetic variation is contained in the surviving individuals. Thus, a bottleneck results in random losses of genetic diversity at a rate of $1/2N_e$ (or conversely, the proportion of variation remaining after one generation of bottlenecking is $1 - [1/2N_e]$).

EXERCISE 5.4

Collaborate on It!

Consider the following examples of N_e: 1, 2, 5, 10, 50, 100. Calculate the amount of genetic variance remaining after one generation at that population bottleneck size, and compile these in a graph that shows the proportion of variation remaining (y-axis) versus the number of individuals, N_e (x-axis). What is the general pattern of loss as a function of population bottleneck size? What are the implications of this pattern for managers?

GENETIC DRIFT. Genetic drift consists of random changes in gene frequency within a population. In a small population, some alleles will not be represented in the next generation by chance alone, and random allele frequency changes and losses of genetic diversity will occur due to chance; the smaller the population, the greater the effect. An analogy with coin tossing will help explain this principle. A coin has two "alleles," heads and tails. If the coin is tossed 50 times (i.e., 25 "individuals" in the population with two alleles each), it is highly unlikely that one of the alleles would be "lost." However, if it is tossed only six times (three individuals in the population), there is a reasonable chance that only heads or only tails will appear; a population of only two tosses (one individual) would make it quite likely that one or the other allele would be lost. If the coin was very unfair and one of the alleles was "rare," appearing, say, only 5% of the time, then it could be lost very quickly in a small population. In a similar way, alleles may be lost in a small population over many generations through this random process. In effect, it simply represents a chronic bottleneck resulting in the repeated loss of diversity in each generation. The proportion of heterozygosity remaining is estimated as

$$[1 - 1/2N_e]^t, \qquad [5.4]$$

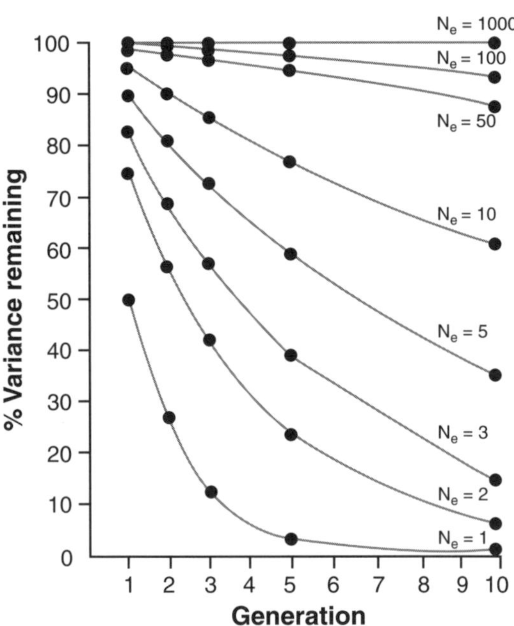

Figure 5.2. *Relative rates of loss of genetic diversity. The losses are demonstrated by the average percentage of genetic variance remaining over ten generations in a theoretical, idealized population at several genetically effective population sizes (N_e). Variation is lost randomly through genetic drift. (From Meffe and Carroll, 1997.)*

where t is time (measured as the number of generations at that population size).

The effect of genetic drift over time varies greatly with population size (Figure 5.2). With 1000 individuals, drift is barely detectable over ten generations. With 50 individuals, the effects of drift are minimal for the first several generations but become apparent after about ten generations. With two individuals, losses from drift are significant even in the first generation and then are very large with each succeeding generation.

INBREEDING. Inbreeding occurs when individuals that are more closely related than by chance alone mate. Small, isolated populations with no dispersal would ensure that inbreeding rates are high because breeding options would be limited. Inbreeding can result in *inbreeding depression*, a reduction in the fitness and vigor of individuals (such as lower fecundity, decreased survivorship, lower growth rates, smaller birth size, and other anomalies) as a consequence of increased homozygosity (Figure 5.3). The level of inbreeding depression depends on the relatedness of the breeding individuals, but the rate of change of the inbreeding coefficient (ΔF) is once again calculated as $1/2N_e$.

Inbreeding effects can be very real in nature. For example, a population crash of song sparrows during a severe winter in British Columbia selectively killed individuals with higher inbreeding coefficients. In an experimental study of purposely inbred white-footed mice in Illinois, inbred individuals had significantly lower survival rates than outbred individuals when released into the wild.

In summary, the rate of loss of genetic variation is a function of effective population size, with the per-generation loss of diversity from a large population estimated as $1/2N_e$. Small populations, those on the order of tens of individuals, stand to possibly lose significant amounts of quantitative genetic variation if they remain at small sizes for multiple generations. A single generation at a small population size, followed by growth to a larger population, probably would not have significant genetic consequences for most species.

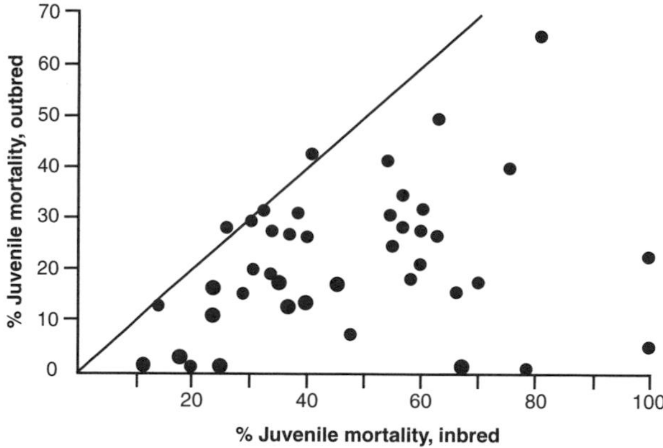

Figure 5.3. *The effects of inbreeding on juvenile mortality in captive populations of mammals. Each point compares the percentage of juvenile mortality for offspring from inbred and noninbred matings. The line indicates equal levels of mortality under the two breeding schemes. Points above the line are higher mortality from noninbred matings, and points below the line are higher mortality from inbred matings. The distance of a point from the line indicates the relative strength of the inbreeding effect. (From Ralls and Ballou, 1986.)*

EXERCISE 5.5

Collaborate on It!

Split into small groups to address the following conservation genetics issue, and be prepared to share your findings with the class. For the ROLE Model scenario, the species is the bog turtle; for the SnowPACT scenario, the species is the yellow-legged frog; and for the PDQ Revival scenario, the species is the red-cockaded woodpecker.

A very comprehensive survey of the species in question by a Master's student at the local university estimated the population at approximately 50 males and 50 females. The next year, after a stressful winter, the same student estimated the population at 10 males and 40 females. What were the original and the resulting N_c and N_e of this population? By how much did each decline? What proportion of genetic variation was retained in the remnant population? Provide management recommendations for this population. Should anything be done? Is there cause for genetic concern at this point? What more information might be needed to take serious action?

EXERCISE 5.6

Collaborate on It!

You are the managers for a grizzly bear population in northern Montana. A consensus of the plan is that grizzlies should have no more than a 5% loss of genetic diversity over the next 100 years. Suppose they have a generation time of 6 years, they occur at a population density of one per 25,000 acres, and their N_e/N_c ratio is 0.2. How much grizzly bear habitat will you need?

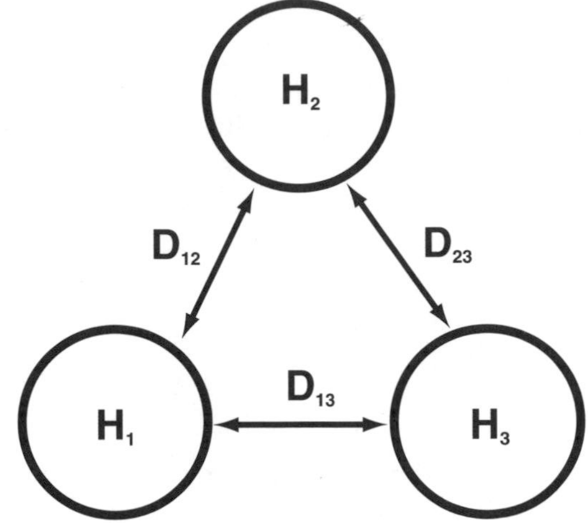

Figure 5.4. *The partitioning of total genetic diversity, H_t, into within-population and among-population variation. This schematic represents a species with three populations, each with some level of within-population heterozygosity (H_1, H_2, and H_3); overall mean population heterozygosity is H_p. Among-population divergences (D_{12}, D_{23}, and D_{13}) are represented by the arrows between the populations; overall mean population divergence is D_{pt}. (From Meffe and Carroll, 1997.)*

CHANGES IN PATTERNS OF GENETIC DIVERSITY AMONG POPULATIONS

The second way that genetic variation may be lost is by changing the natural distribution of genetic diversity among populations, either by artificial isolation of populations through habitat fragmentation or by mixing naturally isolated populations. To understand this, we need to relate genetic variation to the geographic distribution of species, which will also connect genetic considerations of management with landscape-level considerations.

The genetic variation we have discussed so far—heterozygosity—is distributed in some way in physical space. This geographic distribution may be "partitioned" into within-population and among-population components of variation. In other words, if we consider the total amount of genetic variation carried by a species, at least two spatial components contribute to that variation: the average levels of heterozygosity found *within individual populations* and the degree to which *different populations differ from one another* genetically.

Consider, for example, a species consisting only of three populations that occur in three distinctly different areas (Figure 5.4). Each population has a measurable level of population heterozygosity, represented by H_1, H_2, and H_3. But each population also may diverge genetically from every other population, represented by D_{12}, D_{23}, and D_{13}. That is, because of long-term genetic isolation, random events, different selection pressures, or other reasons, different populations may be genetically divergent and thus distinguishable from each other, represented by the D (divergence) components in Figure 5.4. Thus, the total genetic variation of this species (H_t) may be partitioned into the mean within-population variation (H_p) plus the mean among-population variation (D_{pt}). (The H_p notation signifies population-level heterozygosity, whereas the D_{pt} notation signifies the mean divergence among populations across the total range of the species.)

Now let's consider a more complex example. In this case, there are 17 populations of, say, oak trees assayed for genetic variation. This genetic information would allow us to group populations on the basis of their genetic similarities and compare these similarities with their geographic distributions (Figure 5.5). The first level in an eventual hierarchy is simply the individual populations, each of which has a measured heterozygosity level (which, as we saw above, would be partly a function of population size). We may then begin to group populations hierarchically on the basis of

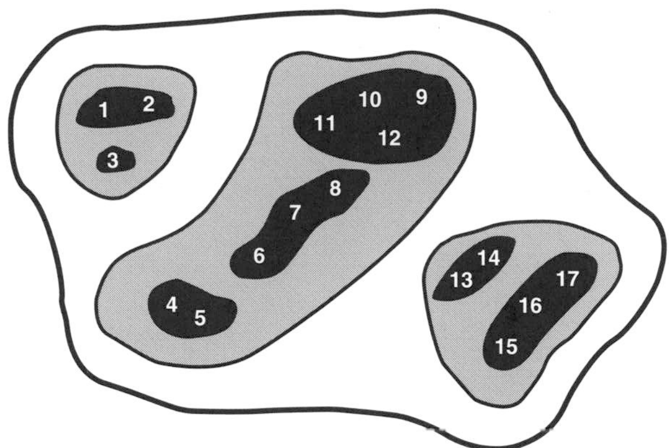

Figure 5.5. *The partitioning of genetic diversity in a more complex and realistic situation. In this hypothetical case, 17 populations of oak trees are distributed across a landscape, and genetic data allow for their clustering based on similarities in genetic content. (From Waples, 1995.)*

their genetic similarities. Clustering level I joins those populations that are most similar, such as 1 and 2, or 4 and 5 (in this case, resulting in seven black groupings). Level II then clumps those seven groups based on their overall similarities, resulting in three main clusters (the three shaded groupings). Level III is simply the entire cluster of 17 populations, which then could be compared with another such group of populations elsewhere.

Why would such information on the geographic partitioning of genetic variation be important to natural resource managers? Quite simply, effective conservation of genetic diversity must consider both within- and among-population genetic variation to retain the highest amount of variation and to maintain a natural population genetic structure. For example, if most of the genetic variation carried by a species was represented by among-population divergence (the D_{pt} component), then it would be more critical to maintain population isolation than it would be if there were little divergence—that is, if most of the genetic variation was represented by within-population heterozygosity (H_p) and most populations were similar to one another. Strong genetic divergence could be an indication of adaptations to local conditions, and movement of individuals among such populations could break up such adaptations. It also could be a result of ancient divergences and potential speciation events in progress, again indicating a need for isolation.

As an example of within- and among-population distribution of genetic variation, let's consider how variation is distributed in plant species with very different geographic ranges (Table 5.2A). Endemic species are those with extremely limited distributions, narrow species are more broadly distributed but still occur within a small range, regional species are those found across a wide region (e.g., southeastern U.S.), and widespread species are those with even more cosmopolitan distributions. When the amount and distribution of genetic variation in a large number of plant species was compiled, some patterns emerged.

First, the total levels of genetic variation (H_t) increased with increasing geographic distribution. This suggests that species with narrow distributions have lesser overall genetic diversity. Second, mean within-population heterozygosity levels (H_p) increased with geographic distribution, implying that narrowly endemic species lose heterozygosity, the point made in the previous section. Third, an estimate of the divergence component D_{pt} (in this case using a statistic called G_{ST}, which is the proportion of total variation due to among-population differences) shows that more narrowly distributed species tend to develop more among-population divergence, probably due to their isolation.

An interesting pattern also emerges when considering the breeding system of plant species (Table 5.2B; for simplicity, we show only the divergence component). Plant species that self-pollinate have very much higher divergence components than those that use mixed strategies or completely cross-pollinate; that is, self-pollinating populations are more genetically isolated because of this mode of reproduction. The species with the least divergent populations are those that reproduce using wind-dispersed pollen, which is the least selective form of movement of pollen (and thus genes) among individuals or populations.

These examples indicate how differences in the flow of genetic material—either because of physical (geographic) isolation or because of different

Table 5.2. *Mean Values of Genetic Diversity and Population Structure Derived from the Literature for Plant Species Categorized by Geographic Range and Breeding System*

	Within Species H_t	Within Populations H_p	Among Populations G_{ST}[1]
A. *Geographic range*			
Endemic	0.096	0.063	0.248
Narrow	0.137	0.105	0.242
Regional	0.150	0.118	0.216
Widespread	0.202	0.159	0.210
B. *Breeding system*[2]			
Selfing			0.510
Mixed, animal			0.216
Mixed, wind			0.100
Outcrossing, animal			0.197
Outcrossing, wind			0.099

[1] G_{ST} is the proportion of the total genetic diversity found among populations (i.e., the divergence component).
[2] Mixed breeding systems involve both selfing and outcrossing; also indicated is whether pollination occurs by wind or by animal vectors.
Source: Data from Hamrick and Godt, 1989.

reproductive styles—can dictate overall patterns of the distribution of genetic variation. Knowledge of such patterns can be useful when making management decisions, especially those that involve the movement of individuals for any of a variety of reasons, including reintroductions, relocations due to habitat destruction, translocations to supplement existing populations, or habitat changes that either remove or reinforce population isolation.

EXERCISE 5.7

Collaborate on It!

In groups, select a species from your scenario that occurs in more than one area within that scenario, and discuss how among-population genetic variance could be affected by different management practices. What life history traits might result in higher or lower levels of within- versus among-population variation? Are there potential dangers from either combining isolated populations or isolating populations that presently experience gene flow? Is there anything that could or should be done proactively to avoid such danger? Share your conclusions with the class.

The Loss of Allelic Richness

What we have discussed thus far is the loss of quantitative variation—that is, the loss of levels or changes in the geographic distribution of heterozygosity. Another, perhaps more serious, problem is the loss of rare alleles in small populations. Rare alleles—those that occur at low frequencies but which may confer an adaptive advantage under stressful situations—tend to be lost at a faster rate than the erosion of heterozygosity.

Here is an analogy that might help clarify this concept. Suppose everyone in a classroom of 30 students was permitted to select two drinks from the back of the room. This would be analogous to the two alleles at a given gene locus (one allele for each hand). The available drinks ("alleles"), 60 in total, are colas (35), orange juice (17), iced tea (6), and spring water (2). The cola and orange juice alleles are thus relatively common in the population, the iced tea allele is somewhat rare, and the spring water allele is extremely rare. At a signal, students are allowed to go to the back of the room and select their two drinks. The most agile student in the room manages to grab both of the two rarest alleles, the spring water. The other alleles are distributed among the other 29 students. Thus, in our population, most students will have common alleles (colas and orange juice), a few will carry a moderately rare allele (iced tea), and one carries two very rare alleles (spring water).

The class, complete with two drinks apiece, now goes on an ecosystem management field trip to a local agroforestry project. Seven of the 30 students wander away on a small trail and become lost. If we consider this movement of drink alleles equivalent to mortality, then these 7 students and their alleles are lost from the population. You can readily see that it is impossible for the common alleles (colas or orange juice) to be lost, even if they were the only alleles carried by these 7 students; there are just too many of those alleles present elsewhere in the population for this level of "mortality" to have extinguished them. However, there is a good chance that the iced tea allele would be lost and an even better chance that the spring water allele would disappear with the "mortality" of 7 of the 30 students.

Moving somewhat closer to reality, a mathematical demonstration of allelic frequencies in a hypothetical population compares relative losses of alleles and the erosion of quantitative genetic variation (Table 5.3). In this case, a population starts with 8 alleles, 7 of which are rare. The original allele frequencies of the 8 alleles are 0.80, 0.07, 0.03, 0.03, 0.02, 0.02, 0.02, and 0.01. The number of alleles, the percentage of original alleles, and the percentage of quantitative genetic variation remaining after one generation at various population sizes is illustrative. At a large population size of 1000, nearly all of the allelic and quantitative variation remains. For smaller populations, the loss of allelic diversity is much greater than that of quantitative variation. Rare alleles are lost at a demonstrably faster rate than is overall heterozygosity.

Now here is a real example from an endangered daisy species (*Rutidosis leptorrhynchoides*) from Australia. Allelic richness measures were collected from populations of very different sizes, ranging from tens to tens of thousands of individuals. Allelic richness (the number of alleles detected) increased significantly with population size (Figure 5.6a), implying that alleles are lost from small populations, as predicted. When rare and common alleles were separated, it was clear that common alleles are present in all sizes of populations, but that rare alleles are in fact the ones that are lost in small populations (Figure 5.6b).

Table 5.3. *A Hypothetical Example of the Loss of Rare Alleles as a Function of Population Size*

N_e	No. Alleles Remaining	% Original Alleles Remaining	% Heterozygosity Remaining
1000	≅ 8.00	≅ 100.0	99.95
100	7.81	97.6	99.5
10	3.86	48.3	95.0
5	2.69	33.6	90.0
1	1.35	16.9	50.0

Why are rare alleles important? Such alleles may not be critical continually in the population, but they could be important during extreme events. For example, an experimental study with a desert fish, *Poeciliopsis monacha*, examined survival under conditions of low oxygen (hypoxia) and cold temperatures as a function of which alleles were carried at a particular gene locus (*Ldh-C*). Individuals homozygous for the common allele (+/+) had 100% survival in hypoxia; homozygotes for the variant allele (*v*/*v*) had survival rates of about 53%, whereas heterozygotes (+/*v*) had intermediate survival of about 91%. Conversely, under cold stress conditions, survival rates were the opposite: 100% for *v*/*v*, 92% for +/*v*, and 81% for +/+, indicating that different alleles can confer differential advantages under stressful conditions. Thus, a loss of rare alleles from a population could be detrimental; at minimum, it reduces the base level of genetic variation of that population.

What does this loss of rare alleles mean for managers? It is another indication that small population sizes are to be avoided if possible. Not only are they more prone to extinction simply because of their small size and potential loss of heterozygosity, but they also are demonstrably prone to a loss of rare alleles.

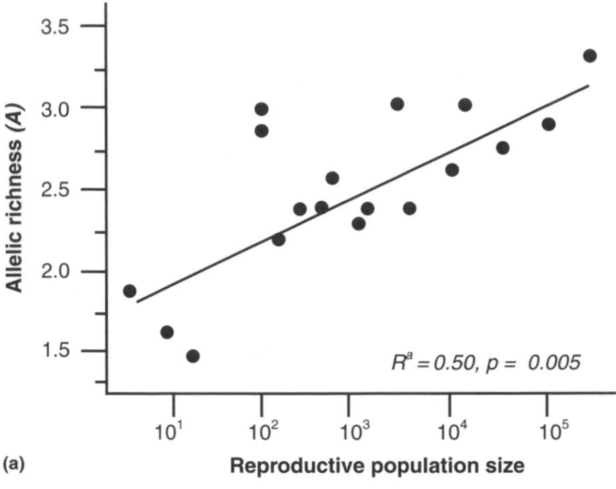

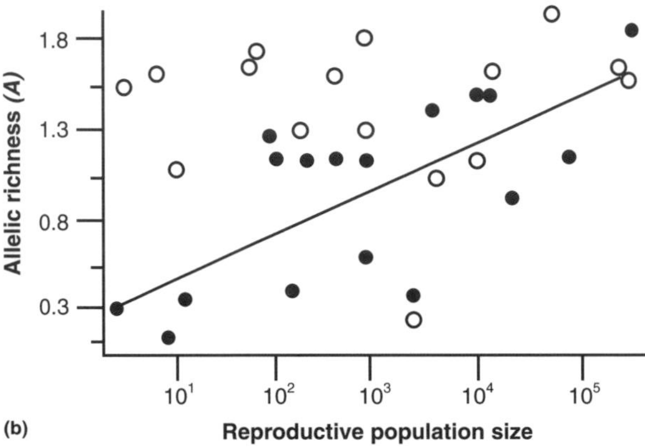

Figure 5.6. *Measured allelic richness in various-sized populations of an endangered daisy,* Rutidosis leptorrhynchoides, *in Australia. (a) The overall relationship between average number of alleles found and population size. (b) The common (open circles) and rare (solid circles) alleles shown separately. Note that common alleles have no relationship with population size; it is only the rare alleles that are reduced (lost) in smaller populations. (From Young et al., 1999.)*

> EXERCISE 5.8
>
> **Talk About It!**
>
> Discuss as a group whether the conclusions you drew and recommendations you made in Exercise 5.5 would be altered at all if you knew that two rare alleles had been lost from the population.

The Role of Genetics in Conservation and Ecosystem Management

Conservation genetics is a set of tools that can be of great utility in conservation in general and ecosystem management in particular. The materials we covered here are but a brief introduction to a very complex, fascinating, and useful knowledge base. The actual roles of genetics in conservation are potentially much larger and more diverse than we have discussed (Box 5.3), and they may be pursued through some of the literature cited at the end of this chapter. We have touched upon mostly the first of these roles—issues of heterozygosity—and a bit of the third role—defining population structure. The many other roles of genetics can be just as critical.

In the larger picture, however, genetic information can only make limitated contributions to conservation, and we must recognize that as well. Genetics is a very important and powerful tool—in certain circumstances. When dealing with small or declining wild populations, when working with captive propagation, when translocating individuals in restoration efforts, when trying to determine historical or contemporary dispersal patterns, when defining taxonomic units, and in many other specific cases, genetics can and should be a prominent component of the toolbox.

But genetics is not fundamentally the issue when dealing with large-scale habitat conversion, or in socioeconomic conflicts, or when institutions and human communications are ineffective or dysfunctional. Genetics cannot provide the answers to restoring natural hydrologic conditions, how grazing should be managed on a public grassland, or how or whether to conduct controlled burns. The point is, it is easy to lose sight of the larger picture and begin to call for genetic studies in any given situation. Genetic analyses are impressive and they involve fascinating technologies; they are scientifically based and produce tangible results that we can debate. Although all of this is very good, it is not always the right approach. We simply stress that the urge to use sophisticated, "high-tech" solutions should be limited to those situations when such an approach can address specific and appropriate questions.

BOX 5.3

Empirical and Conceptual Roles for Genetics in Conservation Biology

In their 1996 book *Conservation Genetics: Case Histories from Nature*, John Avise and James Hamrick discussed five major empirical and conceptual roles that genetics plays in conservation biology, thus illustrating the many and diverse ways that genetic information can assist in conservation decisions. The first role was to address *issues of heterozygosity* (within-population genetic variability), including questions such as: Is genetic variation reduced in endangered species? If so, is this cause for ecological or evolutionary concern? Should populations be managed for increased variation? Is a certain class of genes of special fitness significance? How serious a problem is inbreeding depression?

The second role was to address *issues of parentage and kinship*, including questions such as: Who has bred with whom in captivity and nature? How does this illuminate breeding and social structure? What is the impact of the mating system on effective population size (N_e), inbreeding, and gene flow?

The third role of genetics was to address *issues of population structure and intraspecific phylogeny*. This area includes problems such as the levels of gene flow between populations, whether gender-biased dispersal can be documented, how gene flow relates to demographic connectedness, and whether there are significant historical partitions within species.

Issues involving species boundaries and hybridization phenomena was the fourth role. This area includes questions such as: How genetically distinct are endangered species? Do phylogenetic perspectives alter our species concepts? How widespread are hybridization and introgression in nature? Do such findings affect the legal status of endangered species? What forensic applications do genetic markers provide?

Finally, genetics can address *issues of species phylogenies and macroevolution* to answer questions regarding the phylogenetic relationships of species and higher taxa.

References and Suggested Readings

Allendorf, F.W., and R.F. Leary. 1986. Heterozygosity and fitness in natural populations of animals. Pp. 57–76 in M.E. Soulé (ed.). *Conservation Biology: The Science of Scarcity and Diversity*. Sinauer Associates, Sunderland, MA.

Avise, J.C. 1994. *Molecular Markers, Natural History, and Evolution*. Chapman and Hall, New York.

Avise, J.C., and J.L. Hamrick. 1996. *Conservation Genetics: Case Histories from Nature*. Chapman and Hall, New York.

Billington, H.L. 1991. Effect of population size on genetic variation in a dioecious conifer. Conservation Biology 5:115–119.

Falk, D.A., and K.E. Holsinger. 1991. *Genetics and Conservation of Rare Plants*. Oxford University Press, New York.

Fisher, R.A. 1930. *The Genetical Theory of Natural Selection*. Clarendon Press, Oxford.

Frankham, R. 1995. Effective population size/adult population size ratios in wildlife: A review. Genetical Research 66:95–107.

Hedrick, P.W., and P.S. Miller. 1992. Conservation genetics: Techniques and fundamentals. Ecological Applications 2:30–46.

Jiminez, J.A., K.A. Hughes, G. Alaks, L. Graham, and R.G. Lacy. 1994. An experimental study of inbreeding depression in a natural habitat. Science 266:271–273.

Keller, L.F., P. Arcese, J.N.M. Smith, W.M. Hochachka, and S.C. Stearns. 1994. Selection against inbred song sparrows during a natural bottleneck. Nature 372:356–357.

Meffe, G.K., C.R. Carroll, and Contributors. 1997. *Principles of Conservation Biology*, 2nd ed. Sinauer Associates, Sunderland, MA.

Nunney, L., and D.R. Elam. 1994. Estimating the effective population size of conserved populations. Conservation Biology 8:175–184.

Pope, T.R. 1996. Socioecology, population fragmentation, and patterns of genetic loss in endangered primates. Pp. 119–159 in J.C. Avise and J.L. Hamrick (eds.). *Conservation Genetics: Case Histories from Nature*. Chapman and Hall, New York.

Ralls, K., and J. Ballou. 1986. Captive breeding programs for populations with a small number of founders. Trends in Ecology and Evolution 1:19–22.

Ryman, N., and F. Utter (eds.). 1987. *Population Genetics and Fishery Management*. University of Washington Press, Seattle.

Schonewald-Cox, C.M., S.M. Chambers, B. MacBryde, and L. Thomas (eds.). 1983. *Genetics and Conservation: A Reference for Managing Wild Animal and Plant Populations.* Benjamin/Cummings, Menlo Park, CA.

Stangel, P.W., M.R. Lennartz, and M.H. Smith. 1992. Genetic variation and population structure of red-cockaded woodpeckers. Conservation Biology 6:283–292.

Thornhill, N.W. (ed.). 1993. *The Natural History of Inbreeding and Outbreeding: Theoretical and Empirical Perspectives.* University of Chicago Press, Chicago.

Vucetich, J.A., T.A. Waite, and L. Nunney. 1997. Fluctuating population size and and the ratio of effective to census population size (N_e/N). Evolution 51: 2015–2019.

Waples, R. S. 1995. Evolutionarily significant units and the conservation of biological diversity under the Endangered Species Act. American Fisheries Society Symposium 17:8–27.

Young, A.G., A.H.D. Brown, and F.A. Zich. 1999. Genetic structure of fragmented populations of the endangered daisy (*Rutidosis leptorrynchoides*). Conservation Biology 13:256–265.

Issues Regarding Populations and Species

ALTHOUGH GENES ARE THE FUNDAMENTAL BUILDING blocks of life and protecting genetic diversity can be critical in the right circumstances, people in general do not get very excited about conservation genetics. Genes themselves are not visible, they do not flit among tree branches in the spring and sing pretty songs, nor are they cute and furry with big brown eyes. Most conservation attention occurs, rightly or wrongly, at the level of the organism; thus, we now leave genetics to consider higher levels of biological organization. Populations, and especially species, are the typical focus for many conservation efforts. Species are more tangible entities than genes or ecosystems; they are formally recognized by scientists and protected by law; they have been used and enjoyed and abused and exterminated by humans for millennia. Species will remain a strong conservation focus for the foreseeable future.

The Species

What is a species? This may seem a strange question, for surely most of us can identify many species with ease and confidence. Even the uninformed are not confused in their identifications of blue jays, cardinals, and mourning doves, for example. Yet, the definitions of species, and the particular boundaries that define given species, have been the subjects of scientific debate for centuries. Various species concepts have been and are being used. The **biological species concept**, which defines species on the basis of reproductive (genetic) isolation, was the prevailing species concept for much of the twentieth century. More recently, the **phylogenetic species concept**, which defines species according to patterns of ancestry and descent and shared derived characters, has gained popularity. Other species concepts include evolutionary, ecological, cohesion, nominalist, and pluralist, and serve as the basis for great debates among systematists. Obviously, how we define a "species" can vary a great deal depending on the species concept we use.

The species, as it turns out, is merely a working hypothesis and is not immutable. Particular species identifications change over time, and revisions are common. Systematists who are "lumpers" tend to

EXERCISE 6.1

Talk About It!

Discuss the "species" as a concept. How many different definitions do you know? Can these definitions be applied universally—for plants and animals, for common and uncommon organisms, for sexual and asexual species, for microbes as well as complex organisms? How might the various definitions affect species protection, either legally, practically, or through the eyes of different members of the public?

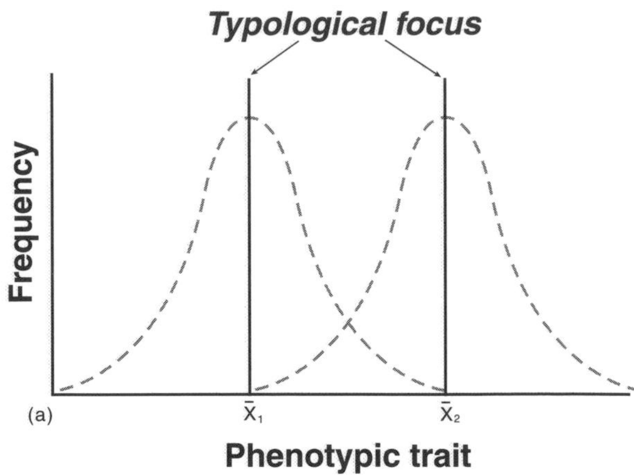

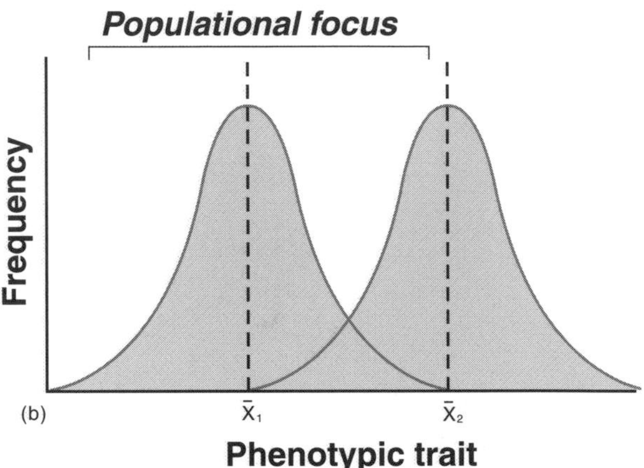

Figure 6.1. *Conceptual models of typological and populational thinking, illustrated by the distribution of individuals of two species for a given phenotypic trait. In each conceptualization, the perceived reality is represented by a solid line; a dotted line represents the perceived less important aspect. (a) In the typological view, the essence of the species is represented by the mean (the "type," $\bar{X}_1$ and $\bar{X}_2$). Variance around the mean is considered "noise" and therefore unimportant. (b) In the populational view, the essence of the species includes the variance of the trait, and the mean ($\bar{X}_1$ or $\bar{X}_2$) is simply one statistical descriptor. There is no "perfect type," and species may even grade into one another for a given trait with no clear gap. (From Meffe and Carroll, 1997.)*

group individuals with small trait differences under the name of one species, whereas systematists who are "splitters" will use those small differences to identify several different species. It is possible for one collection of similar fishes, for example, to be called a single species by one expert and several species by another.

Why is there so much confusion in assigning individual organisms to species? Very simply, *nature does not come in discretely defined (or labeled) units, and there is much variation within species.* In many cases, nature offers more of a continuum of variation than distinct types, and it may be difficult to sort out that continuum into what humans like to identify as discrete units. This variation can be quite important to conservation concerns.

To understand this better, we need to consider the two major ways of viewing variation, including biological variation, in nature: typological and populational thinking. A **typological view** believes that the objects that we see are manifestations of perfect forms, and any variance in those objects is due merely to imperfections—unfortunate deviations from the "type." This view dates back to the Greek philosopher Plato, who promoted this view of the universe with his notion of the *eidos*, or perfect type. In this perspective (Figure 6.1a), the mean or average form (X_1, X_2) is what counts, and any variance around that mean is "noise," largely to be ignored.

In contrast, the **populational** view holds that objects in nature, including individuals that constitute what we perceive of as species, occur in a continuum of variation and that this variation is in fact meaningful. In the populational view (Figure 6.1b), mean or average forms are merely statistical artifacts of a collection of individuals, for any two or more entities can produce a mean. What is really worth attention is the variance around that mean, which can be substantial.

The great twentieth-century evolutionary biologist Ernst Mayr eloquently summarized this difference when he stated: "For the typologist, the type (*eidos*) is real and the variation an illusion, while for the populationist the type (average) is an abstraction and only the variation is real. No two ways of looking at nature could be more different" (1942). Biological classification began with a typological view of the world. Linnaeus—the father of the modern system of classification of organisms—was a typologist, as were many nineteenth-century biologists, who believed that nature was the reflection of a divine plan, and species were created instantaneously in perfect and immutable form.

All evolutionary biologists and systematists of recent times have adopted the populational view and understand that the variation we observe is the raw material for evolutionary change and adaptation; there must be heritable genetic variation among individuals in order for natural selection to operate. This variation, of course, need not all be due to genetic differences; certainly many environmental factors influence the physical traits that we measure, and one of the difficulties in species identification is to determine whether physical differences are genetically based or merely environmentally influenced and subject to change.

EXERCISE 6.2

Talk About It!

As a class, review for a moment the basic process of natural selection. Be sure you understand how variation is generated, how selection operates to act on that variation, and the role of heritability.

Let's return to Figure 6.1b. Suppose the x-axis was the measurement of a particular trait, like the number of lateral line scales of a fish. Suppose many individuals of two species were counted for the trait, and their frequency distributions produced the two curves in this figure. There may be an area where the counts for the two species overlap. Such a pattern of overlap is not unusual, and it demonstrates the continuum nature of biological diversity. Thus, an individual we recognize as a particular species actually can have a trait that is more compatible with that of a different species! This is why species are sometimes difficult to clearly define; nature does not always come in discrete packages.

It is this within-population variation that we wish to emphasize here. Such variation within species is an important resource that is sometimes overlooked with our typical (and typological) focus on the species as a unit. Within-species variation results in among-population diversity that can be critical to conservation action. Recognizing populational diversity and the roles of different populations is an important management consideration for at least three reasons:

1. *Different populations may contain unique genetic diversity; this diversity is a resource that, if lost, cannot be recovered.* An especially pertinent example for human welfare involves the ancestors of agricultural crops. Many of our present-day food crops have a very narrow base of genetic variation and may be vulnerable to diseases, global climate change, or other environmental stresses. The wild ancestors of many crops are still in existence and are reservoirs of genetic diversity that could prove extraordinarily useful to human agricultural endeavors as unique sources of genetic variation.

2. *Different populations may contain unique local adaptations not found elsewhere.* For example, a high-elevation population of a tree frog species would be expected to be better adapted to cold temperatures than a low-elevation population, and vice versa. Similarly, a population of shrubs living on the edge of an estuary would be expected to be more salt-tolerant than one well inland, next to a mountain stream. A conservation focus strictly at the species level could miss such within-species differences.

3. *Perhaps most importantly, organisms perform functions in their local ecosystems.* When they disappear from that system, it may seem to be "merely" a populational extinction if the species exists elsewhere, but that function is no longer performed there. For example, there are wasp and fig tree species in the new world

tropics that have a tight pollination relationship: A given wasp species pollinates only a single species of fig tree. If that wasp disappears from a valley where the fig occurs, the wasp may still exist elsewhere; but unless it is restored to that valley, its biological function is missing and the fig trees will not get pollinated. Species are more than simply entities on a biologist's diversity list; they are composed of many populations that perform functions wherever they occur. *Populational loss means local loss of biological function*, such as predation, herbivory, nutrient cycling, decomposition, a prey base for other species, pollination, evapotranspiration, or any of dozens of other things that living organisms do.

The point is that biological diversity needs to be conserved at all levels, not simply at the species level, and regardless of the species concept used. Although a tremendous amount of biological variation occurs *within* the species, it can easily be overlooked. Politically, most of the focus is at the species level, although the U.S. Endangered Species Act does account for subspecies and even populational variation for some taxa.

EXERCISE 6.3

Collaborate on It!

Split into groups of four or five. Select a species from your scenario, and discuss what its loss from that landscape could mean with respect to the points made above. Is it likely that unique genetic diversity or a unique adaptation could be lost? What about that species' function in the ecosystem? Is it likely to have repercussions for other species or the system in general? How could you best relate such information to stakeholders in the scenario who have no biological background? Share your discussions with the class.

The Roles of Species in Science and Policy

The species, of course, plays a central role in conservation. We have an Endangered *Species* Act, not an Endangered Ecosystem Act or an Endangered Allele Act. Species can have high profiles in the public arena—positive and negative. The spotted owl is both a symbol of disappearing old-growth forests to preservationists and a harbinger of lost jobs and repressive governmental control to some loggers and developers. Reintroduced wolves in Arizona are a call to the wild for wealthy suburban supporters and a perceived threat to some livestock ranchers. The endangered snail darter provided an opportunity to stop what was considered the unneeded and destructive Tellico Dam in Tennessee to some and an annoying block to progress for others. It also became a political and legal football for supporters and opponents of the Endangered Species Act. Most everyone supports conservation of the cuddly panda, but few will go out on a limb for an endangered snail or spider. It is important for a manager of natural resources to understand how species are perceived by the public and consequently used in legislation, policy decisions, and subsequently management. Those perceptions and uses vary tremendously according to the particular public entities and policy makers involved, as well as political moods.

VIEWPOINTS ON SPECIES

To more effectively understand some of these perceptions and deal with them more effectively as managers, it helps to consider some of the ways in which species are viewed by people relative to biological, political, and economic perspectives. These viewpoints on species resonate differently with groups of people holding different social, economic, and institutional interests. Understanding the various perspectives can help us interact more effectively with stakeholders, partners, policy makers, and the public at large.

Here we identify six categories of how species might be perceived ecologically, socioeconomically, and legally. For each, we provide a brief description followed by one or more questions to ask about a species or population to help determine its possible membership in that category.

KEYSTONE SPECIES. A **keystone species** is one whose effect on the structure of a biological community is well out of proportion to its relative bio-

Figure 6.2. *The gopher tortoise (*Gopherus polyphemus*). This inhabitant of the southeastern U.S. coastal plain is thought to be a keystone species because the burrows it digs have been found to harbor many other species of vertebrates and invertebrates. These burrows may be especially important in protecting species from the frequent fires that were historically prevalent in this region. (Photo by G.K. Meffe.)*

mass. The addition or removal of a keystone species has large effects on the richness and relative abundance of many other species. Examples include gopher tortoises (Figure 6.2) and beavers.

- Would ecological functions of the ecosystem (trophic relationships, community structure, hydrological flow, succession patterns, disturbance cycles, and so forth) be significantly altered if the species were absent?
- Would other species increase, decline, or disappear from the ecosystem if this species were eliminated?

INDICATOR SPECIES. An **indicator species** is indicative of particular conditions in a system (ranging from natural to degraded) and used as a surrogate measure for other species or particular conditions. Examples are darters and stoneflies in a stream, which indicate good water quality.

- Does the species have a highly specific niche or a narrow ecological tolerance (e.g., substrate)?
- Is the species tied to a specific biotic community or successional stage?
- Can the species reliably be found under a specific set of circumstances, but not others?
- Is the species typical of either natural or highly degraded conditions?

UMBRELLA SPECIES. An **umbrella species** is a species that, if secure or flourishing, would protect many other species because of its demand for large expanses of habitat. An example is the grizzly bear.

- Does the species require large blocks of relatively natural or unaltered habitat to maintain viable populations?
- Does the species require relatively natural and undegraded conditions?

FLAGSHIP OR CHARISMATIC SPECIES. A **flagship or charismatic species** is a species that elicits emotional feelings from individuals, including a willingness to contribute financially to the species' well-being or otherwise support their protection. Examples are whales, tigers, and elephants.

- Is this a species that people relate to in a positive emotional way (a warm and fuzzy species, or an especially interesting or attractive species) and that would elicit a strong protective response?
- Is this a species symbolic of human values, beliefs, or attitudes (such as "smart," "hard workers," or "faithful to mate and family")?
- Is this a species that has been covered extensively in the media?

VULNERABLE SPECIES. A **vulnerable species** (or population) is particularly susceptible to extinction. Examples are black-footed ferrets and whooping cranes.

- Is the species' overall population size small and/or range limited?
- Is the species' habitat fragmented or are the individual populations highly isolated, with poor dispersal power?
- Does the species have a highly specialized niche?
- Is the species especially vulnerable to human activities?

- Is the species recognized by specific laws (e.g., listed as endangered by federal or state agencies or by the Convention on International Trade in Endangered Species [CITES])?

Economically Important Species. An **economically important species** is a species that has positive or negative consequences for the local, regional, or national economy. Examples are elk, zebra mussels, and Komodo dragons.

- Is the species harvested for a product (meat, hides, fat, pharmaceuticals, trophies, pet-related items)?
- Does the species' presence and behavior result in economic losses (farmers, ranchers, industries)?
- Is the species attractive for ecotourism? Will people pay to travel and observe the species?

Membership of a given species in these categories is not always clear, and a species may fall into several categories. For example, the mountain gorilla certainly is a flagship or charismatic species; it is also vulnerable, as there are few individuals and they live in a small area subject to civil wars. In addition, the species is economically important because people pay large sums of money (when there are not civil wars) to local guides to see gorillas. Finally, they may also be an umbrella species, because the protection of thousands of acres of their mountain habitat means that other species also will be protected.

Understanding how people relate to given species—how their particular interests are served or threatened—is fundamental to being effective in interacting with stakeholder groups, an activity you will pursue in subsequent chapters and throughout your career. A given image of a species will resound positively with one stakeholder group but have no meaning (or generate negative feelings) for another. For example, it is currently being debated whether the prairie dog is a keystone species; let's say for the sake of argument that it is. This would mean that the prairie dog plays an unusually large ecological role in U.S. western prairies, a role that is understood by and meaningful to members of environmental organizations. It may also be a vulnerable species because most of its formerly huge colonies have been eliminated by habitat destruction and intentional poisoning. But its role as a vulnerable, keystone species is largely irrelevant to ranchers who see the prairie dog (whether correctly or not) as a pest and an economic detriment for the cattle business. Both these perspectives must be understood and appreciated before reasonable discussions can take place between different stakeholder groups with very different goals (see Chapter 10). To see only one perspective and be blind to the other is to ensure highly emotional debates, gridlock, positional bargaining, and little progress in problem solving. Understanding different public perceptions with respect to given species roles is an important step in breaking through value differences and making progress in local problem solving.

EXERCISE 6.4

Talk About It!

Either in small groups or as a class discuss the following questions:

- Can a widespread species be vulnerable?
- Can a species be rare in an ecosystem and still act in the role of a keystone species?
- Can a rare species be widespread?
- If secure, would an umbrella species' presence necessarily protect all the other species within its geographic range?

EXERCISE 6.5

Collaborate on It!

As a class, select about a dozen species from your scenario. Then individually consider how you might classify each one with respect to the six categories just discussed. A given species could fit none, one, or several categories. Then compile the results from the whole class on a chart. Do any patterns emerge? Are there surprises? How consistent are you as a group in assessing these species in a biopolitical sense? What might be the consequences of these patterns when dealing with a diverse public?

Connecting Populations and Species to Landscapes

All these genes, populations, and species we have discussed occur on real landscapes and are spatially distributed in various ways. Those spatial distributions have important consequences for how management might proceed. In preparation for the chapters that follow, which pursue a more in-depth consideration of ecosystem management on a landscape scale, we introduce three concepts to describe the biological distribution of native species across landscapes: *alpha*, *beta*, and *gamma richness*. These concepts are of interest because they relate directly to responsibilities for managing species "on the ground."

Alpha richness (α) is the number of species within small areas of fairly uniform habitat. An example is the number of songbirds or herbaceous plant species in a coniferous forest stand.

Gamma richness (γ) is the number of species within a region (i.e., the cumulative number of species observed in all habitats of a region). An example is the number of songbirds or herbaceous plant species on the Colorado Plateau.

Alpha richness is sensitive to area sampled and definition of uniform habitat, and both indices are sensitive to intensity and effectiveness of sampling effort. Poor sampling or sloppy definition of uniform habitat will greatly influence the results. A person who cannot identify birds by their calls or who is not very skilled in the identification of herbaceous plants will not develop a rigorous and reliable species list. Likewise, defining a habitat as a uniform type when in fact sampling encompasses three subtle but distinct habitat types will inflate the assessment of alpha richness.

Beta richness (β) is the amount of change or turnover in species (i.e., species gained and lost) in going from one habitat to the next (the species *difference* between two habitats). A high beta richness means that different habitats are supporting very different suites of species, and the cumulative number of species recorded increases rapidly as additional habitats are censused.

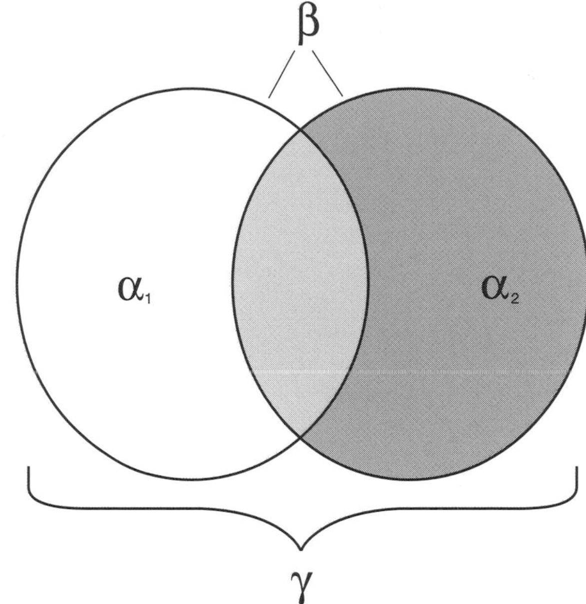

Figure 6.3. *A depiction of alpha, beta, and gamma richness. In this Venn diagram, each circle is a distinct habitat type, and each has its own measured alpha richness. Beta richness is the total of the nonoverlapping species in the two habitats (i.e., the species unique to each given habitat). Gamma richness is the sum total of the species richness across the whole area.*

The relationship between alpha, beta, and gamma richness is well illustrated by a Venn diagram showing two habitats (Figure 6.3). Alpha richness is the number of species found in each habitat, beta richness is the number of species unique to each habitat, and gamma richness is the cumulative total for the two habitats. An example of these terms, using a hypothetical and simple list of bird species found in three habitats (Table 6.1), helps explain the concepts further.

We emphasize that the focus for studies of species richness is *native* species. Invasive exotics certainly can boost alpha richness, but the negative effects are so well known that we do not need to go into them in any depth. Counting a large list of exotic species at the expense of natives simply is a gross misuse of the concepts of alpha, beta, and gamma richness. Thus, we refer only to native biodiversity when using these terms.

What do the concepts of alpha, beta, and

Table 6.1. *A Hypothetical Example of a Bird Survey in Three Distinct Habitats Showing Alpha, Beta, and Gamma Richness*

Species	Habitat 1: Deciduous Forest	Habitat 2: Mixed Pine Forest	Habitat 3: Golf Course
Sandhill crane			X
Canada goose			X
Black-billed cuckoo	X	X	
Olive-sided flycatcher	X		
Eastern bluebird		X	X
Wood thrush	X	X	
White-eyed vireo	X	X	
Yellow-rumped warbler		X	
Ovenbird	X		
Cardinal	X	X	X
Eastern meadowlark			X
Scarlet tanager	X		
Totals			
Alpha richness:	7	6	5
Beta richness:	1 vs. 2, β = 5	2 vs. 3, β = 7	1 vs. 3, β = 10
Gamma richness:		12	

gamma richness mean for the natural resource manager? The way species are distributed across a landscape can have implications for land-use decisions. For example, should you focus on managing for high alpha richness when a refuge is part of a regional network of protected areas? How would you manage differently if the refuge was a highly isolated natural area in a sea of development? Should you maximize and manage for strong differences among habitats in an area to maximize beta richness? Addressing such questions offers valuable perspectives on overall richness across the landscape when making community-based land-use decisions.

Such decisions require not only an understanding of species and populational richness and distributions, as we have discussed here, but a detailed knowledge of how those populations respond to the heterogeneity of real landscapes. It also is necessary to understand techniques that estimate populational persistence under different scenarios and circumstances. We will address such techniques in the next chapter.

EXERCISE 6.6

Talk About It!

1. Consider a variety of different habitats with very little change or turnover in species across them. Is conserving areas with high species richness—if they share the same set of species—of equal value to managing for remnants of ecosystems that share few species (i.e., have a high beta richness)?
2. Can you think of species that are rare at the alpha level but common at the gamma level? What implications might this have for management? Can you think of management actions you could perform that would increase alpha richness at the expense of beta or gamma richness?

EXERCISE 6.7

Collaborate on It!

Working in small groups with your scenario, address one or more of the following issues:

1. Consider that a particular area in your system is an even-aged forest of stands of several different species. Suppose one of these species was of economic value and proposed to be logged out. What would you predict would happen to beta richness of the bird and mammal communities in the forest? Would gamma richness change?
2. Consider two patches of forest in your scenario that are fairly close in proximity but not contiguous. Develop a management action that would increase beta richness between these two forests. Now consider these two forests as composing a regional forest. What management action might increase gamma richness?
3. Consider native species in a riverine system in your scenario. What might the colonization of an invasive exotic species such as zebra mussels, carp, or an aggressive crayfish do to alpha richness of native species? How might it affect beta and gamma richness in the system as a whole?

References and Suggested Readings

Cracraft, J. 1983. Species concepts and speciation analysis. Pp. 159–187 in R.F. Johnston (ed.). *Current Ornithology, Vol. 1.* Plenum Press, New York.

Daily, G.C. (ed.). 1997. *Nature's Services: Societal Dependence on Natural Ecosystems.* Island Press, Washington, D.C.

Donoghue, M.J. 1985. A critique of the biological species concept and recommendations for a phylogenetic alternative. The Bryologist 88:172–181.

Mayr, E. 1942. *Systematics and the Origin of Species.* Columbia University Press, New York.

Mayr, E. 1959. Darwin and the evolutionary theory in biology. Pp. 409–412 in *Evolution and Anthropology: A Centennial Appraisal.* The Anthropological Society of Washington, Washington, D.C.

Otte, D., and J.A. Endler (eds.). 1989. *Speciation and Its Consequences.* Sinauer Associates, Sunderland, MA.

Power, M.E., and L.S. Mills. 1995. The Keystone Cops meet in Hilo. Trends in Ecology and Evolution 10:182–184.

Power, M.E., et al. 1996. Challenges in the quest for keystones. BioScience 46:609–620.

Rohlf, D.J. 1989. *The Endangered Species Act: A Guide to Its Protections and Implementation.* Stanford Environmental Law Society, Stanford, CA.

Rojas, M. 1992. The species problem and conservation: What are we protecting? Conservation Biology 6:170–178.

Schultze, E.-D., and H.A. Mooney (eds.). 1994. *Biodiversity and Ecosystem Function.* Springer-Verlag, Berlin.

Wiley, E.O. 1978. The evolutionary species concept reconsidered. Systematic Zoology 27:17–26.

Experiences in Ecosystem Management: The Copper River Watershed Project

Riki Ott and Kristin Smith

ALASKA IS A GREAT LAND, BUT IT IS ALSO CAUGHT UP in a great myth. It is not all pristine, public, and protected. In fact, it resembles something a little like Swiss cheese. Although it is true that 88% of Alaska's 370 million acres is public land, the remaining 12% is almost all held by native corporations, which were created by the Alaska Native Claims Settlement Act (1971). ANCSA extinguished all aboriginal claims to Alaskan lands and in exchange created 13 native regional corporations and 226 native village corporations, which were deeded title to nearly 44 million acres of land and given $962 million as "start-up" capital. Native corporations selected much of the prime timber and mineral lands, and prime accessible scenic real estate—mostly as in-holdings in public lands. Native corporations are guaranteed access to their in-holdings and, in many cases, the guarantee comes with an exemption to National Environmental Policy Act review.

The problems created by this pattern of land ownership are threefold. First, it exacerbates a hostility of some residents, especially in rural areas that border the public lands, toward the federal and state governments. Second, with so much land perceived as "protected," there is a frontier ethic to develop what is available. Third, the economic needs of the native corporations, largely satisfied currently by unsustainable resource extraction, are quite different than the social, cultural, and economic needs of the native villages and the communities near the in-holdings, and the conservation needs of the public land managers. These conflicting needs have resulted in divisiveness, polarization, and lawsuits.

Competing resource needs are also created by the pattern of land settlement. Half the state's population lives in Anchorage, but the resources are extracted from sparsely populated rural areas such as the Copper River watershed. This region, east of Anchorage, is one-sixth the size of California with only 5600 people, half of whom live in Cordova, the only incorporated town among the 20 settlements in the area. Rural regions and unincorporated communities lack political clout and find their needs largely ignored by urban and corporate interests.

In 1994, an ad hoc group of Cordova citizens met to determine ways to protect our way and quality of life, while diversifying and rebuilding our town's economy, which had been devastated by the *Exxon Valdez* oil spill 5 years earlier. The state of Alaska was considering oil and gas lease sales and infrastructure for industrial-scale tourism in our region, while native corporations were starting clear-cut logging operations. We perceived these development options as threats to the long-term health of salmon habitat and thus to our way of life: salmon are key to both our economy and our subsistence culture. Arguments over resource use had polarized and paralyzed the community. Sensing that our needs would not be met by the state government or large corporations, two dozen individuals came together and agreed to listen to one another.

Three years later, this group incorporated as the

Copper River Watershed Project. Our mission is to work with residents of the Copper River watershed in diversifying the region's economy while sustaining its resource base and its cultural heritage. We represent an effort to coalesce residents along the 287-mile long Copper River in a citizen-driven group that defines its priorities for quality of life and future development. We made three decisions that ultimately defined our leadership role in Alaska for finding ways to integrate conservation and development.

First, there was a need for public visibility to increase political leverage. We started working with the Netherlands-based Artists for Nature Foundation, which uses art as a medium to advocate for sustainable development. As a result of this collaborative effort, we produced a book and a traveling art show. In 3 years, the show has been seen by nearly half a million people in seven cities.

Second, we decided to expand the geographic focus to include the entire Copper River watershed, the vast upper valley, and the river delta. Forests in the valley were being clear-cut, and to protect the Copper River salmon fishery, the upriver fish habitat needed protection. The best way to do that was to "put a better business plan on the table," to show businesses and landowners ways to make money through conservation-based economic development.

Third, we decided to focus on three areas of our economy—forestry, fisheries, and tourism—by conducting resource assessments to determine options for sustainable economic development. The report from this $250,000 effort forms the foundation of our current work. Projects are prioritized on the basis of meeting our criteria of increasing at least one form of social, economic, or natural capital without decreasing the other two and of finding local project partners and a local project leader.

Projects

Our ongoing work on community values has helped people define why they like living in the Copper River watershed. People discovered that most everyone valued our neighborly and safe rural areas, our wilderness (for recreation, spiritual renewal, and subsistence activities), our cultural heritage, visiting with friends, and sharing our values with children. When residents realized that we shared so much in common, it became possible to work together to protect these values by thoughtfully integrating development and conservation—that is, by proactively encouraging value-based jobs.

Our fisheries research found that additional processing of fish, beyond the primary stage of heading and gutting, was key to maximizing the value of this resource locally. While processors produce new lines of product such as fillets, smoked fish, and canned fish, we are working with fishermen's organizations to secure a licensed certification mark for Copper River salmon to further increase the products' market value. We are also working with processors, the city, and the native village of Eyak to develop a fish waste-processing plant. This effort would create a second harvest by turning the trash stream into a cash flow through products like organic fertilizer, bone meal, and nutritional supplements.

We have two forestry projects, one in the Copper River valley and one in the delta. Both involve native landowners. In the valley, we are exploring how sustainable forestry management practices can be applied to replace clear-cutting. Our report found that sustainable forestry can increase the overall value of the forest and the number of jobs, while decreasing the harvesting of timber and maintaining the recreational, cultural, and subsistence value of the forest.

Our forestry project on the delta involves restoring fish habitat on a private, 4000-acre clear-cut inholding within the Chugach National Forest. Trees left in the buffer zone had blown down in places, and, where the logging road bridges were removed, the road material was eroding and clogging salmon- and trout-rearing streams (Figure A). We worked with the Eyak Corporation (one of the 226 native village corporations created under ANCSA) to stabilize key areas that were identified by the state and federal project partners. The ground crew used logs left from the bridges, dug trenches for the logs, then planted fast-growing native alders along the logs to stabilize stream banks

Figure A. *A former logging road crossing of a salmon stream in the Copper River delta. Road material is eroding into the streambed behind a leftover bridge stringer log. (Photo by P. Swartzbart.)*

to slow erosion in these areas (Figure B). Together we improved about 20 miles of riparian habitat in our first summer, and we hope to expand this work into a fish habitat restoration corps with youth participation.

Cordova leads the state in community-based tourism planning. Our report served as a catalyst for residents to define what kind of tourism industry we wanted to foster and to draft a long-term plan to achieve that vision. Our goal is to promote small-scale eco- and cultural tourism in the face of efforts by the state, commercial tourism, and native corporations to open up the Copper River watershed and nearby Prince William Sound to industrial-scale tourism.

To get people working together, we took small steps initially. With the Cordova Chamber of Commerce, we produced a brochure that promotes the wild nature of our town and setting, and a shopping guide designed to reach independent travelers. With the chamber and city, we conducted visitor surveys that helped convince residents that independent travelers were spending more money in town than visitors from the large cruise ships.

In the upper valley, we are planning to work with McCarthy and Chitina residents to promote small-scale tourism in the historic McCarthy/Chitina/Cordova corridor. This will include planning with state agencies for tourist amenities and infrastructure needs (R.V. facilities, solid waste), as well as considering the protection of sensitive habitat. We are also encouraging members and the public to help shape the future of tourism in this region by attending state and federal planning meetings.

Successes and Failures

Although we have successfully attracted local native corporations and public land managers as project partners, we have not yet been able to work with native corporations that are based in urban areas, but that own and want to develop land within our region. Polarization over resource

Figure B. *A physically stabilized streambank, with a worker replanting native alder trees, which fix nitrogen and help to further stabilize the streambank with vegetation. (Photo by P. Swartzbart.)*

use between local residents (native and non-native) and "absentee" landowners who have no stake in the future of the community still occurs.

When we learned to match our work approach to our community's infrastructure, we had more success in building ties among community leaders than our initial approach of holding large-scale community meetings. The large meeting format smacked of government-style, top-down decision making, whereas small get-togethers, going to people's business offices for casual conversations, and simply getting to know each other as individuals ultimately built more lasting relationships.

We succeeded in getting a core of diverse interest groups and leaders to work together, but our work is still not widely understood throughout the region. It takes time to earn and build trust among community organizations and residents as the foundation for future work. It is simply not possible to produce substantive, tangible projects without established trusting relationships.

It will take time, the building of trust, and education to shift communities to sustainable ecosystem management. As more communities become interested in value-based development, we are hopeful that our efforts will merge into a sustainable future where social, environmental, and economic capitals are integrated into long-term wealth for the entire state. We realize that this shift cannot happen until we learn to listen with the goal of understanding, to accept each other's truths as valid, and to integrate other ways of knowing into a new model accepted and used by all.

Discussion Questions

1. Resource extraction, such as clear-cut logging, is prone to boom-and-bust cycles, depending on the pulp markets. In the Copper River watershed, when pulp markets dropped, so did the landowners' interest in pursing sustainable forestry. Our concern is that when the pulp market increases, there will not be enough lead time to develop sustainable forestry projects, including funding through grants, and clear-cut logging will prevail. How can we keep the landowners' attention on proactive and sustainable forestry, regardless of pulp market cycles?
2. How can we engage the regional native corporations in undertaking sustainable development of their resources, which are often in rural regions distant from where the majority of their shareholders live?
3. Ghandi said there are two parts to change: blocking the old and creating the new. Old paradigms are usually underpinned by a collection of laws, economic forces, and even scientific findings. People working to create change at the community level are often overcome by the sheer momentum of the old paradigm. What can people do to hasten change in attitudes and actions when old paradigms or ways of doing business support corporate (economic) needs, but not community (social and environmental) needs?
4. How can communities that want to preserve their rural character quantify or factor the "intangibles"—accessible wilderness, watchable wildlife, quality recreational opportunities—into local and regional planning efforts?
5. What can a rural community with a small land base, faced with large-scale development project(s) near town, do to preserve the characteristics that make it special?

Populations and Communities at the Landscape Level

HISTORICALLY, MANAGING POPULATIONS OF SINGLE species has been the primary focus of natural resource professionals. Indeed, the practice of natural resources management arose from efforts to (1) recover species that had been overharvested, (2) control or eradicate overabundant populations that threatened local economies, and (3) manage species with economic value for a sustainable harvest. For example, Douglas fir trees were harvested and white-tailed deer were hunted because they generated economic value. Wolves and prairie dogs were exterminated and insect outbreaks on national forests were treated with pesticides because they threatened the livelihoods of ranchers and loggers.

Today, single-species management is still important. Under the rubric of ecosystem management, the managed species are more likely than not to be those with declining populations, rather than those of economic interest. This chapter will consider how conservation biologists manage for species whose persistence is threatened. The focus on declining populations of single species has, however, brought an increased interest in managing for other species. Ecosystem management now takes into account the value of biodiversity and, therefore, seeks to also manage for species' communities. This is a wise course that will be beneficial in the long run, because as ecosystems continue to be fragmented, degraded, and converted, more and more species will become imperiled. Resource managers would be overwhelmed if they continued to manage for all species, one at a time. The management of collections of species, or species communities, will be considered in the latter part of this chapter.

Single-Species Management

Ecosystem managers increasingly find themselves dealing with declining populations of single species. Addressing these concerns requires an understanding of populations, factors that lead to population declines, and how ecologists estimate viable population sizes.

Often both a proximate and an ultimate factor can result in a population decline. The **proximate**

factor is the immediate cause for the decline and is usually related to a decreased birth rate, an increased death rate, or both. If uncorrected, these effects may cause a population to decline to the point where extinction is inevitable unless drastic action is taken. The **ultimate factor** is what leads to the increase in death rate or decrease in birth rate. A conservation biologist must understand the ultimate factor(s), which are usually associated with the species' environment, in order to reverse a species' decline. For example, the proximate factor for the decline of the peregrine falcon in North America was a decrease in birth rate (Figure 7.1). The ultimate factor, however, was pervasive pesticide contamination throughout much of the species' range. Falcons accumulated organochlorines by eating birds that had these toxins in their body fat by feeding on invertebrates, which in turn had taken up DDT from the soil.

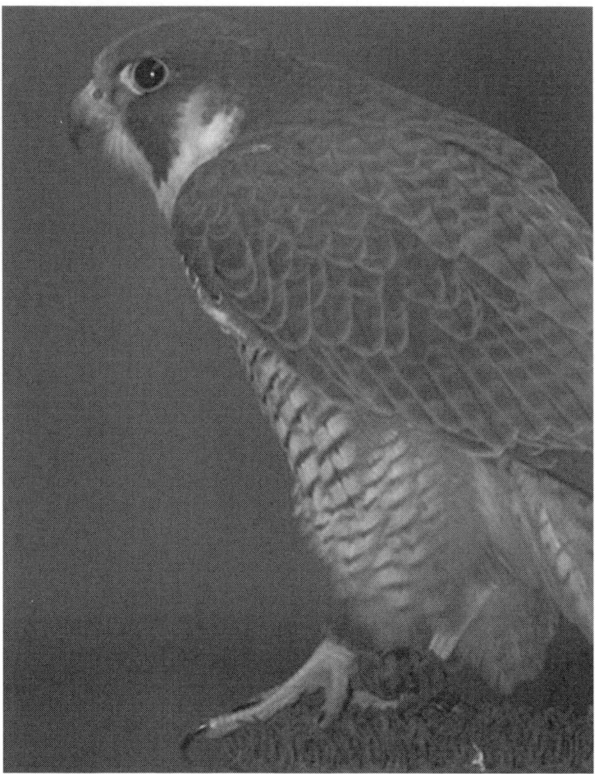

Figure 7.1. *A peregrine falcon. Historically, peregrines experienced reduced birth rates as a result of pesticide contamination. (Photo by Peter S. Weber.)*

Recovering the peregrine falcon from the brink of extinction required understanding both the proximate and ultimate factors. Biologists dealt with the proximate factor, the decrease in birth rate, by introducing young peregrines into the wild, thereby supplementing the reduced birth rate. The ultimate factor required action by the U.S. Congress, which eventually banned the widescale use of DDT in the United States. Over time the peregrine falcon's environment has gradually become less contaminated with DDT, allowing to successfully reproduce once again. Today, peregrine falcons have reoccupied much of their historic range.

Once the proximate and ultimate factors responsible for a species' decline have been determined, which ones should be dealt with first? Perhaps a human analogy will help. If you came upon an automobile accident and found a victim bleeding, what should you do first—stop the bleeding or treat for shock? Both can kill a person, but if the bleeding is not stopped immediately, you will not have time to treat for shock. Stopping the bleeding in a human is analogous to dealing with the proximate factor in a species. If birth and death rates are not dealt with immediately, the population might go extinct before one could deal with the environmental or ultimate factor.

EXTINCTIONS FROM DETERMINISTIC AND STOCHASTIC FORCES

Two kinds of factors may lead to the actual extinction of a population: deterministic and stochastic. **Deterministic forces** result from factors that, unless checked, will unquestionably result in the disappearance of a population. The persistent logging of old-growth forests in the Pacific Northwest is a deterministic force in the decline of species dependent upon large-diameter trees. Similarly, in the Sonoran Desert of the American Southwest, habitat continues to be converted to housing and commercial development at a rapid pace (Figure 7.2), and many species that depend on this ecosystem will experience reduced populations. In both cases, the outcome is predictable: If the deforestation of old-growth forests and the conversion of deserts to

Figure 7.2. *Sonoran Desert ecosystems are rapidly being converted to housing and commercial development, resulting in predictable losses of native populations in a deterministic fashion.*

houses continues unabated, there will be an ever-increasing list of species whose fate is in jeopardy.

Stochastic forces are the result of random events such as demographic changes, the loss of genetic diversity, or unusual environmental factors like an extremely cold winter, a wet spring, or a dry summer (see Chapter 3). Stochastic forces may also result from catastrophic events, either human-caused or natural. Importantly, stochastic extinctions usually occur for populations that are small or already reduced by deterministic forces. If logging reduces an old-growth forest to smaller, isolated stands, populations dependent upon contiguous habitat with large trees are at greater risk from stochastic events.

In other words, stochastic extinctions are a function of both population size and the magnitude of the stochastic or catastrophic event. The probability of a population going extinct is inversely related to population size. Deterministic forces set the stage for stochastic extinctions by reducing population size. Reduced populations are in turn predisposed to being influenced by stochastic forces, either reduced genetic diversity or an unusually harsh environmental event.

EXERCISE 7.1

Collaborate on It!

Split into small groups. In each group, and for the scenario you are using, examine species that are threatened with extinction. For each species, determine the proximate and ultimate factors that are responsible for the population decline. What are the deterministic and stochastic forces that have led to their declines? Each group should share their results with the class. Can you see any patterns among those factors that are proximate and those that are ultimate?

PVA AND MVP

The concepts of population viability analysis and minimum viable population are the responses of conservation biologists to estimating the population size that would be resistant to stochastic extinctions under various conditions. Knowing approximately how large a population should be to minimize the chance of a stochastic extinction is often critical information in ecosystem management projects.

Population viability analysis (PVA) is the science of model development to estimate extinction risk and closely related parameters, such as N_e. A **minimum viable population (MVP)** is the smallest spatially discrete population having a certain probability (e.g., 99%) of remaining extant (not going extinct) for a certain period of time (e.g., 1000 years), despite the effects of demographic, environmental, genetic, and catastrophic events. The actual probability of remaining extant and the specific time scale are determined by the conservation planners when performing a PVA.

A variety of factors must be considered when determining the time period and probability of not going extinct chosen for the model. For example, age at breeding, life span, the level of genetic diversity in the population, and the likelihood of environmental and catastrophic events all should be taken into account when one conducts a PVA. Consider a mouse and an elephant. Whereas a deer mouse may breed in its first year of life and live for less than a year, an elephant may not breed until 10 years of age and live for a half century. The time period for a PVA for the mouse might be 100 years, whereas the time period for an elephant population might be 1000 years or more. One size and one time frame does not fit all; to think otherwise would be a denial of the great diversity of life history strategies shown by species.

APPROACHES TO MVP ESTIMATION

Three approaches are used to estimate an MVP: experimental, observational, and modeling.

THE EXPERIMENTAL APPROACH. In the **experimental approach**, the practitioner experimentally isolates different-sized patches of suitable habitat that contain the species of interest. Population size is assumed to be directly related to patch size. Populations living in these patches are then monitored over time, and an empirical estimate of the minimum viable population size is made based on how long these different populations persist. On smaller patches, populations may quickly go extinct, suggesting these patches did not support populations that would qualify as a minimum viable population.

The best example of this approach was initiated by Thomas Lovejoy and his colleagues in the Amazon rain forests of Brazil. Working with the Brazilian government and timber companies, the scientists monitor a wide range of different species, from butterflies to primates, in different-sized patches of forest in a large experimental array. Now in their third decade, they continue to monitor population persistence on large tracts and extinctions on the smaller tracts. The results from this work are providing managers with information about patch sizes that support populations large enough to ensure that species of conservation concern will persist over time.

The experimental approach to MVP estimation is the most rigorous and least used method of estimating a minimum viable population. This is surprising, considering how much land use occurs on America's public lands. It is unclear why public land agencies that regulate logging, livestock grazing, outdoor recreation, and other land uses have not worked with conservation biologists in viewing these land uses as adaptive experiments. These are opportunities waiting to happen!

THE OBSERVATIONAL APPROACH. The **observational approach** examines populations of a species over time in habitats of different sizes, but not habitats that were created experimentally, as in the experimental approach. If populations inhabit patches of different sizes and have different persistence times, one can use this information to deduce a viable population. The observational approach assumes that populations are at equilibrium and that population size is solely a function of patch size and not due to differences in other patch attributes, both unlikely assumptions. The observational approach is similar to the experimental approach in that it monitors real populations of different sizes over time, yet it differs in that it measures populations in naturally occurring habitat patches, not on experimentally delineated ones.

This method was used to estimate the minimum population size of bighorn sheep necessary to persist over time. Bighorn sheep populations occurring on different-sized mountain ranges in California, Colorado, Nevada, New Mexico, and Texas have been monitored by state wildlife agencies for decades. By plotting herd sizes over the time periods they have been surveyed, one readily gets an idea of what constitutes a minimum viable population (Figure 7.3). Herds of bighorn sheep that exceed 100 animals have a very high likelihood of

EXERCISE 7.2

Collaborate on It!

In small groups, and for your scenario, choose two species. For one species, design an observational approach to estimate the MVP, and for the other species, develop an experimental approach. Be specific in how you would design these studies, and share your results with the class. What problems are associated with these two methods? List at least one reason agencies seldom use these approaches and one reason they might use them more often. Do your ideas illustrate the use of passive adaptive management?

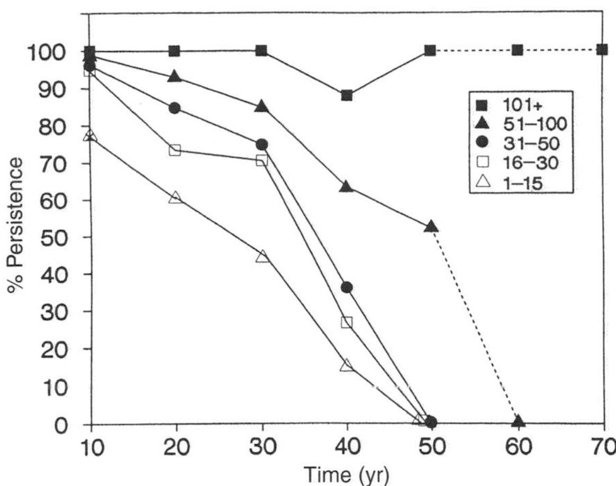

Figure 7.3. *The persistence of different-sized bighorn sheep herds in the American West over time. Different initial herd sizes were monitored for up to more than 50 years. Dashed lines are extrapolations based on trends. (From Berger, 1990.)*

persisting for 60 years or more, whereas those smaller than 100 have a reduced probability of persisting. This suggests an MVP for this species of at least 100 individuals.

THE MODELING APPROACH. The most commonly used approach to assess the extinction risk of species, as well as population viability, is the **modeling** approach. But first it is important to understand a statistic, *lambda* (λ), that is frequently used when modeling MVPs.

Lambda (λ) provides an easily understood projection of how well a population is doing because it is an estimate of the population's rate of change. Consider a population of cutthroat trout that this year contains 150 individuals. Last year your survey found that the population contained 100 trout. Intuitively you know the population grew by 50%. Lambda for this population captures that rate of change: 150 animals this year divided by 100 animals last year = 1.5. A λ value greater than 1 indicates a population is increasing; a λ value less than 1 indicates the population is decreasing (100 trout this year and 90 trout a year later: 90 ÷ 100 = 0.90). Finally, a λ value of 1 tells us the population is static and did not change during the time interval between two surveys (100 ÷ 100 = 1).

Conservation modelers use λ to determine whether or not the population they are modeling is increasing. If λ is not greater than 1, then their model suggests the population is declining. The two general categories of models used to estimate an MVP are deterministic and stochastic.

Deterministic MVP models contain fixed birth and death rates for different age classes, based on the averages of actual birth and death rates measured in the field from wild populations. In deterministic models, the only parameter one usually varies is the initial population size. These models give a population estimate that is then compared with the population estimate at some earlier time, and λ is determined. A λ value greater than 1 indicates the population is increasing. These models can be run for several generations, evaluating λ over time. If λ is greater than 1, the population is increasing, and the birth and death rate averages used in the model are those for which a manager can plan.

Deterministic models are called "deterministic" because they always give the same population estimate, unless you change the initial population size or if you change the birth and death rates used in the model. If, instead, you use the variation embedded in the data, you are using an approach called stochastic modeling.

Stochastic models are similar to deterministic models but they incorporate uncertainty into the model. By using average birth and death rates and by not including genetic and environmental stochasticity, deterministic models fail to acknowledge the variation that is part of any real population existing in nature. Stochastic models enable you to incorporate uncertainty; that is, they allow the natural variation in the data to express itself. For example, Stan Temple and John Cary from the University of Wisconsin built a population model for forest songbirds in Wisconsin. They had determined from studies that forest birds nesting close to forest edges would experience reduced birth rates, through nest predation and cowbird parasitism. If the Wisconsin countryside consisted of woodlots that were increasingly fragmented, would this bode poorly for the songbirds nesting

Table 7.1. *Selected Parameters in a Stochastic Model of Population Dynamics for a Forest Songbird Community in Wisconsin*

Model Parameter	Value or Description[1]
Clutch size	4 eggs (a constant)
Average fecundity (<100 m from forest edge)	0.7 young/nest (stochastic, c.v. = 20%)
Average fecundity (100–200 m from forest edge)	2.3 young/nest (stochastic, c.v. = 20%)
Average fecundity (>200 m from forest edge)	2.8 young/nest (stochastic, c.v. = 20%)
Adult annual survival	62% (stochastic, c.v. = 30%)
Subadult annual survival	31% (stochastic, c.v. = 30%)

[1]Stochastic variation is put in the model by using the c.v. (coefficient of variation) for fecundity and survival.
Source: Modified from Temple and Cary, 1988.

in them? What minimum populations would be necessary to keep these populations viable?

Temple and Cary's model was stochastic in that they did not use fixed values for many of the model parameters. Instead of using a fixed annual survival rate for adult birds, they allowed for stochastic variation, in this case a coefficient of variation (the standard deviation divided by the mean) of 30% around the mean survival rate each year (Table 7.1). Temple and Cary also allowed fecundity to vary in their model, with a coefficient of variation of 20%. The size of a population varies from year to year; seldom will it be the same size it was the year before. Birth and death rates, along with other population and environmental parameters, cause this annual variation. Models like this give different population estimates each time they are run; after all, the parameters that compose the model vary. Rather than a single answer as expected from a deterministic model, stochastic models present a range of projected population estimates over successive years (Figure 7.4a).

These various estimates of population size, reflecting variation in the data, can be used to estimate an MVP. Model projections in Figure 7.4b showed that 5% of the population estimates at 15 years were zero; in other words, there was a 5% chance the population would go extinct within 15 years. The definition of an MVP is the probability of remaining extant during a predetermined time period, so one can determine there is a 95% probability of the

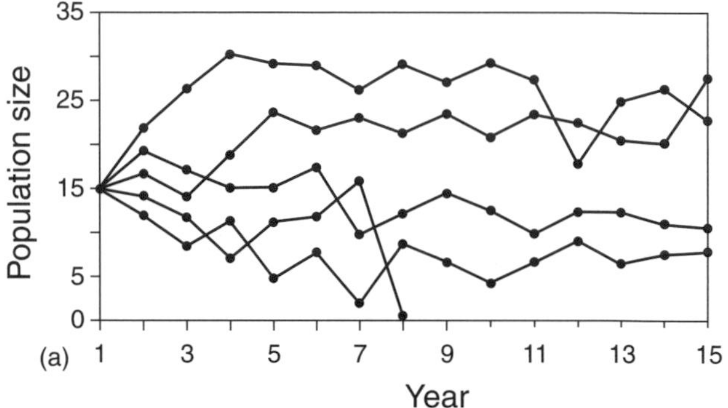

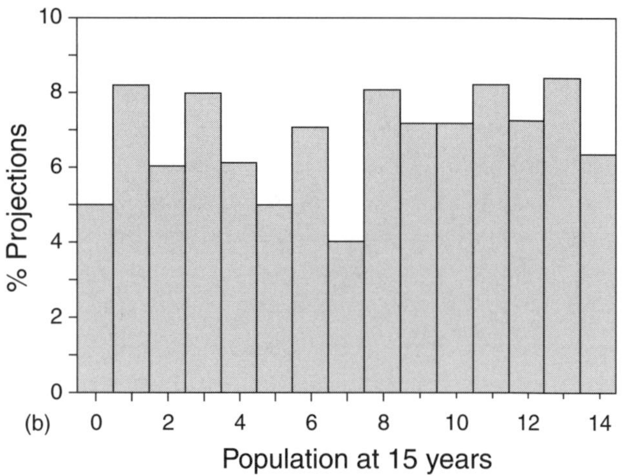

Figure 7.4. *Results from a stochastic model for estimating an MVP. (a) Five different population projections are shown for a simulation of 15 years. Note that one population in this case goes extinct at 8 years. (b) The percentage of population projections at the end of 15 years for numerous runs of the model. Note that 5% of the time, projected populations went extinct.*

population remaining extant for a 15-year period, given the model parameters used.

SOME THOUGHTS ON PVA AND MVP ESTIMATION

PVA and MVP efforts can be useful in conservation planning, particularly if the models are actually *validated* by matching the model output to a new set of data that was not used to develop the model. Validation can be done through manipulative field experiments or by sensitivity tests. Sensitivity tests systematically vary model functions and parameters over a range of conditions, while holding all other aspects of the model constant.

Having said this, however, keep in mind that the A in PVA is not for "answer"; it stands for "analysis." These efforts should be placed in that context. It would be naive to think one should manage for a single, magical number and when that number is achieved the population is secure. If undue attention is given to a single population number, it can divert attention away from the ecological processes and landscape attributes that are responsible for the persistence of a population.

Moreover, remember that MVP estimates for one species do not transfer to another species; most likely, they will even be different for different populations of the same species. Finally, MVP estimates do not consider spatial factors because a PVA is often done without considering real landscapes. The next section on metapopulations begins to incorporate spatial reality.

EXERCISE 7.3

Talk About It!

An MVP was determined for the northern spotted owl, using both the deterministic and the stochastic modeling approaches. The two methods yielded similar results. Should this surprise you? (Hint: Both models used the same data set, one using averages while the other used the actual variation in birth and death rates. Data were obtained from monitoring owls for 4 years.)

METAPOPULATIONS

In stochastic extinctions, the probability of extinction is inversely related to population size; the smaller a population, the greater the likelihood the population will go extinct. The larger the population, the greater the chance the population will persist. Conservation biologists have long been aware that local populations of small size commonly go extinct and, at some time later, reappear. How does this happen? After all, haven't we been told that "extinction is forever"? Extinction *is* forever when there are no other populations nearby from which individuals can disperse and recolonize an area that has lost its last individuals. But recolonization via dispersal does occur, and this phenomenon may allow for either the persistence of a population over time or the reappearance of populations that have disappeared. From these observations came the concept of a metapopulation.

A **metapopulation** is a regional population consisting of a number of spatially discrete populations distributed among habitat fragments and connected via dispersal (Figure 7.5). Unlike a single population in a largely uniform environment, a metapopulation is affected to a large degree by birth and death rates within the spatially discrete populations, as well as by how often dispersal occurs among them. Birth and death rates, in turn, are affected by the habitat quality of the patches supporting these discrete populations.

In Figure 7.5, for example, notice the outermost circle in the upper-left part of the metapopulation. This circle represents a spatially discrete habitat supporting a population to which individuals occasionally disperse. But notice that individuals do not disperse from it. What is occurring here? Conservation biologists consider a habitat supporting a spatially discrete population in which mortality exceeds productivity to be a **sink population** (Figure 7.6). But what about the largest circle in Figure 7.5, at the bottom of the figure? The population is large, as suggested by the size of the circle, and individuals consistently disperse from it, suggesting excellent habitat because the birth rate exceeds the death rate. This habitat supports a population called a source (Figure 7.6). A **source population** is a population

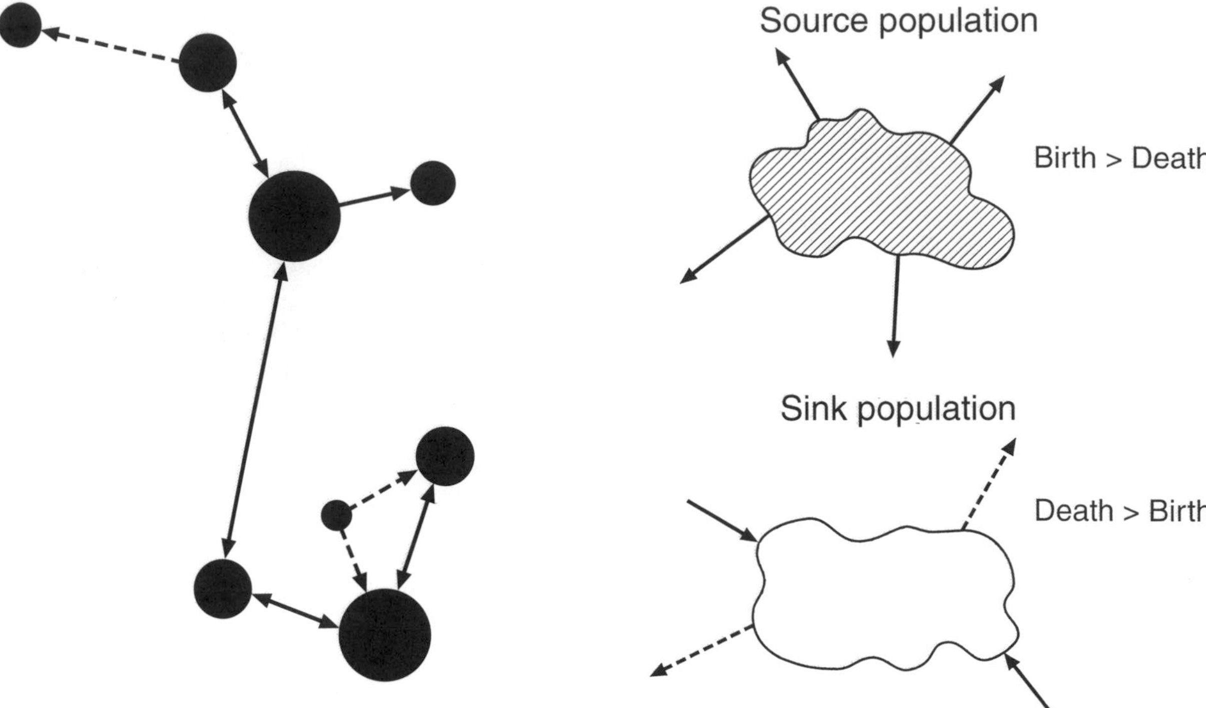

Figure 7.5. *A metapopulation showing different populations that vary in size (size of the circles) and that are connected via dispersal. Solid arrows indicate regular dispersal; dashed arrows indicate that dispersal occurs intermittently. Arrows indicate direction of dispersal. (From Meffe and Carroll, 1997.)*

Figure 7.6. *Metapopulations consist of source populations and sink populations. Source populations are spatially discrete populations in which birth rate exceeds death rate; sink populations are populations in which death rate exceeds birth rate. Sinks will go extinct without management intervention or individuals dispersing into the population; the latter is known as the rescue effect.*

in which productivity exceeds mortality—in other words, a self-sustaining population capable of producing individuals that may disperse to other populations.

The process of source populations bolstering sink populations through dispersing individuals, known as the **rescue effect**, is at the heart of the idea of metapopulations. The fact that metapopulations consist of spatially discrete patches containing populations that may be kept from going locally extinct by dispersing individuals is of great importance for conservation managers. The realization that metapopulations exist encourages managers to be aware of individual populations and the quality of habitat supporting them.

Because small populations are likely to go extinct repeatedly from the stochastic and catastrophic forces that shape habitat quality, it is essential to know whether populations are part of a larger metapopulation, connected via dispersing individuals. Because a metapopulation is shaped by the presence of multiple populations, a metapopulation is the sum of the characteristics of its constituent populations, each of which has its own birth and death rates shaped by habitat quality. Importantly, metapopulations incorporate a degree of spatial realism. Dispersal is mediated by the landscape between the populations. Whether this landscape is friendly, benign, or deadly will largely determine whether successful dispersal occurs. Although the metapopulation concept acknowledges dispersal, it says little about the actual matrix across which individuals disperse. Models that specifically address these details are called spatially explicit.

EXERCISE 7.4

Collaborate on It!

For the scenario you are using, pick a species that exists as a metapopulation and discuss the following questions. What management actions would you take if you had several populations that were sinks? Would you forget about them, knowing that they are not self-sustaining, or would you attempt to improve the habitat so that birth rates exceeded death rates? Or, still yet, might you focus your management to ensure high habitat quality of adjacent source populations, believing that dispersing individuals will maintain the adjacent sink populations?

Figure 7.7. *The Florida scrub jay's numbers are declining because of fire suppression and urbanization. A spatially explicit model is helping conservationists plan an appropriate response to ensure that this species persists. (Photo by G.K. Meffe.)*

SPATIALLY EXPLICIT MODELS

Whereas metapopulations are conceptual and do not include the complexities of real landscapes, they are nevertheless useful to conservation planners. Keep in mind, however, that general models yield general insights. Most conservation work is connected to real landscapes, with their complexities of diverse ownership and uses, various habitat types and conditions, and intact and altered ecological processes, as well as different densities of human populations. **Spatially explicit models** incorporate the actual locations of organisms and suitable patches of habitat. They also explicitly consider the movement of organisms among such patches across real landscapes.

Spatially explicit models are dependent upon geographic information systems (GIS). They can be either grid-cell-based or individual-based. In grid-cell-based models, cells are polygons of the same size and contain information such as ecosystem type, ownership, and population size. The information within these cells is then tracked over time, and the dynamic conditions of a population are determined. The individual-based models track the location of individuals across the landscape, and the fate of the species is determined by fitness attributes (probability of surviving and reproducing) based on the patches they occupy.

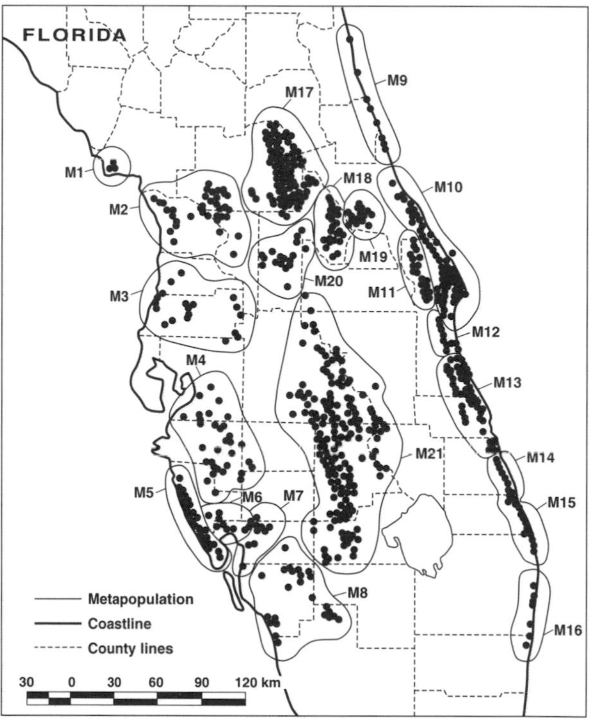

Figure 7.8. *The delineation of Florida's 21 metapopulations of scrub jays, based on 1992–1993 surveys. (From Stith, 1999.)*

An example of how spatially explicit models enable managers to protect a species at risk is illustrated by work done on the Florida scrub jay

Table 7.2. *Results from a Spatially Explicit Model for Scrub Jay Persistence of the Cedar Key Population in Florida*[1]

	Original	Restoration, No Acquisition	Restoration: 30% Acquisition	Restoration: 70% Acquisition
Population size (pairs)	4	17	34	54
Extinction risk (%)	100	10	0.0	0.0
Quasi-extinction risk[2]	—	82	5	0.0

[1]All simulations were run for a duration of 60 years.

[2]Quasi-extinction risk is having fewer than 10 pairs during any time during the simulations.

Source: Data from Stith, 1999.

(Figure 7.7). Florida scrub jay populations have declined precipitously through the suppression of natural fires and rapid urbanization of their native habitat. Statewide mapping efforts found that there were 21 metapopulations thought to be demographically isolated from one another (Figure 7.8). In response to the continuing decline of scrub jays, a land acquisition program is gradually adding more jay populations to the network of protected areas. The spatial variability where these populations occur, however, makes it difficult to decide which populations are adequately protected and where further land acquisition should be directed. Deciding which jay populations have the greatest need for land acquisition is further complicated by the fact that currently protected lands have not been properly managed and could support more jays if restored to optimal condition. If habitat in these protected areas were restored and jay populations recovered to normal densities, how viable would these populations be, with and without further land acquisition? And which patches should have the highest priority for acquisition?

These questions were addressed using a spatially explicit, individual-based population model that incorporated demographic and environmental stochasticity, catastrophic events, and other details of jay biology. The model simulated jay population dynamics on realistic landscapes developed from satellite imagery and aerial photographs. Actual jay territory locations were used in the model, as well as demographic data collected over many years of field research. A series of simulations was run for each metapopulation based on different reserve design scenarios, ranging from only protected areas to an ideal configuration consisting of all critical patches being acquired through acquisition.

For example, the Cedar Key population (M1 in Figure 7.8) was predicted to be extremely vulnerable without ecological restoration and land acquisition (Table 7.2). With restoration but no acquisition, there is a 10% liklihood of extinction and an 82% chance of quasi-extinction (population dropping below 10 pairs). With restoration *and* land acquisition, it is apparent that the chance of the population persisting is very high. The output from the model enables managers to identify critical unprotected habitat and protected areas that need restoration. This type of modeling effort would not have been possible without the detailed research conducted on the Florida scrub jay over many decades, as well as the landscape-level data collected remotely and incorporated into a GIS study.

INFORMATION NEEDS FOR MVP, METAPOPULATION, AND SPATIALLY EXPLICIT MODELS

PVA models depend on good demographic data, including fecundity and survival rate estimates for different-aged individuals that compose a population. PVA models with stochasticity built into them require even more knowledge about a population, such as variation in fecundity and survival rates, carrying capacity, catastrophic events, and other stochastic phenomena that might affect a species (Table 7.3). Indeed, data requirements for stochastic models are at least twice as great as for

Table 7.3. *Data Required for Constructing PVA, Metapopulation, and Spatially Explicit Population Models*

		Model Type			
		PVA			
Data Type	**Data Needs**	**Deterministic**	**Stochastic**	**Metapopulation**[1]	**Spatially Explicit**[1]
Demographic	Age class	X	X	X	X
	Age of first breeding	X	X	X	X
	Mean fecundity for each age class	X	X	P	P
	Mean survival for each age class	X	X	P	P
	Variance in fecundity		X	X	X
	Variance in survival		X	X	X
	Carrying capacity and density dependence		X	P	P
	Variance in carrying capacity		X	X	X
	Frequency and magnitude of catastrophes		X	X	X
	Covariance in demographic rates		X	X	X
	Spatial covariance in rates			P	P
Landscape	Patch types			X	X
	Area of patches			X	X
	Location of patches				X
	Transitions among patch types				X
	Matrix types				X
Dispersal	Number individuals dispersing			P	P
	Age class and timing of dispersal			X	X
	Density-dependent or independent dispersal			X	X
	Dispersal-related mortality			X	X
	Number of individuals immigrating			P	P
	Movement rules				X

[1]"P" means information needs to be patch specific.
Source: Modified from Beissinger and Westphal, 1998.

deterministic models. Furthermore, obtaining good estimates for the variation around fecundity and survival rates requires that measurements be made over many years; otherwise the full range of environmental variation will not be incorporated into the data.

Not surprisingly, these categories of data are known for few populations. Indeed, species whose populations are declining are often poorly understood, secretive, exist in low densities, or are protected by strict laws, collectively making collecting such data difficult. Sometimes resource managers substitute data from more common or closely related species; however, using information from such surrogate species may not be biologically appropriate. Because PVAs are so dependent upon demographic data and probabilities of stochastic events, they should be used with extreme caution when this information does not exist, is based on biological intuition, or is taken from a taxonomically similar species.

Metapopulations expand the spatial realism from single populations to spatially discrete habitats and their constituent populations, which are connected via dispersing individuals across landscapes. Spatially explicit models are similar to metapopulation models, but they contain spatially explicit information regarding habitat and demographic data, both within patches where the species occurs and across the landscape where the species disperses. Both modeling approaches differ from PVA models by acknowledging dispersal.

Whereas metapopulation models require data on discrete habitat patches, spatially explicit models require much more detail about the patches and the intervening lands across which the dispersing individuals move (see Table 7.3). Suffice it to say that such detailed information exists for few populations in the wild.

The use of PVA, metapopulation, and spatially explicit models is in vogue, and there is no indication that they will suddenly become unpopular. This is somewhat surprising considering how much information is required to correctly use such models. Are models constructed with inadequate data dangerous? Yes and no. Poorly constructed models that give numbers viewed as "truth" are indeed dangerous. But models that attempt to reduce complex systems to something more manageable can be useful. Indeed, models that enable managers to ask questions about a system and then see how the system responds have great power to teach. Whether to develop a model and implement its recommendations depend on the quality of the data and our understanding of the system.

Models are the most successful when used in conjunction with field studies and adaptive management, focusing on limiting factors and examining different management scenarios. PVA models were extremely useful in helping guide management actions for the northern spotted owl and the grizzly bear because the models were preceded and followed by extensive field investigations designed in an adaptive management approach. This process is quite different from simply putting numbers in a canned software program and then believing the A in PVA stands for "answer" rather than "analysis." In summary, models are tools conservation planners can use in developing successful, flexible approaches to maintaining viable populations of single species.

EXERCISE 7.5

Collaborate on it!

For your scenario, pick a species that exists as a metapopulation. Break into groups and develop a management scenario that incorporates the MVP, metapopulation, or spatially explicit approach to the management of that species. Then, in a three-way discussion, present the strengths and weaknesses of each approach. Be careful to list the assumptions you made when evaluating the approach you chose.

Managing for Species Communities

The previous material has addressed how managers deal with single species of conservation concern at the landscape scale. It is useful to understand how the alteration of ecosystems affects species, particularly those that are imperiled. However, managers using the ecosystem management approach increasingly find themselves addressing more than single-species issues. Indeed, ecosystem management is a far more inclusive approach than simply managing for rare and declining species. If concern over threatened species were the central focus of ecosystem management, there would be little need for this approach, because fishery and wildlife biologists, forest managers, and other professionals in natural resources have long been occupied with managing single species, including threatened ones. Ecosystem management, in addition to its concern for single species, also acknowledges managing for collections of species that comprise important parts of all ecosystems.

This is not to say that managing for species communities will be easy, only that it may be more efficient. There are many difficulties in managing for communities of species that occupy an ecosystem. Consider a few:

- By far the largest number of species are neither threatened nor endangered. For most of these species, ecologists know little about their biology and requirements. Single species that are declining often receive attention from managers because there is legislation protecting them, such as the Endangered Species Act (ESA). But what about biological communities? Do we have a national Endangered Ecosystem Act? No, and even though the ESA has provisions that allow for managing collections of species with similar ecological requirements, no single piece of legislation focuses on species communities.

- Will it be any less expensive managing for communities than single species? Perhaps not, considering how many species are in this broad group.
- Few natural resource managers have received training in community-level management. Historically, colleges and universities that produce land managers have offered courses that dealt with single species; community-level and landscape-level management courses are far less common.
- Will diverse publics understand management that focuses on ecological processes and landscape matrices to manage for collections of species? They are willing to lend their support to conserving single species, whether it be whooping cranes or pandas, but what will they think of conservation actions that promote communities of species, of which many are poorly known and seldom seen?

These are legitimate questions to ask, none of which have simple answers. In any case, there is an encouraging trend in ecosystem management for managers to focus on communities of species rather than single species. Therefore, it is appropriate to consider ways to manage for collections of species. There are at least three approaches, none of which is mutually exclusive, to managing for species communities: the species approach, the ecological process approach, and the landscape approach.

THE SPECIES APPROACH

The **species approach** concentrates on the manipulation of demographic variables (such as birth and death rates) of a single species that affects a broader collection of species. For example, this approach might attempt to increase the fecundity of a community of forest songbirds by reducing (increasing the death rates of) brown-headed cowbird populations to decrease nest parasitism. Cowbirds have historically not been in contact with forest songbirds, but increasing human densities and the conversion of natural ecosystems to agricultural use and urbanization have expanded both the numbers and the distribution of cowbirds. Cowbirds are able to enter forest edges and seek out nests of songbirds in order to parasitize. Because forest songbirds have only recently come into contact with cowbirds, they have no defenses against this highly effective strategy. By reducing cowbird numbers near forest edges, managers can increase the birth rates of forest songbirds and help ensure their persistence.

Other examples include salmon and prairie dogs. Pacific salmon die in glacier-fed rivers following spawning. Their carcasses provide a rich food source for a wide array of scavenging species, such as bald eagles, bears, crows, gulls, and ravens. As native salmon runs decline or are replaced by hatchery-reared salmon, the community of scavenging species is no longer sustained and experiences a decline in its members. Hatchery-reared salmon are not available to scavengers because they return to the hatchery to spawn and the carcasses are not returned to the river.

Likewise, a wide array of vertebrates, plants, and invertebrates thrive in prairie dog colonies. As prairie dog populations decline across the American West, so do the species that depend upon them (Figure 7.9). Both prairie dogs and salmon are considered by many to be keystone species in that they have a disproportionate impact on communities of other species, relative to their own biomass. By managing for this single species, a manager may promote the welfare of many species whose fitness is enhanced in prairie dog colonies.

The species approach relies heavily on management concepts that have emerged from fishery and wildlife management, as these fields have traditionally focused on managing single species. An important difference, however, is that the species approach manages a single species—not for that species' sake, but for the myriad other species that depend on or are affected by it.

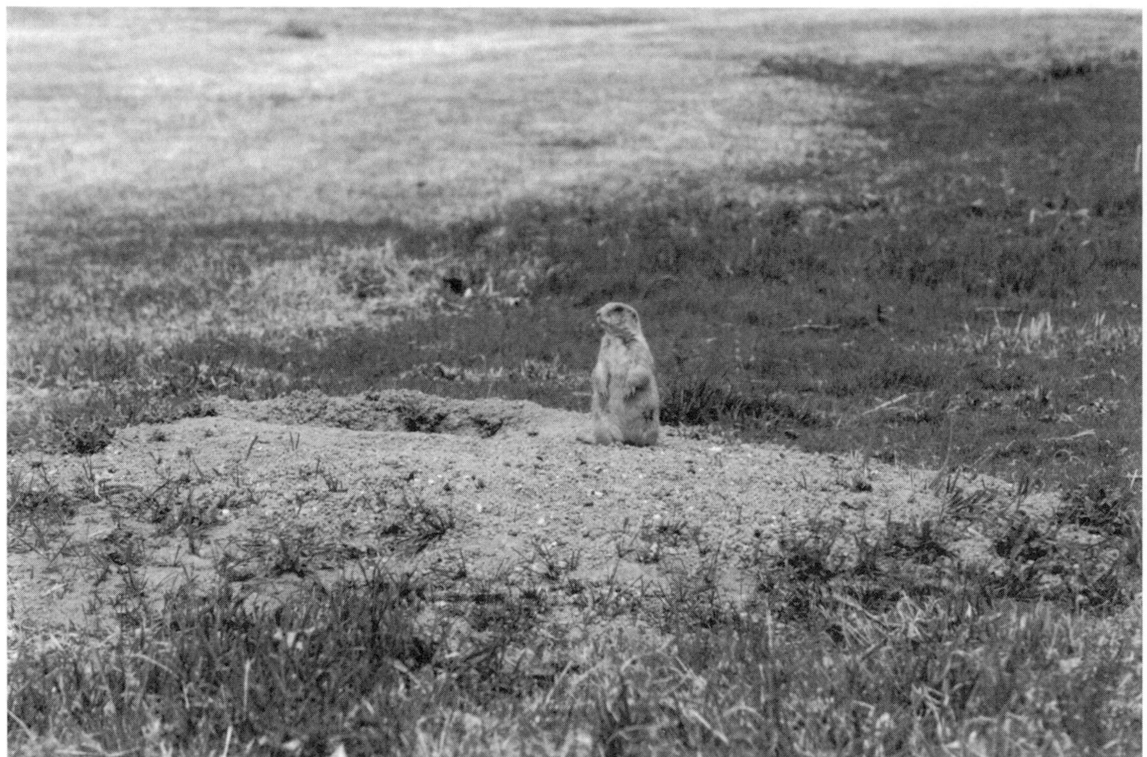

Figure 7.9. *Black-tailed prairie dog colonies support a vast array of vertebrate, plant, and insect communities. This is an example of the species approach to managing for species communities. (Photo by Wendell Gilgert.)*

THE ECOLOGICAL PROCESS APPROACH

The **ecological process approach** to managing for species communities draws on ecosystem science for its inspiration. Natural resource managers applying this strategy focus on ecological processes such as flooding, fire, herbivory, and predator-prey dynamics. This approach assumes that if ecological processes are occurring within their historic range of spatial and temporal variability, then the naturally occurring biological diversity will benefit. Perhaps the most prominent example of this approach is the reinstatement of fire to fire-dependent ecosystems (Figure 7.10). Land management agencies, from The Nature Conservancy to the U.S. Forest Service, are once more allowing fires to burn or are starting prescribed fires to mimic the historical fire regime of ecosystems. This requires knowing the fire interval, time of year when fires naturally occur, and average fire-patch size.

The same principles apply to flooding and herbivory. Riparian areas require flooding to maintain the natural abundance of their respective species communities, whereas grassland ecosystems depend upon grazing by herbivores. In the absence of these ecological processes, plant and animal communities change over time to something different from what historically defined the ecosystem. The U.S. Department of the Interior recently used flooding in the Grand Canyon to reestablish riverine geomorphology and biological parameters below a dam by releasing large amounts of water to mimic a natural flood event.

THE LANDSCAPE APPROACH

A third method of managing for species communities is the **landscape approach**, which draws on the discipline of landscape ecology for its stimulus. This approach, which focuses on landscape pat-

Figure 7.10. *Prescribed fires are an essential ecological process for ensuring that species communities persist in fire-dependent ecosystems. This is an example of the ecological process approach to managing for species communities. (Photo by Heather A. L. Knight.)*

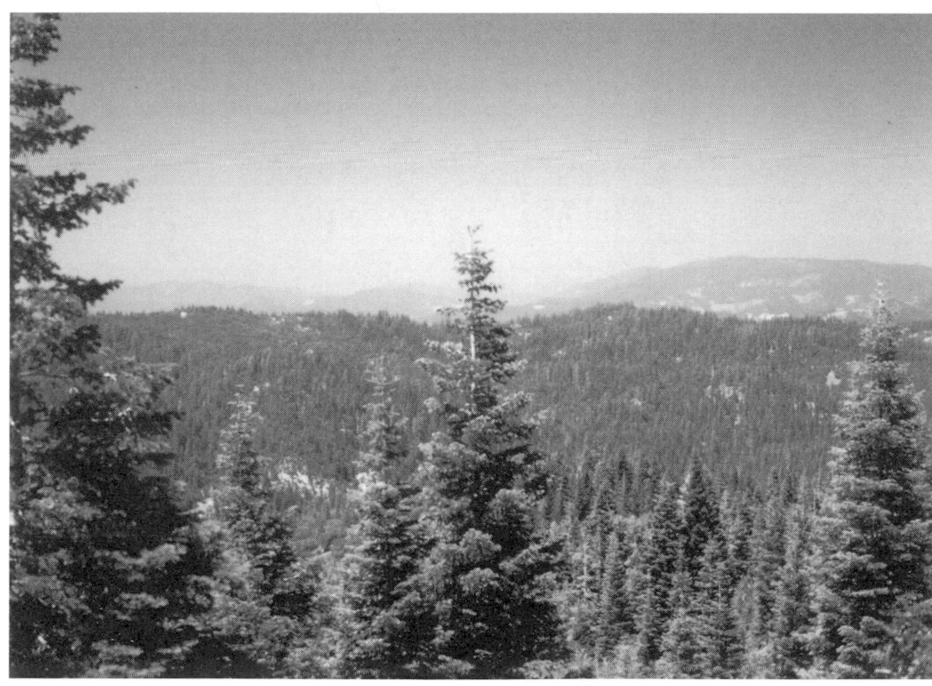

Figure 7.11. *A mixture of old-growth forest stands that are large, less edgy, and more connected is an example of the landscape approach to managing for species communities. (Photo by Wendell Gilgert.)*

terns rather than processes, manages landscape elements in such a way as to collectively influence groups of species in a desired direction. To maintain wildlife communities that depend on late-successional-stage forests, for example, forest stands could be managed for old age, large area, and minimal edge (Figure 7.11). To promote communities of species that are dispersal-sensitive, landscapes could be managed so that ecosystem remnants are more proximal or are connected by movement corridors. This approach assumes that by managing a landscape for its components, the naturally occurring species will persist.

EXERCISE 7.6

Collaborate on It!

For the scenario you are using, devise a species, ecological process, and landscape approach to manage for a collection of species mentioned in the scenario. Which of the three approaches seems most difficult to envision? Which of the three approaches do you think would be most attainable? What obstacles stand in the way?

THE ROLE OF MONITORING IN EACH APPROACH

The species, ecological process, and landscape approaches can be quite effective in ensuring that viable species communities exist across landscapes. It is important, however, that each approach be monitored to ensure the desirable outcome of management actions. Managing in the absence of monitoring is wasteful and can potentially worsen a situation if managers do not anticipate undesirable outcomes.

It is important to remember that these three approaches were devised to effectively manage scores of species simultaneously. Each method assumes that the community of species benefits by the focus on one or more of the three approaches. For example, in the ecological process approach, managers assume that by reinstating fire in the appropriate spatial and temporal scales, the fire-dependent community of plants and animals prospers. Using the ecological process approach not only benefits many species; it is more efficient than managing for species one by one.

To monitor the species approach, one might monitor the population of the critical species. For example, if managing for reductions in cowbird populations to benefit songbird communities affected by cowbird parasitism, one could monitor cowbird populations. Alternatively, a monitoring scheme could be devised that actually checks on the species community of interest—in this case, the number of nests that are annually parasitized. In such instances, one would hope to see either the cowbird population decrease or reduced nest parasitism (Figure 7.12a).

To assess the effectiveness of the ecological process approach, a manager may monitor the process itself. For example, if fire is prescribed to ensure fire-dependent species, then keeping track of average fire size and total areas burned each year might be appropriate (Figure 7.12b). Alternatively, one could manage some aspect of the community being benefited, such as species composition or the relative abundance of species that comprise the community of interest.

Finally, a manager might evaluate whether the landscape approach is working based on monitoring attributes of the landscape elements. If wildlife communities dependent upon old-growth forests were of management concern, one might measure patch size of late-successional-stage forests over time. If patch size increased, one might assume that communities depending on old-growth forests benefit (Figure 7.12c).

It is encouraging that land managers are using each of these three approaches to ensure the maintenance of native communities. The U.S. Forest Service and Bureau of Land Management, for example, commonly focus management on key species, ecological processes, and landscape features. Nongovernmental organizations such as The Nature Conservancy do so as well, when managing bioregions. Both government agencies and NGOs are increasingly attempting to manage landscapes that have complex administrative boundaries. By finding ways to focus on ecosystems, even though

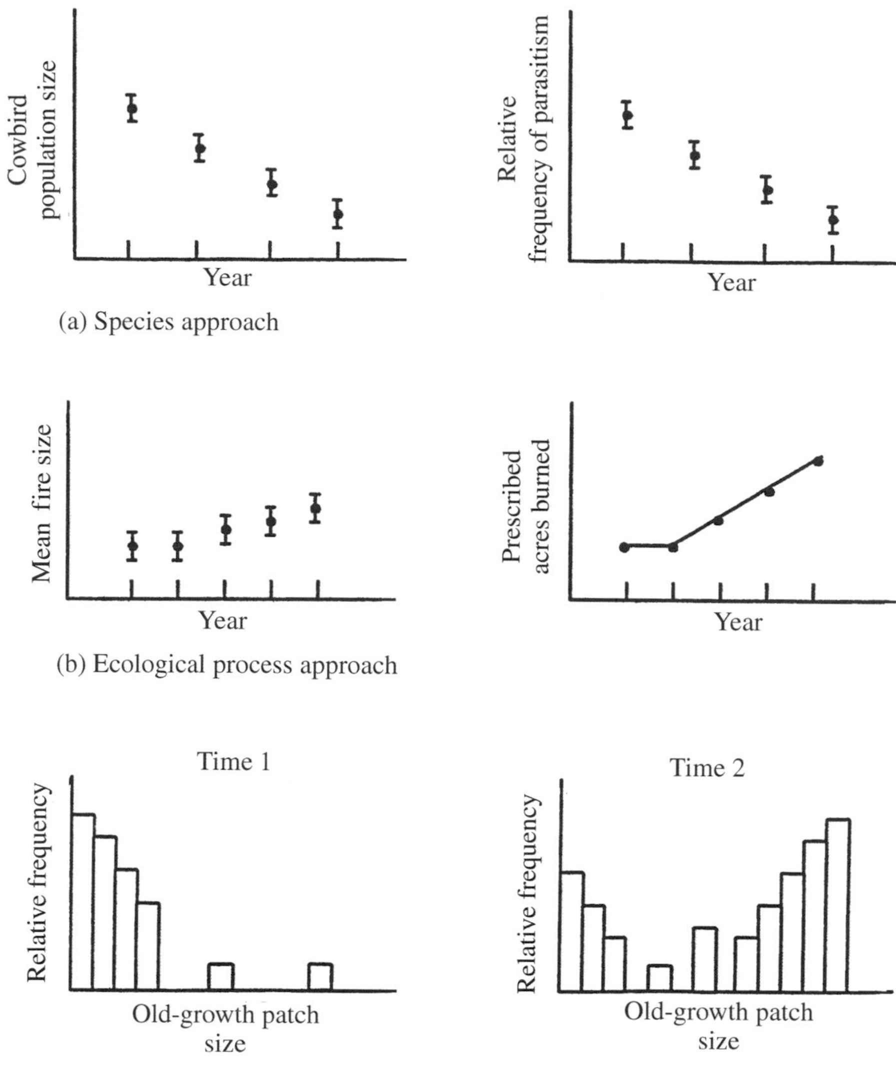

Figure 7.12. *Examples of how to monitor the three different approaches to managing for species communities. (a) Managers might assess whether the species approach is working by monitoring cowbird populations to increase the productivity of forest songbird communities whose nests are parasitized by cowbirds. On the left, cowbird populations are declining each year, whereas on the right, the frequency of songbird nests parasitized is decreasing over time. (b) An example of monitoring an ecological process to manage for species communities. In this case, prescribed fires are being reinstated on a landscape. On the left, the average fire patch size over time is increasing, suggesting that prescribed fires are getting larger, supposedly to mimic the historic fire regime. On the right, the number of acres burned every year increases, showing an increase in prescribed fires. (c) An example of how the landscape approach to manage for species communities can be monitored. The bar graphs show that the frequency of large patches of old-growth forest are increasing over time. This would suggest that wildlife communities that depend on older forests are benefiting.*

they encompass diverse ownership boundaries, managers are achieving greater success. By integrating the three approaches, a more appropriate balance results—neither unduly stressing nor leaving out species, ecological processes, or landscape features. Today, land managers are using all three methods in integrated approaches to ensure the maintenance of biologically diverse ecosystems.

References and Suggested Readings

Berger, J. 1990. Persistence of different-sized populations: An empirical assessment of rapid extinctions in bighorn sheep. Conservation Biology 4:91–98.

Beissinger, S.R., and M.I. Westphal. 1998. On the use of demographic models of population viability in endangered species management. Journal of Wildlife Management 62:821–841.

Boyce, M.S. 1992. Population viability analysis: Adaptive management for threatened and endangered species. Transactions of the North American Wildlife and Natural Resources Conference 58:520–527.

Franklin, J.F. 1993. Preserving biodiversity: Species, ecosystems, or landscapes? Ecological Applications 3:202–205.

Hunter, M.L., Jr. 1996. *Fundamentals of Conservation Biology*. Blackwell Science, Cambridge, MA.

Knight, R.L. 1998. Ecosystem management and conservation biology. Landscape and Urban Planning 40:41–45.

Knight, R.L., and T.L. George. 1995. New approaches, new tools. Pp. 279–295 in R.L. Knight and S.F. Bates (eds.). *A New Century for Natural Resources Management*. Island Press, Washington, D.C.

Liu, J., J.B. Dunning, and H.R. Pulliam. 1995. Potential effects of a forest management plan on Bachman's sparrow (*Aimiphila aestivalis*): Linking a spatially explicit model with GIS. Conservation Biology 9:62–75.

McCullough, D. (ed.). 1996. *Metapopulations and Wildlife Conservation Management*. Island Press, Washington, D.C.

Reed, J.M., L.S. Mills, J.B. Dunning, Jr., E.S. Menges, K.S. McKelvey, R. Frye, S.R. Beissinger, M. Anstett, and P. Miller. 2002. Emerging issues in population viability analysis. Conservation Biology 16:7–19.

Ruckelshaus, M., C. Hartway, and P. Kareiva. 1997. Assessing the data requirements of spatially explicit dispersal models. Conservation Biology 11: 1298–1306.

Stith, B.M. 1999. Metapopulation viability analysis of the Florida scrub-jay (*Aphelocoma coelurescens*): A statewide assessment. USFWS report 1448-40181-98-M324. Jacksonville, FL.

Temple, S.A., and J.R. Cary. 1988. Modeling dynamics of habitat-interior bird populations in fragmented landscapes. Conservation Biology 2:340–347.

Woolfenden, G.E.W., and J.W. Fitzpatrick. 1984. *The Florida Scrub Jay: Demography of a Cooperatively Breeding Bird*. Princeton University Press, Princeton, NJ.

Experiences in Ecosystem Management:

The Winyah Bay Focus Area

Roger L. Banks

SINCE 1995, THE U.S. FISH AND WILDLIFE SERVICE has gone through profound organizational and structural adjustments in an attempt to convert to an ecosystem approach that protects fish and wildlife resources for Americans. Basically, the concept is straightforward: When carrying out their role of managing and protecting fish and wildlife resources of national significance, service employees are encouraged to (1) think on a landscape scale; (2) work across program lines; and (3) when appropriate, work closely with stakeholders. The goal of the USFWS Directorate has been to instill this philosophical approach to resource management in the heads of all USFWS employees, to the point where it becomes second nature. Although results to date have been spotty, across the country there are places where USFWS personnel have contributed to the protection of fish and wildlife resources by adopting the ecosystem approach, working with partners within and outside of government. When the approach has been successful, there is unanimous agreement that the value of achievements through successful partnering was far greater than what would have occurred had individuals worked separately.

The Winyah Bay Focus Area (WBFA) is a good example of a partnership that has been successful and has resulted in the long-term protection of thousands of acres of valuable coastal habitats. In addition to exemplifying the value of partnering, this effort also demonstrates why thinking on a landscape scale is so important. To fully understand what the WBFA partnership is all about, it is important to recognize how it fits into South Carolina's coastwide resource protection programs.

In 1988, coastal South Carolina represented the southernmost extension of the Atlantic Coast Joint Venture of the North American Waterfowl Management Plan (NAWMP). During this time, a group of five individuals representing a cross section of the public and private sectors joined together to form a landscape-scale, long-term land protection effort that was coined the ACE Basin Project. ACE is an acronym of the three river systems—Ashepoo, Combahee, and Edisto—that compose the basin. The ACE Basin Task Force included a private landowner, The Nature Conservancy, Ducks Unlimited, the South Carolina Department of Natural Resources, and the USFWS. The goal of the ACE Basin Project was to accomplish long-term habitat protection in the focus area through the use of conservation easements, acquisition, and management. Unlike many of the land protection efforts in vogue at the time, which were more restrictive and regulatory oriented, the approach for the ACE Basin was to be proactive and landowner-driven.

The timing could not have been better. The heavy-handed, regulatory approach to meaningful resource protection was just not working. Although scattered and small areas of important wetlands were being protected through the wetlands permitting program, they were seldom part of a planned, landscape-scale protection effort. Moreover, the resource value of what was protected was often negligible.

The ACE Basin, more than any other location in

coastal South Carolina, was the ideal area within which to undertake such a large-scale, relatively unconventional conservation effort. Key reasons were the remaining land ownership patterns and that so many of the landowners were somewhat conservation-minded and were excited about keeping the basin relatively undeveloped. Within the entire ACE Basin Project area, there were only about 50–60 individual landowners to deal with. Fortunately, many of the antebellum-era plantations remained virtually intact, including the old ricefields, making this one of the most diverse and productive natural resource areas in the state. Environmental and developmental threats to the basin are virtually nonexistent. Unlike Savannah to the south, and Charleston and the Myrtle Beach Grand Strand to the north, the ACE Basin is not immediately threatened by urban sprawl, industry, or resort and residential development. This is not to imply that implementation of this landscape-scale land protection effort was easy.

Initially, the goal of the ACE Basin Task Force was to establish long-term protection on 90,000 acres of important habitats in this 350,000-acre focus area. Voluntary conservation easements were intended to be the primary land protection device. Within the first 2 years, efforts had been so successful that the group elected to increase their project goal from 90,000 to 200,000 acres. Based on this success, a decision was made in South Carolina to increase the number of coastal focus areas from one to five. Each new focus area was to be set up, organized, and patterned after the ACE Basin. For example, each one had a chairman from the private sector; each was limited to five to seven members, most of whom were from the private sector; and each area had its own set of threats and opportunities. For all the newly established focus areas, however, the goal was the same: long-term protection of the remaining important habitats through conservation easements and other proactive approaches.

In 1990, the Winyah Bay Focus Area was born (Figure A). Situated between Georgetown to the south and the Myrtle Beach Grand Strand to the north, the WBFA encompasses about 525,000 acres of some of the most diverse habitats in coastal South Carolina. Beginning with the expansive intertidal salt marshes to the east, the habitats progress through wide bands of longleaf pine ecosystems, tidal blackwater rivers lined with diverse stands of forested wetlands (Figure B), and thousands of acres of remnant ricefield impoundments dating back over 200 years. In sharp contrast to the ACE Basin Focus Area, however, the remaining natural resources of the WBFA are under a high degree of threat. Myrtle Beach—the fastest-growing area in coastal South Carolina—is a hotbed of development activity, ranging from highway projects to residential complexes and golf course resorts. The population of the Myrtle Beach area is expected to double by 2020, and the number of tourists is expected to reach more than 1 million people per day. In the face of such pressures, the task of pursuing meaningful, long-term, landscape-scale habitat protection is made increasingly difficult.

Fortunately, the ACE Basin Project had demonstrated that large-scale areas could be protected if partners were willing to work together. From the beginning, it was recognized that for the WBFA to be successful, it would take a somewhat different composition of partners, as well as a more diverse slate of protection approaches. Through the insight of the chairman of the WBFA effort, a task force was established that brought together a mix of agency and private-sector individuals who had a dismal history of being able to work together. In light of this mixed bag of partners, the first rule of order was that everyone involved would have to leave their baggage at the door. Without such an understanding, the effort would have failed from the outset. Despite the mix of federal regulators, resort developers, industry representatives, state natural resource biologists, and private conservation groups, the WBFA Task Force did meld together as an extremely effective partnership. It was clear from the beginning that all the members of the task force, regardless of their different backgrounds, shared the same desire to achieve meaningful, long-term protection in the WBFA.

The WBFA Task Force established a goal to protect at least 85,000 acres of important habitats within the core of the focus area. In light of the

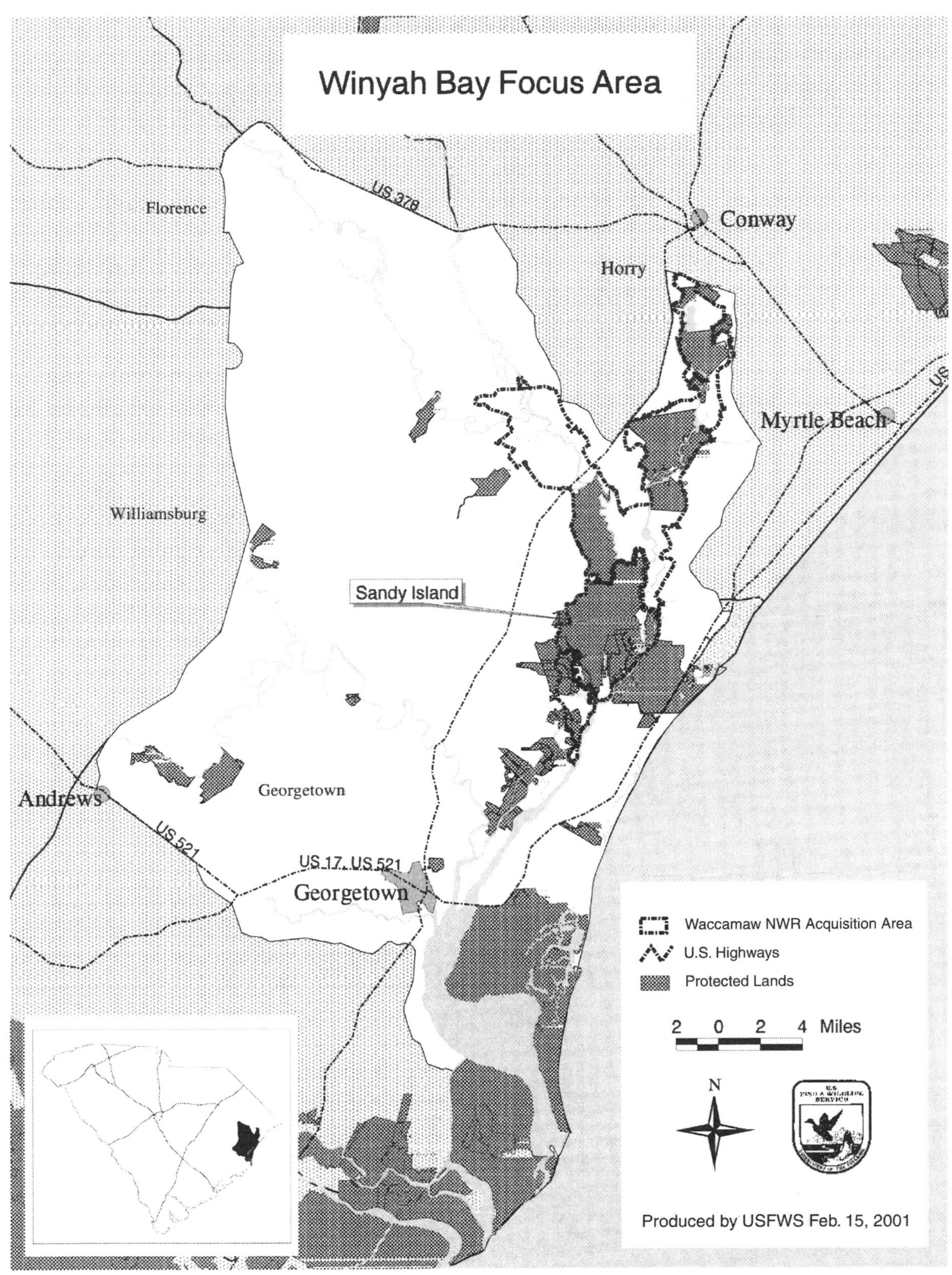

Figure A. *A map of the Winyah Bay Focus Area. The inset shows its location within South Carolina.*

Figure B. *An example of the tidal forested wetlands being protected within the Winyah Bay Focus Area. This is the Little Pee Dee River within the Waccamaw National Wildlife Refuge. (Photo by David H. Gordon.)*

ongoing threats of urban sprawl and runaway golf course resort development, the task force knew it had a window of opportunity of about 5 years within which to protect the most pivotally important properties in the focus area. Recognizing the time constraint, it jointly developed a strategy that would give the effort the best chance of success. Briefly, the Task Force developed a strategy that included:

- Collectively establishing the boundaries of the focus area.
- Determining what important properties in the focus area were already under some type of protection.
- Establishing a "core" area of about 85,000 acres, within which most of their initial efforts would be directed.
- Identifying which properties within the core area were in most imminent need of protection.
- Determining which protection mechanisms would be most appropriate to pursue for the important properties identified (conservation easements, acquisition, or management).
- Agreeing on a plan of action.

As a result of this strategic planning, the task force identified at least eight properties within the core area that would have to be protected for the effort to be successful. The highest-priority focal property for protection was a place called Sandy Island. Everyone felt that Sandy Island was pivotal, and if it was not protected, adjacent landowners would be reluctant to subject their properties to perpetual protection under conservation easements for fear of the development that might follow.

Having established its long-term protection goal, and identifying key properties in need of protection, the task force then identified three approaches it would pursue:

1. Establishing a 50,000-acre National Wildlife Refuge.
2. Placing voluntary conservation easements on as many private properties as possible within the core of the focus area.
3. Facilitating the management and enhancement of key properties whenever possible.

Recognizing that a landscape-scale approach to the long-term protection of important resources was still not generally recognized, the WBFA Task Force members carried their message to key agencies. The idea of basing regulatory decisions on landscape-scale planning, rather than on project-specific bases, was not widely accepted by the U.S. Corps of Engineers, the Environmental Protection Agency, or the state permit issuing agencies. The task force members called on these agencies, explained what the WBFA was trying to accomplish on a landscape scale, and asked if they could accommodate these efforts. Numerous tours around Sandy Island and other properties were provided to any agency personnel interested in learning more about the effort and wishing to see the quality of resources to be protected.

A major hurdle was to achieve credibility with the state and federal regulatory agencies. Traditionally, agencies do not give much consideration to resource protection efforts originating with the private sector; this is where the value of agency and private-sector partnering manifests itself. Agency representatives on the task force were able to demonstrate to the regulatory agencies that the WBFA Task Force goals were also the goals of the resource agencies. In addition, they were also able to demonstrate that agencies would gain insight simply through input from private partners.

The chairman requested that the USFWS consider developing a new National Wildlife Refuge in the WBFA. The proposal was accepted, and the result was approval to initiate planning for the 50,000-acre Waccamaw Refuge. All task force members, both agency and private sector, played key roles during the environmental planning and public hearing stages of the effort. The end result was overwhelming support from Washington and from the local community for the refuge. Approval for the refuge was granted in 1995, and initial acquisition funding was allocated the first year.

During this same time period, the owners of Sandy Island had suffered a setback in their efforts to acquire a permit to access the island for development and decided to offer it for sale. Recognizing the opportunity to acquire wetland mitigation credits for at least three major highway projects, the state Department of Transportation, in conjunction with The Nature Conservancy, was able to acquire the island and develop the Sandy Island Mitigation Bank. In the end the Department of Transportation acquired a total of 17,000 acres of critical habitat within the core of the WBFA in exchange for what will probably not exceed 500 acres total wetlands impact. The Sandy Island Mitigation Bank has been recognized nationally as the most cost-effective mitigation bank known in the country. Recognizing that this habitat protection would not have occurred without input from a number of entities, it is safe to conclude that the initial role of the WBFA was critical for setting the stage to make the acquisition a reality.

It is important to remember that the two major approaches to achieving the desired level of resource protection in the WBFA were acquisition and voluntary conservation easements. While the local media were giving tremendous coverage to the acquisition-related success stories, like the establishment of the Waccamaw Refuge and the acquisition of Sandy Island, little was being said about protections being afforded through voluntary conservation easements, which have now been placed on more than 7000 acres of quality coastal properties.

The decision on the part of private landowners to place an easement on their property in perpetuity is not taken lightly. Whatever decision is made—whether who should hold the easement or how much conservation protection to build into it—they know that the commitment is forever. Consequently, a landowner wants to feel comfortable that if he or she commits to protecting the land forever, other property owners in the area are willing to make the same commitment. What they do not want is to end up being an island of protection in a sea of development.

Ecologically, conservation easements represent the "ultimate" in terms of the kinds of protection that can be afforded private property. Most importantly, protection via a conservation easement is a positive, proactive effort in that it is a voluntary agreement made between the landowner and the easement holder, and it is permanent. In a way, the landowner is in a position to determine the destiny of the property, or literally manage from the grave, because he or she determines the level of conservation protection, as well as any other details thought appropriate to ensure the property is protected. This is in sharp contrast to most regulatory forms of protection, in which the agencies dictate the terms of protective covenants.

Moreover, the American public benefits because properties under easement remain in private ownership and stay on the tax rolls, and traditional uses, such as hunting and fishing, farming, and certain silviculture activities, are still permitted. Unlike some forms of protection that apply only to wetlands or some other segment, in most cases conservation easements are applied to the entire property. This is important because easements protect habitats for the smaller, less economically important species, as well as the more charismatic, larger species, many of which have economic value.

An important value of a landscape-scale land protection effort like the WBFA is that, when local citizens begin to see resource protection taking place around them, the effort becomes contagious. Seeing the positive impact of these kinds of decisions on their quality of life, more people want to become a part of it. Nowhere is this more clearly demonstrated than in the area of conservation easements.

The partnering activities within the WBFA Task Force continue to be the cornerstone of success. In partnerships like the WBFA, the actual partnering effort takes various forms. Not all members of the WBFA Task Force are involved all the time, or with every issue. The establishment of the Waccamaw Refuge and the acquisition of Sandy Island, for example, required a great deal of involvement from the chairman and the agency representatives on the task force, whereas the partners most heavily involved in the efforts to acquire conservation easements are those from the private sector. Many landowners are inherently suspicious of state and federal agencies and would not feel comfortable negotiating with them on the long-term fate of their properties.

In light of what the WBFA Task Force has accomplished to date, one could conclude that is has been a successful conservation partnership. But not everything we tried has worked, and we have learned some important lessons:

- Partnership goals should be clear and achievable.
- To be successful, all partners must believe in and support the same goals.
- Partnering is extremely time-consuming. One must be willing to commit the time and energy necessary to see the effort through.
- To be a successful partner, one must be willing to relinquish his or her own authority.
- A successful partnership is one in which nobody can detect who is in charge.
- Partnerships designed to achieve acceptance by the general public should be chaired by someone from the private sector.
- Project areas should be of manageable size; in areas that are too large, it is difficult for people to have a sense of ownership.
- All partners must be willing to leave their "baggage" at the door; this is critical for partnerships involving individuals with differing interests.

Discussion Questions

1. The WBFA is a proactive, private-sector-led resource protection effort. Why has the private factor been so valuable in the success of this effort? How does it differ from more traditional resource protection efforts?
2. Recognizing that the setting of the WBFA is somewhat unique and could probably not be duplicated elsewhere, what features of this land protection partnership could be implemented in other locations?
3. What are some advantages of achieving resource protection through the placement of voluntary conservation easements versus the achievement of protection via regulatory means?
4. Explain the value of identifying or delineating focus areas within which to pursue meaningful resource protection.

CHAPTER 8

Landscape-Level Conservation

IN THE PREVIOUS CHAPTER, WE EXAMINED METHODS of assessing species populations with various degrees of spatial realism. As we learned, PVAs have little spatial realism, metapopulation models have some spatial realism, and spatially explicit models include considerable spatial detail. We also discussed the need for managing species communities. Considering how many species there are and the increasing difficulty of managing them one species at a time, it is not surprising that managers are increasingly exploring innovative ways to manage for entire assemblages of species. In this chapter, we will build on previous material and examine the broader role that landscapes play on species populations, as well as on collections of species that share ecosystems. We must always remember that landscapes and ecosystems determine the health of species populations; it is seldom the other way around. (However, recall the discussion of keystone species in Chapter 6.)

At the landscape level, species persistence is threatened by three primary factors (Figure 8.1):

1. Habitat loss, the conversion of natural ecosystems to human uses.
2. Habitat fragmentation, the dissection of natural habitat by human activities.
3. Matrix quality, the overall integrity of the landscape in which natural areas are embedded.

These three factors are closely interrelated. As a landscape is converted to some human use (habitat loss), the level of fragmentation increases and the quality of the surrounding landscape is degraded.

Imagine an area in the northeastern part of America that is a blend of deciduous and coniferous forests. Over time, the land use in this area changes from timber production to residential development. As the landscape is converted from forest to homes and commercial development, there is a loss of forest habitat, an increased level of habitat fragmentation from roads and homes, and an overall decline in the quality of remaining habitat because of more dogs, cats, automobiles, non-native species, and human activities.

Clearly, one cannot approach the conservation

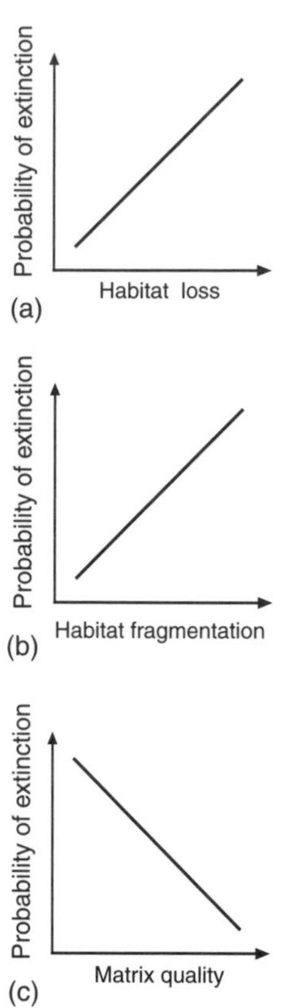

Figure 8.1. *Three landscape-level threats to the persistence of a species population. The probability of extinction increases with (a) increasing habitat loss, (b) increasing habitat fragmentation, and (c) increasing degradation of landscape-matrix quality.*

of biodiversity at the scale of landscapes without an appreciation of habitat loss, habitat fragmentation, and habitat quality. The realization that habitat is being lost—that is, ecosystems are being converted from native vegetation to some form of human use—is of chief concern for all those interested in healthy natural communities. Habitat loss has been the universal rallying call of conservationists for over a century. The fact that habitat fragmentation and the quality of the matrix can also be important factors shaping biodiversity has become recognized more recently. Ecosystem management, in addition to its emphasis on the human dimensions, also stresses the interplay of humans and their effects on biodiversity. Accordingly, it is as important to understand the complexity of landscapes as it is to appreciate the dynamics of human communities. Here we begin to explore the ideas of habitat fragmentation and habitat, or matrix, quality.

EXERCISE 8.1

Think About It!

For the landscape where you are attending college, itemize human activities that (1) result in habitat loss, (2) fragment the habitat, and (3) degrade the quality of the landscape. Is each activity distinct, or do some activities result in habitat loss, fragmentation, and decrease in matrix quality? Can you see any similarities and differences among these activities? What was the landscape during the time of the first Americans?

Habitat Fragmentation

Habitat fragmentation, as much as any other human-induced landscape activity, has helped define conservation biology. There is very little humans do on landscapes that does not increase fragmentation (try to think of some)—road construction, commercial development, agriculture, building rural subdivisions, and so on. Therefore, it is useful for natural resource managers to be aware of the concepts associated with habitat fragmentation. This is because the ecological effects do seem to be very real and because habitat fragmentation is increasing at least as rapidly as the human population.

Habitat fragmentation is the conversion of a contiguous area of native vegetation into remnant patches of native vegetation for human use (Figure 8.2a). An example would be the landscape-wide conversion of native grasslands to agricultural crops. Over time, grasslands are converted to agricultural lands; the grassland ecosystem is fragmented, leaving only grassland patches.

Another type of habitat fragmentation is called **perforation**, in which human uses alter small patches within an area of natural vegetation (Figure 8.2b). A housing development beyond the city limits where lot sizes are measured in acres

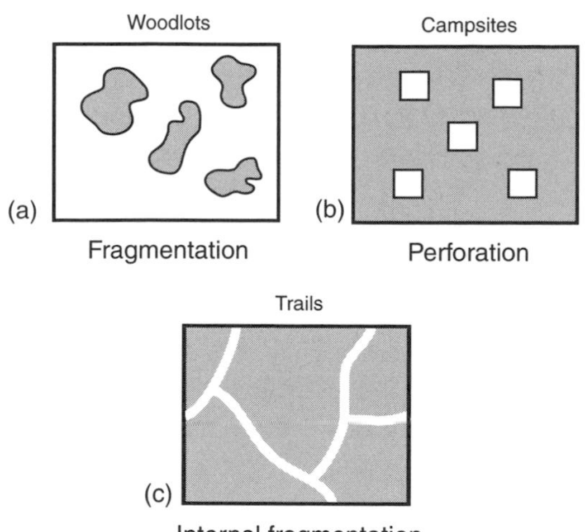

Figure 8.2. *Three types of habitat fragmentation. (a) Traditional fragmentation occurs when one habitat nearly replaces another. (b) Perforation occurs when small openings (e.g., home sites, forest clear-cuts) are created within otherwise contiguous habitat. (c) Internal fragmentation occurs when linear rights-of-way (e.g., power lines, roads) dissect a landscape.*

perforates the landscape because homes replace only small portions of the native vegetation; that is, the homes *perforate* the landscape rather than *dominate* it. Whereas forest cutting may truly fragment a landscape—replacing forests with forest clear-cuts and leaving only forest patches—housing developments in rural areas leave much of the natural vegetation intact beyond the house edges.

Another type of fragmentation is called **internal fragmentation**, which occurs when linear corridors (e.g., roads, power lines) dissect an area (Figure 8.2c). The rural housing development shown in Figure 8.3 also illustrates internal fragmentation. Each home has an access road so the owners can commute to nearby towns. These roads internally fragment, or dissect, the landscape into smaller parcels.

Three landscape-level consequences following fragmentation of a landscape are the loss of area, an increase in edge, and increased isolation of the remaining patches.

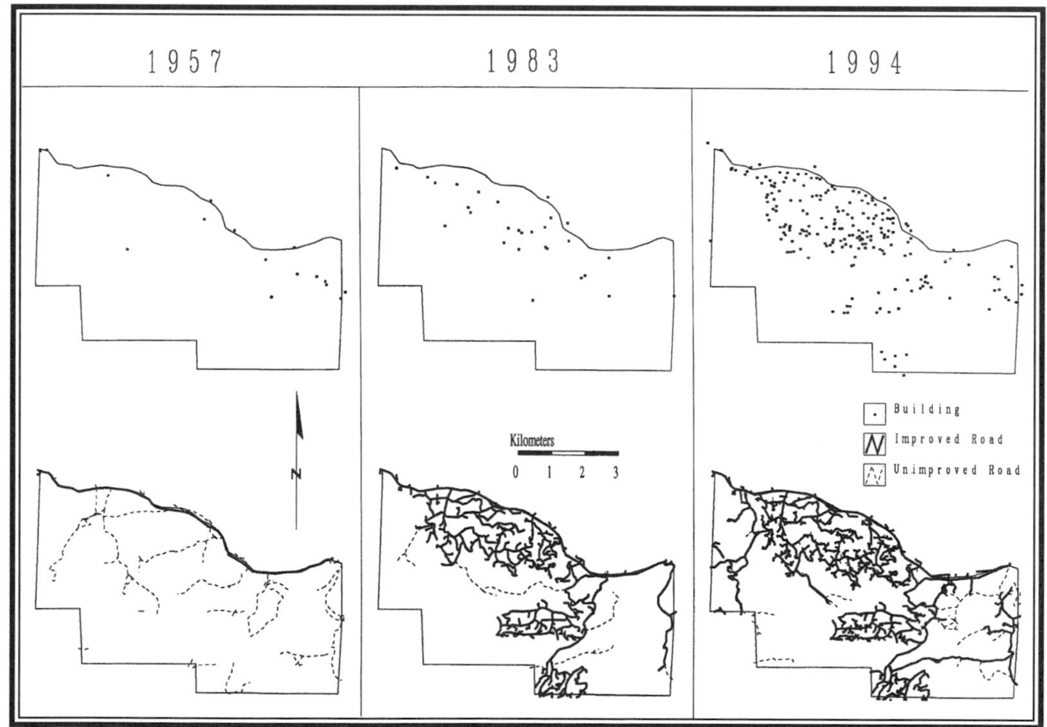

Figure 8.3. *A former cattle ranch in Larimer County, Colorado, that was sold and developed. Notice the increase in homes and roads since development began in the 1950s. (From Knight, 1997.)*

THE LOSS OF AREA

It is inevitable that the area of an ecosystem decreases following fragmentation, at least within a certain time frame. For example, after a forest is logged, the landscape is fragmented, resulting in isolated patches of remnant forests. These patches are smaller in size than before logging activities. This division of a larger area into smaller, isolated patches can affect the number of species that may persist in that ecosystem. This observation is based on what ecologists call the **species-area relationship**, which states that the number of species increases as area of an ecosystem increases. This generalization holds for most types of species—plants, insects, and vertebrates—and an example for birds in Chilean forests illustrates the point. The number of bird species, regardless of season, increases with increasing size of the forest fragments (Figure 8.4). Interestingly, even the largest forest fragments were still missing certain bird species that would normally be present if the landscape were less fragmented. The missing species were characterized by large body size, poor dispersal ability, or habitat specialization—all characteristics that predispose a species to be sensitive to habitat fragmentation.

More species are encountered in ecosystem remnants of increasing size for a variety of reasons. With increasing area, there is a greater likelihood of ecosystem heterogeneity, topographic diversity, and different habitat types (thus increasing beta

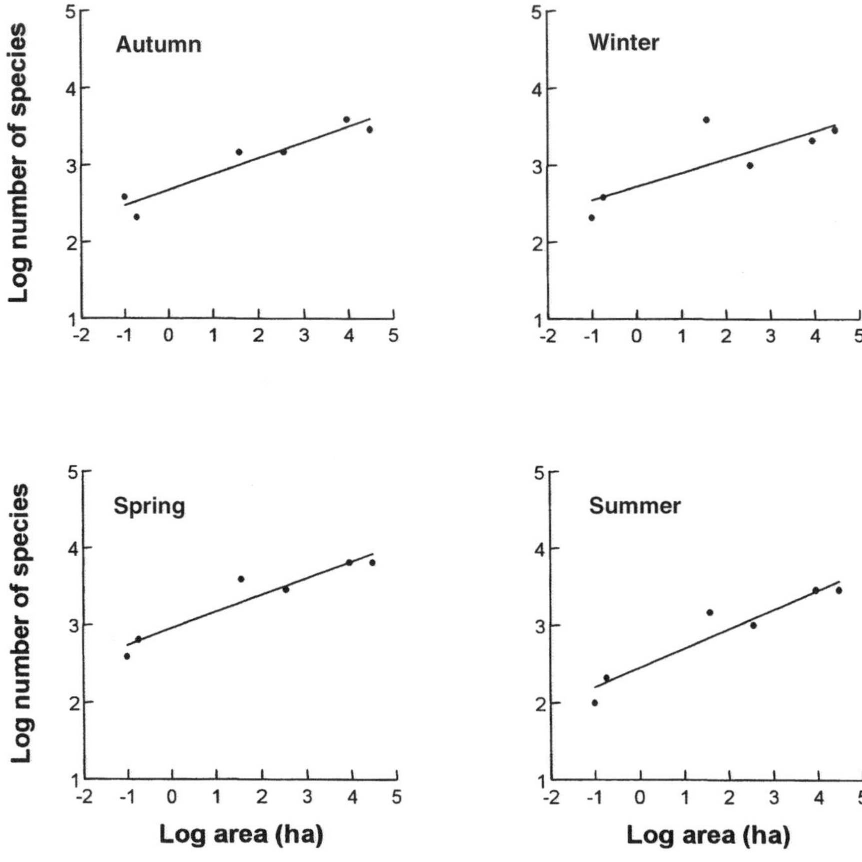

Figure 8.4. *Seasonal variation in the species-area relationship for birds present in different-sized forest fragments of Fray Jorge National Park, Chile. Forest patch size varied from 0.5 ha to 22.5 ha. The axes have been logarithmically transformed. The number of breeding bird species increases with increasing plot size. (From Cornelius et al., 2000.)*

richness). Increasing heterogeneity is accompanied a better chance of more species because there are more niches available. Likewise, with increasing patch size, there is an increasing likelihood of dispersing individuals encountering the patch. In addition, larger patches can support larger populations of a species. Larger population size is accompanied by an increased likelihood of the species persisting over time in that patch—that is, a greater chance of its having a minimum viable population (MVP). Larger patches are also more likely to have species that are area-sensitive.

Area-sensitive species require large areas to survive and reproduce, for several reasons. A species may have a restricted ecological niche; for example, carnivores have a specialized diet that requires large areas, simply because of their food requirements. Mountain lions have large home ranges because of their dependence on adequate populations of deer, their primary prey. Other species may be area-sensitive because of seasonal movements. Elk in the Rocky Mountains have seasonal movements from summers spent at high elevations to winters in lower elevations. By necessity they are area-sensitive because their winter and summer ranges comprise extensive areas. Other species are area-sensitive simply because of their large body size. Black bears are a good example; they eat a broad range of foods and occur in a wide range of habitats. They are area-sensitive because they roam over large areas, commensurate with their body size.

A study of rain forest fragments of different sizes near Manaus, Brazil, further illustrates area sensitivity. Nine species of insect-eating birds were surveyed in forest fragments that varied in size from 2.5 acres to 250 acres. In addition, similar-sized plots of forest in large, intact rain forests also were surveyed as a control. The researchers documented 55 cases of local extinction in the fragments, beginning 1 year after isolation up to the time of their final surveys 14 years later. This corresponded to 74% extinction of the local populations in these fragments. The most dramatic effect of fragmentation on extinction was in the 2.5-acre fragments. All the species recorded in these fragments before isolation were absent in later surveys. Alarmingly, these extinctions occurred despite the second-growth connection of some fragments to continuous forest as close as 225 feet away. This work suggests that these species were all area-sensitive, being able to persist in the short term only in the largest fragments.

It is important to remember that the mere presence of an area-sensitive species in an ecosystem remnant does not ensure its long-term persistence (see Chapter 7 and MVP). Because large species are also long-lived, one might encounter an individual and falsely believe that the population is viable when in fact it is near extinction. In addition, the simple presence or absence of a species does not differentiate between species that occur in both small and large fragments but that differ in some fitness attribute among different-sized fragments.

Consider the results of researchers studying birds nesting on different-sized grassland remnants in Missouri. The greater prairie chicken was a typical area-sensitive species in that it was absent from smaller fragments, occurring only on remnants greater than 325 acres. Two other species, the Henslow's sparrow and the dickcissel, however, illustrate how the mere presence of a species does not necessarily demonstrate whether it is area-sensitive. Both species are small migratory songbirds, and although they were present in all-sized grassland patches, the Henslow's sparrow occurred in reduced densities on the small remnants. The density of dickcissels did not vary with patch size, but this species experienced lower nesting success on the smaller remnants when compared with the

EXERCISE 8.2

Collaborate on It!

In groups, visit the Species in Parks: Flora and Fauna Databases Web site (http://ice.ucdavis.edu/nps/) and compile species lists from National Parks of different sizes. Plot the number of species (on the vertical axis) versus park size (horizontal axis), and note whether the species-area relationship holds. Do this for different taxa (plants, mammals, etc.), and see if the trend holds. Can you deduce any pattern from the shape of the curves (linear, curvilinear)? Does the range of park size influence the shape of the curve?

larger ones. This example illustrates that you cannot rely solely on census data to determine whether a species is area-sensitive. It may be necessary to have density estimates as well as demographic data, such as fecundity and survival estimates.

AN INCREASE IN EDGE

Habitat fragmentation is accompanied by an increase in edge and an **edge effect**, a phenomenon whereby some species are negatively affected near habitat edges (Figure 8.5). Not all species evolved in landscapes that were naturally edgy or heterogeneous; many, in fact, require large areas of unfragmented habitats. What happens to those species when, over time, their ecosystems become increasingly fragmented by human activities? Often the result is a reduction in their ability to survive and reproduce. Indeed, sufficient studies exist on the responses of certain species to increased edge that they are known as edge-sensitive species. An **edge-sensitive species** is one whose fitness is reduced near habitat edges. Three general categories of edge effects decrease the fitness of species sensitive to ecosystem edges: abiotic effects, biotic effects, and human effects.

ABIOTIC EFFECTS. The abiotic effects that may impinge on edge-sensitive species include sharp gradients across edges in temperature, relative humidity, solar radiation, and moisture. For example, certain plant species evolved in contiguous patches of forest. They thrived under a relatively narrow range of ambient temperature, wind speeds, and relative humidity. When ecosystems are fragmented, individuals of these species often find themselves in close proximity to an edge. Because temperature, moisture, wind speed, and other abiotic conditions differ along habitat edges when compared with habitat interiors, the fitness of these species is often reduced. Likewise, many species of insects, such as butterflies, thrive only in forest interiors, where temperature and wind speed are more uniform than those found along forest edges.

BIOTIC EFFECTS. Just as there are some species that did not evolve in association with ecosystem edges, there are other species that did. These species do best when they are associated with edges, and their ability to survive and reproduce are enhanced. These are **edge-generalist species,** and they can exert biotic effects on other species. For example, blue jays and American crows are quite effective nest predators of forest songbirds, whereas brown-headed cowbirds are efficient nest parasites of songbirds, not building nests of their own but instead laying

Table 8.1. *Edge-Generalist Species in Wisconsin Pine Barrens*

	Frequency Detected	
Species	**Edge Transects**	**Interior Transects**
Blue jay	70%	0%
American crow	15%	0%
Brown-headed cowbird	55%	10%

Source: Niemuth and Boyce, 1997.

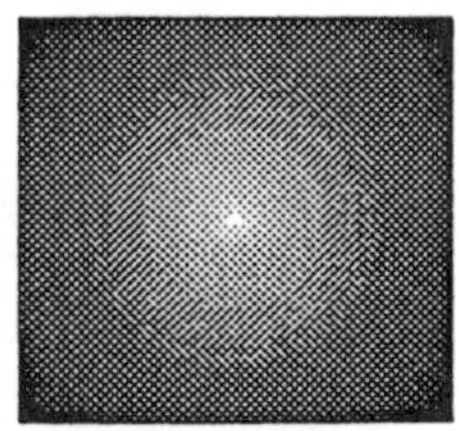

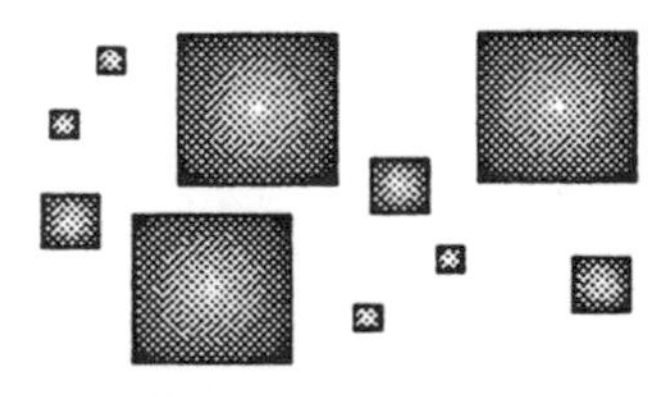

Figure 8.5. *An ecosystem that is 640 acres (1 square mile) in size, before and after it is fragmented. Notice that following fragmentation, the habitat remnants have over 50% more edge than when the ecosystem was intact and much less interior habitat.*

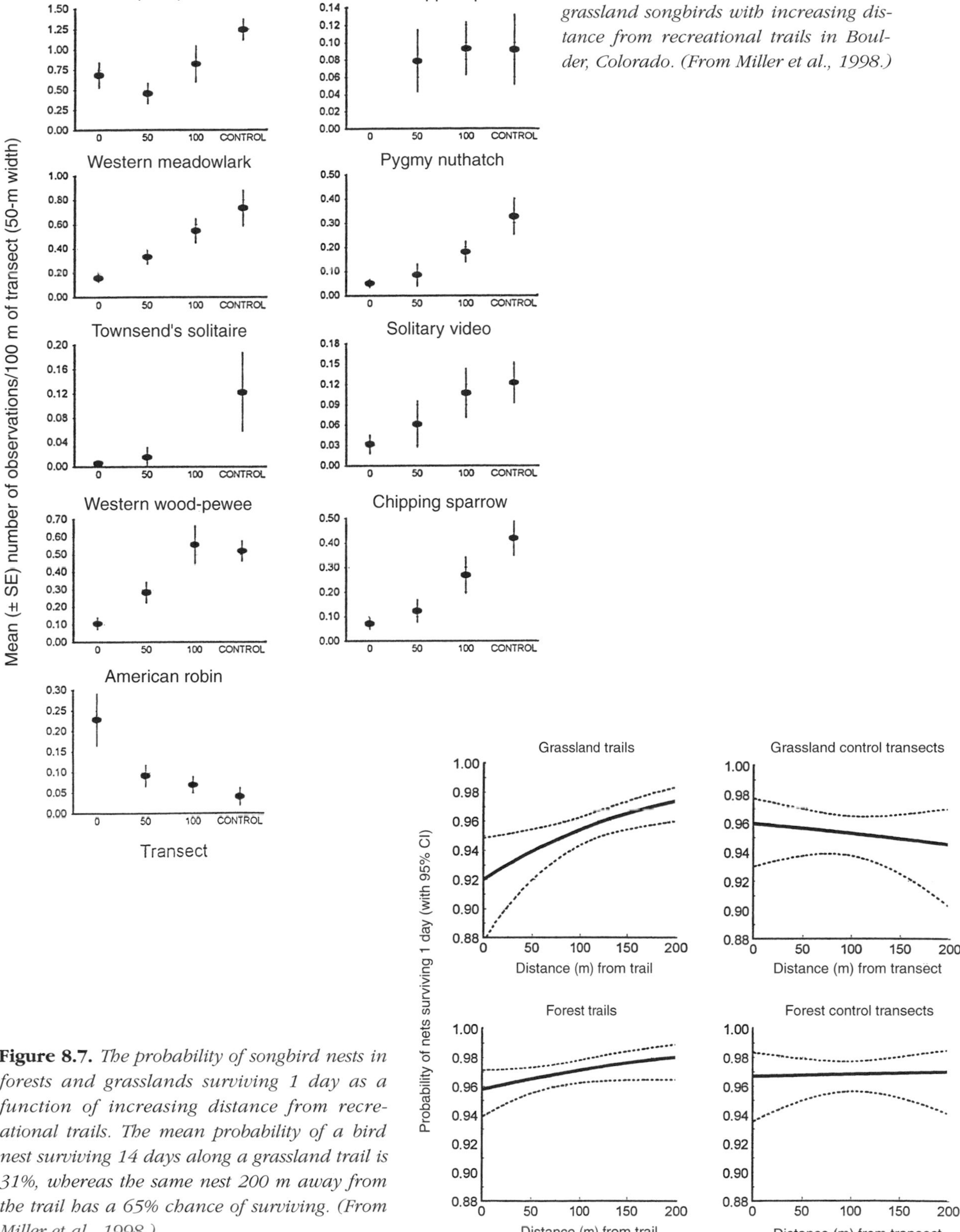

Figure 8.6. *Populations of forest and grassland songbirds with increasing distance from recreational trails in Boulder, Colorado. (From Miller et al., 1998.)*

Figure 8.7. *The probability of songbird nests in forests and grasslands surviving 1 day as a function of increasing distance from recreational trails. The mean probability of a bird nest surviving 14 days along a grassland trail is 31%, whereas the same nest 200 m away from the trail has a 65% chance of surviving. (From Miller et al., 1998.)*

their eggs in the nests of other species. Jays, crows, and cowbirds qualify as edge-generalist species because they are more likely to be encountered along an ecosystem edge than in its interior. In Wisconsin, researchers found that all three species were more likely to be encountered along the edges of pine barrens than within the forest interiors (Table 8.1).

With increased edge following fragmentation, there is an increased likelihood that edge-sensitive species will come into contact with edge-generalist species. Often these generalist species are effective competitors, predators, or nest parasites, such as the jays, crows, and cowbirds found along Wisconsin forest edges. The result may be lower survival and reproduction of edge-sensitive species, resulting in reduced populations. Recreational trails that internally fragment ecosystems illustrate this point. Studies in ponderosa pine forests and grasslands of Colorado have shown that populations of some songbirds are reduced in association with these edges. Notice in Figure 8.6 that eight species of songbirds had lower densities along trail edges, in both forests and prairies, than away from trails. By contrast, the American robin, a classic edge-generalist species, had its highest populations near trail edges. Nesting success also was reduced. The probability of a nest surviving a single day increased with increasing distance from trails, in both forests and grasslands (Figure 8.7).

Human Effects. Certainly human effects associated with edges following fragmentation could be lumped under biotic effects. On the other hand, considering the diversity and magnitude of human activities that are associated with edges, they probably warrant separate consideration. Human effects that affect edge-sensitive species include all conditions associated with human activities along edges, such as pets, weapons, and disturbances from human activities and structures. Because most of what we do fragments landscapes, it is not surprising that humans have a tendency to occur at high densities in association with edges. Even superficial examinations of human activities support the idea that humans tend to be an edgy species.

Earlier we indicated that some human activities—such as home building—perforate landscapes, a form of habitat fragmentation. Houses have edge effects that may extend hundreds of yards away from a house into a natural ecosystem. These human effects include pets, yard lights, activities such as play, lawn mowing, and altering backyard habitat through birdbaths, feeders, and nest boxes. When homes are built within natural ecosystems, beyond city limits, they may have profound effects on biodiversity. For example, the number of predators and competitors on songbirds may increase dramatically in association with the edge created by the house and its occupants.

Some of these aspects were studied near rural homes in Pitkin County, Colorado (Figure 8.8). American robins and black-billed magpies were more numerous close to homes (Figure 8.9). Robins are effective nest competitors with other songbirds, being larger, aggressive, and more likely to be present year-round. Magpies commonly prey on the eggs and nestlings of other songbirds. Perhaps these were contributing factors, along with increased numbers of dogs and cats, for the reduced populations of other songbirds close to homes (Figure 8.10). Interestingly, however, house wrens and bluebirds were more numerous near homes than away from homes, apparently because of nesting boxes (see Figure 8.9).

EXERCISE 8.3

Talk About It!

In groups, examine the different land uses that abut the protected areas in the scenario you are working on. Which ones might have the greatest likelihood to cause an edge effect? Which ones may have the least likelihood? Which species are most likely to be affected, and why? What management steps might you take to minimize these effects?

Figure 8.8. *Rural homes in Pitkin County, Colorado. These homes occur in a matrix of oak and shrub forests west of the Continental Divide. (Photo by Eric Odell.)*

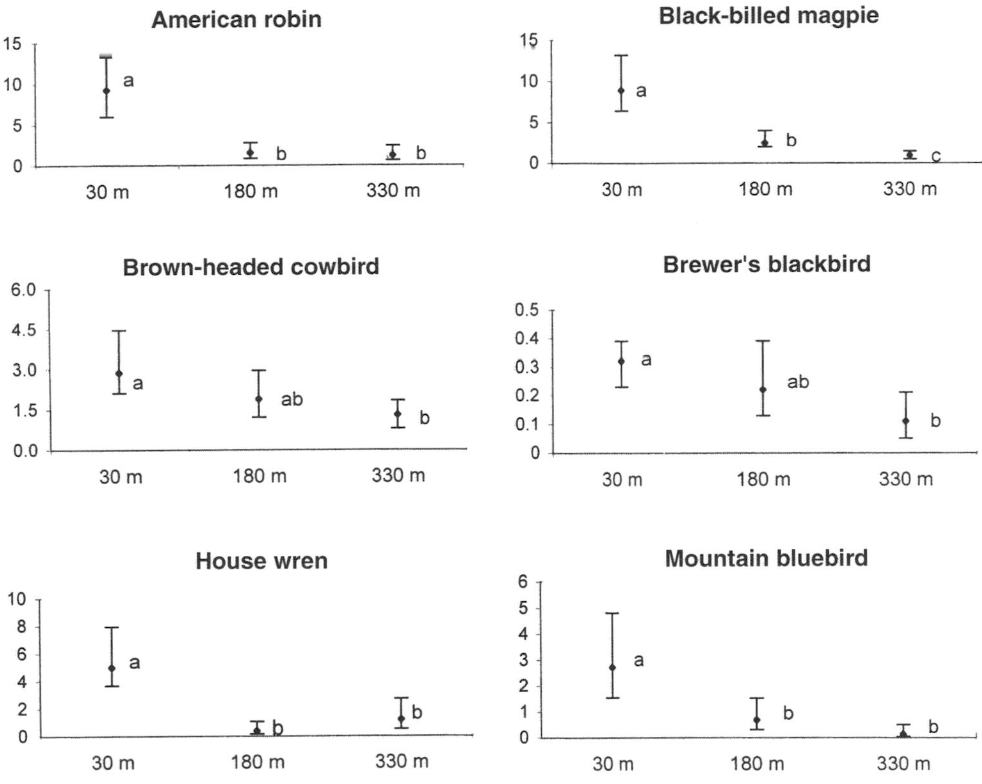

Figure 8.9. *Population estimates of six species of songbirds in the vicinity of homes in Pitkin County, Colorado. These species were more numerous near homes than away from homes. (From Odell and Knight, 2001.)*

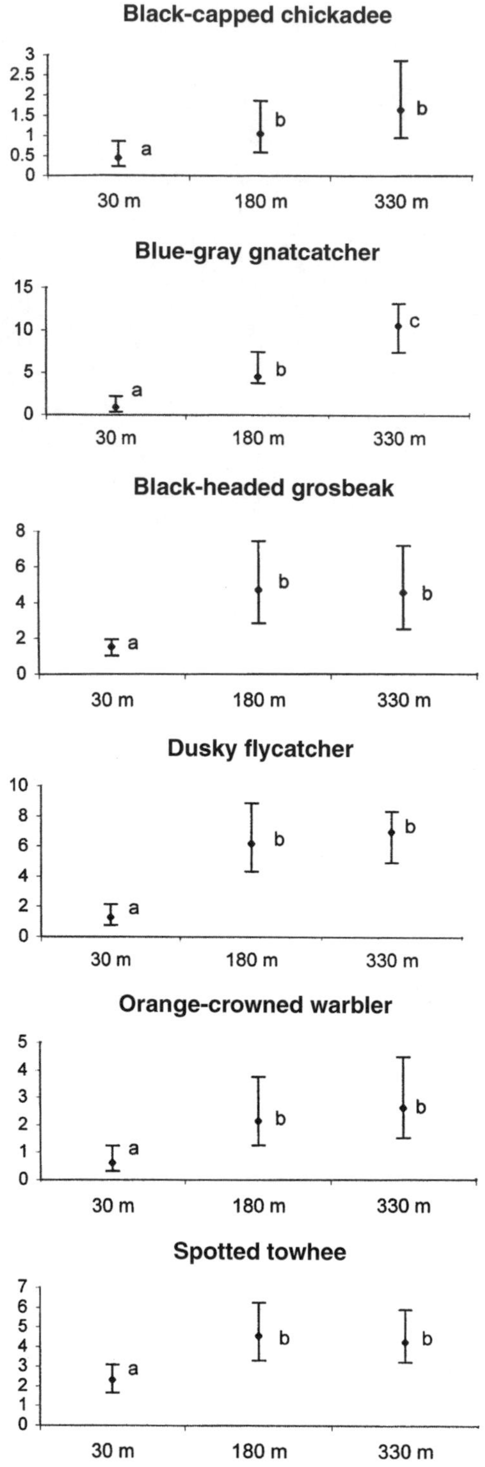

Figure 8.10. *Population estimates of six species of songbirds in the vicinity of homes in Pitkin County, Colorado. Notice that each species is less abundant in the vicinity of homes. (From Odell and Knight, 2001.)*

EDGE EFFECT DISTANCES AND PATCH SHAPE. Edge effects have impacts at variable distances from edges into ecosystem remnants. For example, the depth-of-edge effects of raccoon predation is different from the depth-of-edge effect of brown-headed cowbird parasitism. Likewise, the depth-of-edge effect of temperature at a forest edge differs from that of relative humidity. Humans have different depth-of-edge effects, depending on their activities. Poachers are willing to go farther away from an edge to illegally kill wildlife than someone who wishes to illegally remove plants or cut firewood.

It is difficult to generalize about depth-of-edge effects, other than to say they need to be considered site by site and example by example. The idea of variable distances of different types of edge effects, however, is of critical importance in conjunction with the shape of a patch following fragmentation. Patches of equal area may vary considerably in the amount of area exposed to edge effects. Consider two ecosystem patches that are similar in area but different in shape (Figure 8.11). The circular patch has less edge per unit area than the rectangular patch. Now superimpose a fixed-distance edge effect on both figures. Notice that the rectangular patch has no area immune from the edge effect, whereas the circle has a small area that would not be affected.

A study of songbirds in woodlots of southern Wisconsin illustrates this point. Researchers examined whether songbirds sensitive to forest fragmentation occurred in patches simply based on patch size or the portion of the patch that was beyond a 100-meter edge effect. They used a 100-meter depth-of-edge effect based on how far into a woodlot cowbirds would go in search of nests to parasitize. The shapes of similar-sized patches determined how much of each patch was immune from this edge effect (Figure 8.12). For every species considered, the core area model (the area of forest more than 100 meters from an edge) was a better predictor than total area for whether birds sensitive to fragmentation were detected in surveys. Consider the two similar-sized forest patches in Figure 8.12. When a depth-of-edge

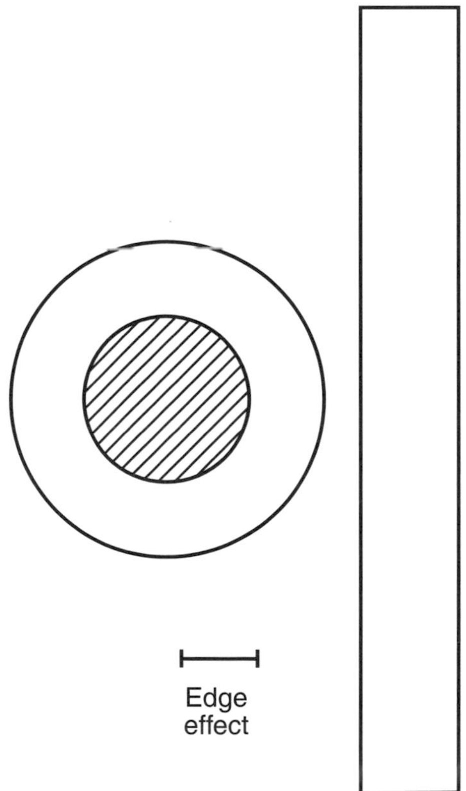

Figure 8.11. *Two ecosystem remnants of the same size but of different shapes. Notice that the circle has less edge per unit area than the rectangle. When an edge effect is placed on both patches, the rectangular patch has no area not affected by the edge effect, whereas the circular patch does have some "non-edge habitat."*

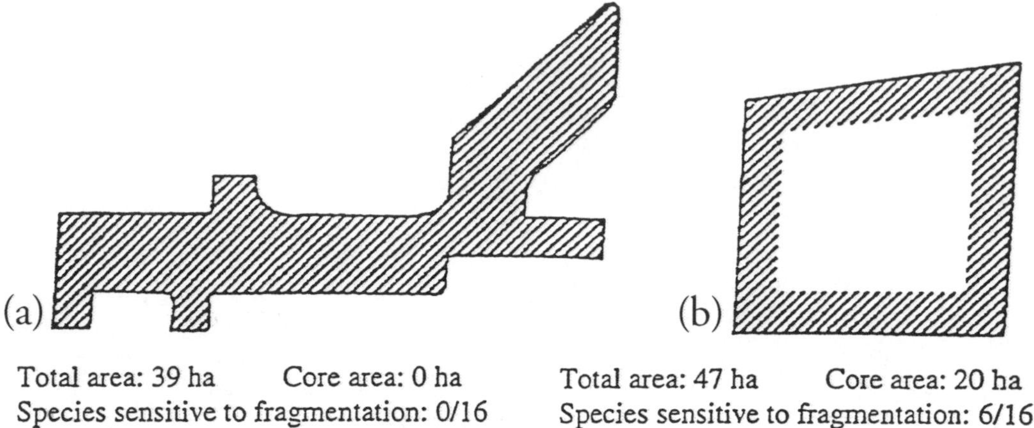

Figure 8.12. *Two forest remnants in Wisconsin, approximately the same size but considerably different in shape. Due to the depth of the edge effect and the shapes of the fragments, the two forest remnants have quite different areas not subject to an edge effect. (a) This fragment is entirely edge habitat. (b) This more square-shaped fragment contains 20 ha of core area. Of 16 species known to be sensitive to fragmentation, none bred in fragment a, whereas 6 bred in fragment b. (From Temple, 1986.)*

effect was placed over each patch, shape determined how much unaffected area each patch contained. In Figure 8.12a, there was no area not affected by the edge, and no birds sensitive to fragmentation were encountered. In Figure 8.12b, there was a core area unaffected by the edge effect, and six species sensitive to edge effects were detected.

Considering the shapes of ecosystem remnants following fragmentation is important because it can profoundly determine how much of the fragment is beyond the depth-of-edge effects. Sometimes ecologists try to capture the complexity of patch area, patch perimeter (edge), and shape by computing the ratio of patch perimeter to patch area. For any regular shape (circle, square, rectangle), arithmetic increases in perimeter are associated with geometric increases in area. In particular, perimeter is directly proportional to the square root of the area. Accordingly, a circle will have the least amount of perimeter relative to area. Real ecosystem patches, however, are much more varied in shape, so the amount of perimeter becomes somewhat independent of area. Researchers studying birds in woodlots in the vicinity of Toronto, Canada, found that irregularly shaped woodlots resulted in perimeter-to-area ratios that were quite different from those found for regular shapes, such

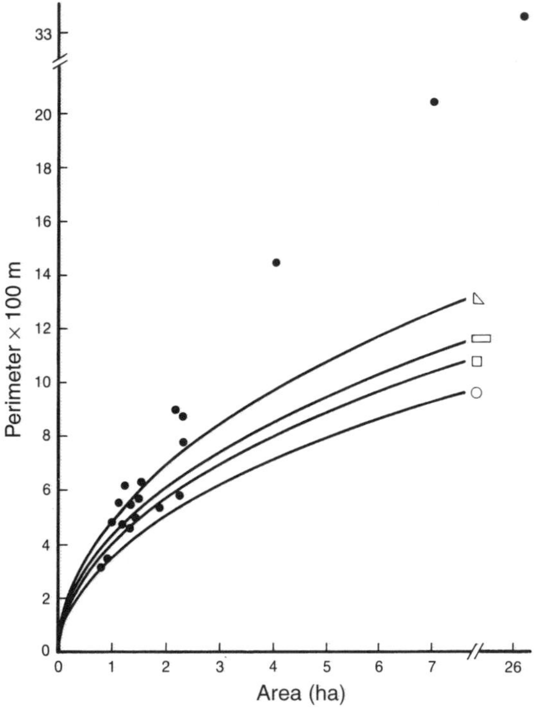

Figure 8.13. *Perimeter-to-area curves for regular shapes (circle, square, rectangle, triangle) compared with remnant woodlots (solid dots) in the vicinity of Toronto, Canada. Notice how the perimeter-to-area ratio for woodlots (solid dots) diverged from the ratios for regular shapes of triangles, rectangles, squares, and circles. (From Gotfryd and Hansell, 1986.)*

as circles, squares, and rectangles (Figure 8.13). This suggests that shape of a patch may be more useful than the perimeter-to-area ratio commonly computed when examining amounts of edge and area in fragmented patches.

> EXERCISE 8.4
>
> **Talk About It!**
>
> For the scenario you are using, examine the areas that are protected, and evaluate them in terms of suitability for edge-sensitive species. Include area (approximate, because there is no scale), edge, and shape in your discussion.

INCREASED ISOLATION

The third major change following fragmentation, in addition to a decrease in area and an increase in edge, is increased isolation of ecosystem remnants (Figure 8.14). Increased isolation between patches affects **dispersal-sensitive species**, which, for a variety of reasons, have a limited success in dispersing between isolated patches. Some species are limited in their dispersal ability by their morphology. For example, plants whose seeds are tiny and thus wind-dispersed might successfully disperse from one patch to another. But what about plants whose seeds are larger and dependent upon insects, such as ants, for dispersal? Can they effectively disperse across an alien landscape that now separates patches? Not likely. Morphological constraints also affect animals. Although rapidly moving animals, such as coyotes, might easily cross a two-lane road and avoid being hit by an automobile, what are the chances that a desert tortoise can do the same? Indeed, ecologists consider paved roads to be effective barriers for the successful dispersal of desert tortoises and other species that are morphologically constrained.

Other species are physiologically limited in their ability to disperse from one patch to another. Eastern chipmunks are diurnal and live in forests. If a forest is fragmented into isolated patches, chipmunks may have difficulty dispersing during the day through the intervening open spaces because of heat stress. Physiologically, they are capable of only moving limited distances in direct sunlight.

Some species have behavioral limitations that preclude successful dispersal between isolated patches. An often-cited example involves ant-following bird species in lowland neotropical rain forests. Following fragmentation, understory bird

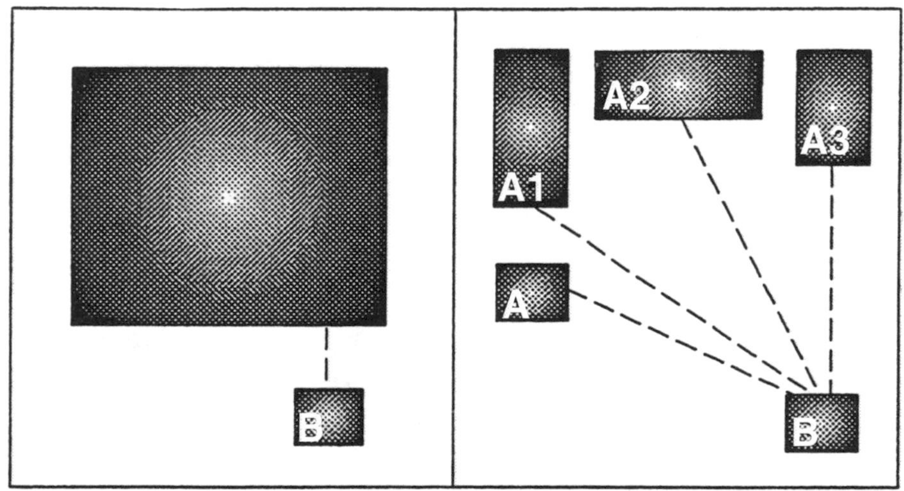

Figure 8.14. *When habitat patch A is fragmented, patch B is more isolated, limiting movement between A and B for dispersal-sensitive species.*

species that normally forage on prey flushed by raiding army ants disappeared from fragments separated from the parent forest by as little as 80 yards. Although these bird species were quite capable anatomically and physiologically of crossing over to the larger forest, they were behaviorally constrained. Many Amazonian forest species have had little need to cross open or nonforested areas in their evolutionary history, so an unwillingness to leave forest cover may be an innate behavioral response. Because these species do not migrate, there is no mechanism for recolonization of isolated fragments.

Researchers in Ohio found that common forest-dwelling birds of North America will cross openings of a limited size and that larger birds were more likely than smaller ones to cross gaps, especially large gaps. These results, when applied to a variety of landscape scenarios following fragmentation, illustrate how dispersal-sensitive species may be precluded from occupying otherwise suitable habitat.

Consider the white-breasted nuthatch, a small forest songbird that nests in cavities and forages for insects under loose bark. Ecologists found that the maximum width of an opening this species would cross within their home range was about 200 yards. Nuthatches have, on average, a home range of 10 acres. Now examine how nuthatches do in four landscape scenarios, each of which has the same total area of woodland habitat (Figure 8.15). The number of nuthatches each can support is limited by the species' ability to cross openings. In each case there is a total of 40 acres of good habitat, but the number of nuthatches that can be supported differs because of the spatial relationships of the habitats and their degree of isolation. This work illustrates how otherwise suitable habitat may be unsuitable for a species that is restricted in its ability to cross open areas.

Whereas some species are morphologically, physiologically, or behaviorally constrained from dispersing, many species are not. Instead, they are fully capable of dispersing but experience elevated mortality when crossing increasingly human-dominated landscapes. Our most immediate experience with these species is seeing their remains

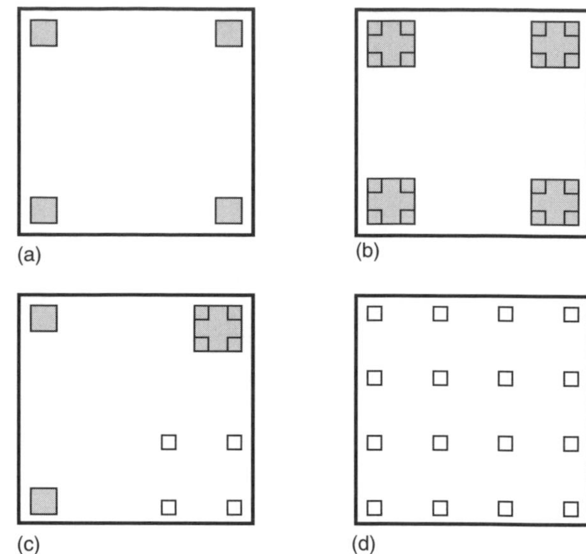

Figure 8.15. *Patterns of occurrence of white-breasted nuthatches as a function of woodland fragment size and distance between patches. Woodland fragments (squares) total 40 acres in each of the four large landscapes. (a) Each woodlot of 10 acres is large enough to constitute a home range (shaded) for one nuthatch. (b) Four 2.5-acre woodlots are close enough for nuthatches to cross, giving habitat suitable for four nuthatches. (c) One set of four 2.5-acre woodlots is spread too far apart to furnish a home range (lower right). (d) None of the 16 2.5-acre woodlots is close enough to be suitable nuthatch habitat. (From Grubb and Doherty, 1999.)*

EXERCISE 8.5

Collaborate on It!

For your scenario, devise a list of actions that would (1) increase the area of patches, (2) decrease the amount of edge of patches, and (3) decrease the isolation of patches. Because it will take time to implement your actions, develop a time line in which these actions will occur. Do not focus exclusively on public lands; include private lands in your thinking. Be specific in listing strategies to implement your vision of a landscape that is less fragmented than when you found it.

along roadsides. Species such as white-tailed deer are not morphologically, physiologically, or behaviorally constrained from moving across alien landscapes. But the deer are more likely to encounter humans, their pets, and automobiles, as well as human activities that result in their death. Indeed, the recognition that many species experience elevated mortality rates when dispersing has led conservation biologists to emphasize the importance of movement corridors (see Chapter 9).

Mosaic and Matrix

We began this chapter with a discussion of how habitat loss, habitat fragmentation, and decreasing matrix quality were issues affecting biodiversity. It was useful to discuss habitat fragmentation before introducing mosaic and matrix because the end results of increasing fragmentation are altered mosaics and matrices. Mosaic and matrix are essential concepts in applying the ecosystem approach for managing biodiversity at the scale of landscapes.

THE LANDSCAPE MOSAIC

The landscape **mosaic** includes the spatial characteristics of all the natural and human-created aspects of the environment. Consider the components of any area familiar to you. In your mind's eye, notice the roads, towns, neighborhoods, city parks, woodlots and prairie remnants, lakes and reservoirs, agricultural fields, shopping malls, parks, and recreation areas. These, collectively, compose the mosaic of your own region.

When evaluating the mosaic of an area, it is important to remember two things. First, a mosaic is not static; it is not fixed in time of the way you remember it. Second, the components of the mosaic are not disconnected and isolated from each other. Because we discussed fragmentation, let's use that concept to illustrate the point that a mosaic is not static. After fragmenting a forest by clear-cutting, are the forest clear-cuts and forest remnants permanent? No, because forests grow back following logging. Indeed, a forest following logging is much more dynamic in its regrowth than a downhill ski run following logging. In the former, trees begin growing almost immediately following cutting. In the latter, any tree attempting to grow in a ski run will be pruned back.

Now consider that the components of a mosaic are connected. Organisms, materials, and human influences can flow between mosaic components, even when they have distinct boundaries. For example, wind-dispersed seeds may be able to cross openings, ensuring plant dispersal from one ecosystem remnant to another. Likewise, domestic cats may range out from the housing developments onto the prairies and hunt native songbirds. People can move virtually anywhere, often taking with them their gasoline engines, pets, and rifles. Therefore, whether these connections that tie together components of a mosaic are good or bad, they certainly occur and must be considered when implementing ecosystem management.

It is clear that mosaics are neither static nor rigidly isolated by impervious boundaries; they are dynamic and interconnected. Indeed, the recognition that mosaics are dynamic led ecologists to coin the phrase "shifting mosaic." To effectively implement ecosystem management, managers need to be aware that administrative boundaries are permeable, not equally for everything, but leaky for a variety of materials and species. In some cases, managers may want to encourage movement across boundaries, to reconnect fragmented ecosystems. In other cases, they may strive to minimize the movement of invasive species or substances. Successful ecosystem management requires understanding the dynamic nature and interplay of an area's mosaic.

EXERCISE 8.6

Collaborate on It!

Break into small groups. For the scenario you are using, prepare a table listing the elements of the mosaic that compose the scenario. What elements are most numerous and which compose the greatest area?

THE LANDSCAPE MATRIX

A landscape **matrix** is the most extensive, most connected, or most influential landscape element of an area. You can think of it as a landscape's general type. For example, in much of the American Midwest, the matrix is agricultural; in Orange County, California, the matrix is urban; and in the center of Yellowstone National Park, the matrix is coniferous forest. The matrix is important in that it can influence ecological processes that may affect biodiversity. Nest parasitism by brown-headed cowbirds, for instance, is common along forest edges where the matrix is agricultural (Figure 8.16a) but uncommon along forest edges where the matrix is forest (Figure 8.16b). Knowing the matrix can help an ecosystem manager determine what conservation issues are important.

Consider for a moment two different matrices, one urban and the other forest. If you were interested in protecting species sensitive to elevated human densities, which of these two would be more challenging? If you were designing a system of protected areas in both matrices, in which one would the protected areas have to be larger, in order to buffer the harmful effects of surrounding land uses? Clearly, in both cases, managing for biodiversity in an urban matrix will be far more challenging and will require larger protected areas to minimize the human effects occurring beyond the boundaries of the protected area.

In the Pacific Northwest, conservation planners were very much aware of matrix quality in conserving the northern spotted owl. The Interagency Scientific Committee designed the "50-11-40 rule" to improve prospects for the successful dispersal of owls between protected areas. This rule required that 50% of the matrix be maintained in trees 11 inches in diameter or larger, and with a 40% canopy cover. It was apparent to the planners that the condition of the matrix surrounding protected areas for spotted owls was critical to overall landscape connectivity. Human activities can produce a matrix in which protected areas are embedded in an alien sea of land use, where the chances of survival are doubtful. Working with stakeholders and other interests, conservation planners can also promote the ability of a matrix to allow successful movement and survival of organisms that cross over the boundaries of protected areas.

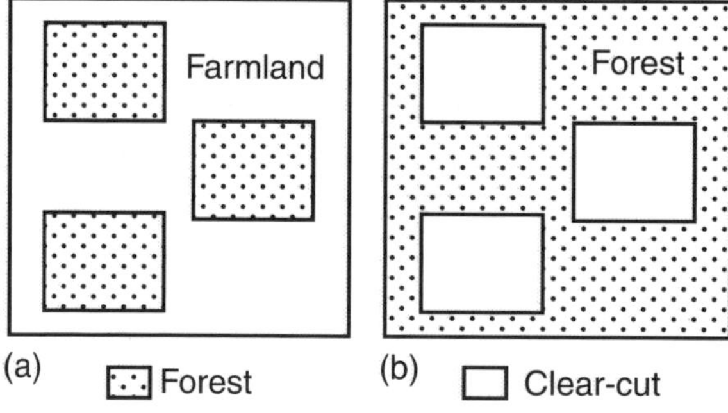

Figure 8.16. *Two different landscape matrices.(a) This matrix is agricultural land, with remnant patches of forest embedded within it. (b) This matrix is forestland, with small openings from forest clear-cuts embedded within it. Although they both have forest edges, one might expect quite different ecological processes and species communities in these two settings.*

> EXERCISE 8.7
>
> **Talk About It!**
>
> What is the matrix where you live? Where you vacation? What is the matrix of the scenario you are using? Is it difficult to determine? Does knowing the mosaic help in determining the matrix?

FRAGMENTATION AND THE LANDSCAPE MATRIX

Once fragmentation begins to alter an ecosystem, the area of that ecosystem decreases, the amount of edge increases, and isolation among ecosystem fragments increases. As an ecosystem becomes increasingly fragmented and area, edge, and isolation change, one eventually modifies a landscape to the point where a different kind of ecosystem dominates, usually a human-created landscape. When the area of this new ecosystem increases, these patches of the *new* ecosystem become larger and more connected. And as humans replace one ecosystem with another through fragmentation and habitat loss, many species that thrived in the original landscape are replaced by species that are better adapted in the new landscape. The natural heritage of increasingly human-dominated matrices contain more weedy, generalist, human-adapted species and fewer specialist species that do not thrive with elevated human densities. Along the Colorado Front Range, as the grass and shrublands are converted to rural housing and commercial development, there will be more black-billed magpies, garter snakes, and raccoons, and fewer lark buntings, prairie rattlesnakes, and bobcats. By implementing ecosystem management at the landscape level, being fully aware of changes in the mosaic and matrix, you may be able to minimize the alteration of species communities that require intact, functioning ecosystems.

It is exceedingly difficult to track the individual fates of all species affected by a landscape-wide conversion of one ecosystem to another. This is one reason why tracking landscape indicators—such as area, edge, and isolation of ecosystem remnants—serves as an effective shorthand to landscape-level changes. There is no universally accepted index to capture these landscape parameters, so it may be preferable to separately measure and evaluate different aspects of a mosaic. This approach would be analogous in humans to evaluating health parameters independently rather than trying to combine them into a single index of a person's overall health. For example, heart rate, blood cholesterol level, and body temperature are better evaluated by your family physician separately than merged into a single number. In any case, ecosystem remnants following fragmentation can never be viewed isolated from the mosaic and matrix in which they occur.

EXERCISE 8.8

Collaborate on It!

For your scenario, compile a list of land uses that have resulted in fragmentation of the landscape. Considering the mosaic and matrix of your scenario, and the species that may be sensitive to fragmentation, devise a comprehensive approach that will initiate a long-term effort to minimize the harmful effects of past and future fragmentation. This will require you to use the concepts presented in this chapter. It will also require you to think across administrative boundaries, the economies that may be affected by your actions, and the societal aspects of your scenario that need careful consideration of the human dimensions affected. Think big, but also think about the long term in your deliberations.

References and Suggested Readings

Andrén, H. 1994. Effects of habitat fragmentation on birds and mammals in landscapes with different proportions of suitable habitat: A review. Oikos 71:355–365.

Bierregaard, R.O., Jr., T.E. Lovejoy, V. Kapos, A.A. Dos Santos, and R.W. Hutchings. 1992. The biological dynamics of tropical rainforest fragments. BioScience 42:859–866.

Cornelius, C., H. Cofre, and P.A. Marquet. 2000. Effects of habitat fragmentation on bird species in a relict temperate forest in semiarid Chile. Conservation Biology 14:534–543.

Franklin, J.F. 1993. Preserving biodiversity: Species, ecosystems, or landscapes? Ecological Applications 3:202–205.

Gotfryd, A., and R.I.C. Hansell. 1986. Prediction of bird-community metrics in urban woodlots. Pp. 321–326 in J. Verner, M. Morrison, and C.J. Ralph (eds.). *Wildlife 2000: Modeling Habitat Relationships of Terrestrial Vertebrates.* University of Wisconsin Press, Madison.

Grubb, T.C., Jr., and P.F. Doherty, Jr. 1999. On home-range gap-crossing. Auk 116:618–628.

Haila, Y. 2002. A conceptual genealogy of fragmentation research: From island biogeography to landscape ecology. Ecological Applications 9:1448–1458.

Helzer, C.J., and D.E. Jelinski. 1999. The relative importance of patch area and perimeter-area ratio to grassland breeding birds. Ecological Applications 9:1448–1458.

Knight, R.L. 1997. Field report from the new American West. Pp. 181–200 in C. Meine (ed.). *Wallace Stegner and the Continental Vision.* Island Press, Washington, D.C.

Knight, R.L. 1998. Ecosystem management and conservation biology. Landscape and Urban Planning 40:41–45.

Knight, R.L, F. W. Smith, S. W. Buskirk, W. H. Romme, and W. L. Baker. 2001. *Forest Fragmentation in the Southern Rocky Mountains.* University Press of Colorado, Niwot.

Miller, S.G., R.L. Knight, and C.K. Miller. 1998. Influence of recreational trails on breeding bird communities. Ecological Applications 8:162–169.

Niemuth, N.D., and M.S. Boyce. 1997. Edge-related nest losses in Wisconsin pine barrens. Journal of Wildlife Management 61:1234–1239.

Odell, E.A., and R.L. Knight. 2001. Songbird and medium-sized mammal communities associated with exurban development in Pitkin County, Colorado. Conservation Biology 15:1143–1150.

Paton, P.W.C. 1994. The effect of edge on avian nest success: How strong is the evidence? Conservation Biology 8:17–26.

Stratford, J.A., and P.C. Stouffer. 1999. Local extinctions of terrestrial insectivorous birds in a fragmented landscape near Manaus, Brazil. Conservation Biology 13:1416–1423.

Temple, S. 1986. Predicting impacts of habitat fragmentation on forest birds: A comparison of two models. Pp. 301–304 in J. Verner, M. Morrison, and C.J. Ralph (eds.). *Wildlife 2000: Modeling Habitat Relationships of Terrestrial Vertebrates.* University of Wisconsin Press, Madison.

Villard, M-A. 1998. On forest-interior species, edge avoidance, area sensitivity, and dogmas in avian conservation. Auk 115:801–805.

Winter, M., and J. Faaborg. 1999. Patterns of area sensitivity in grassland-nesting birds. Conservation Biology 13:1424–1436.

Experiences in Ecosystem Management:

Southern California Natural Community Conservation Planning

Michael O'Connell

PERHAPS NOWHERE ARE THE COMPLEXITIES OF ADAPtive, community-based conservation more evident than in urbanizing settings. Urban areas are generally characterized by rapidly growing human populations, dense development, high land values, complex sociopolitical structures, and possibly the greatest challenge of all to conservation—impacts to biodiversity that are for the most part irreversible. In a managed landscape, such as for timber or water delivery, impacts are more easily adapted over time: harvest rotations can be changed, valves can be turned on and off, and effects can be adjusted based on feedback from a management monitoring program.

But in urbanizing areas, most impacts are permanent; indeed, they are both cumulative and synergistic. They come in the form of pavement, houses, roadways, and elevated human densities. Once conservation systems such as reserves are designed and implemented, it is nearly impossible to change them. Adaptive management is limited in such cases to adjusting management prescriptions within reserves, or perhaps changing boundaries slightly as reserves are acquired and protected. Habitat fragmentation, while an issue for all adaptive management, is particularly acute in urbanizing settings because land uses outside protected areas are antithetical to the protection of the resources inside. Islands in a sea of urbanization is the norm, making connectivity among reserves an essential element of urban adaptive conservation.

The Southern California Case

If urbanizing areas are the most complex challenges for community-based conservation, coastal southern California is the archetypal location. The 6000 square mile region (Figure A) contains nearly 15 million people—one out of every 17 U.S. residents—and is expected to add 50% to this total over the next 20 years (California Department of Finance, 2000). It has some of the highest undeveloped land values in the world. A single buildable acre in Orange County overlooking the coast routinely commands in excess of $1 million. Further complicating the situation, the region's conservative political views are legendary, with many of the most active Congressional conservatives hailing from the five-county area.

Southern California is also among the most biologically diverse ecoregions on Earth. Recent ecoregional diversity analyses (Ricketts et al., 1999; Chaplin et al., 2000) place the region high among globally significant ecosystems, both in taxonomic diversity of endemic plants, birds, invertebrates, and reptiles, and in degree of threat. Noss and Peters (1995) cited coastal sage scrub, the dominant terrestrial system in the region (Figure B), as the ninth-most imperiled ecosystem in the United States; nearly 90% of this habitat type has been lost over the last 100 years (Noss and Peters, 1995). As a result, almost two-dozen species have been listed as threatened or endangered by state and federal governments. Another 75 species await consideration for listing.

Until the early 1990s, most conservation action in

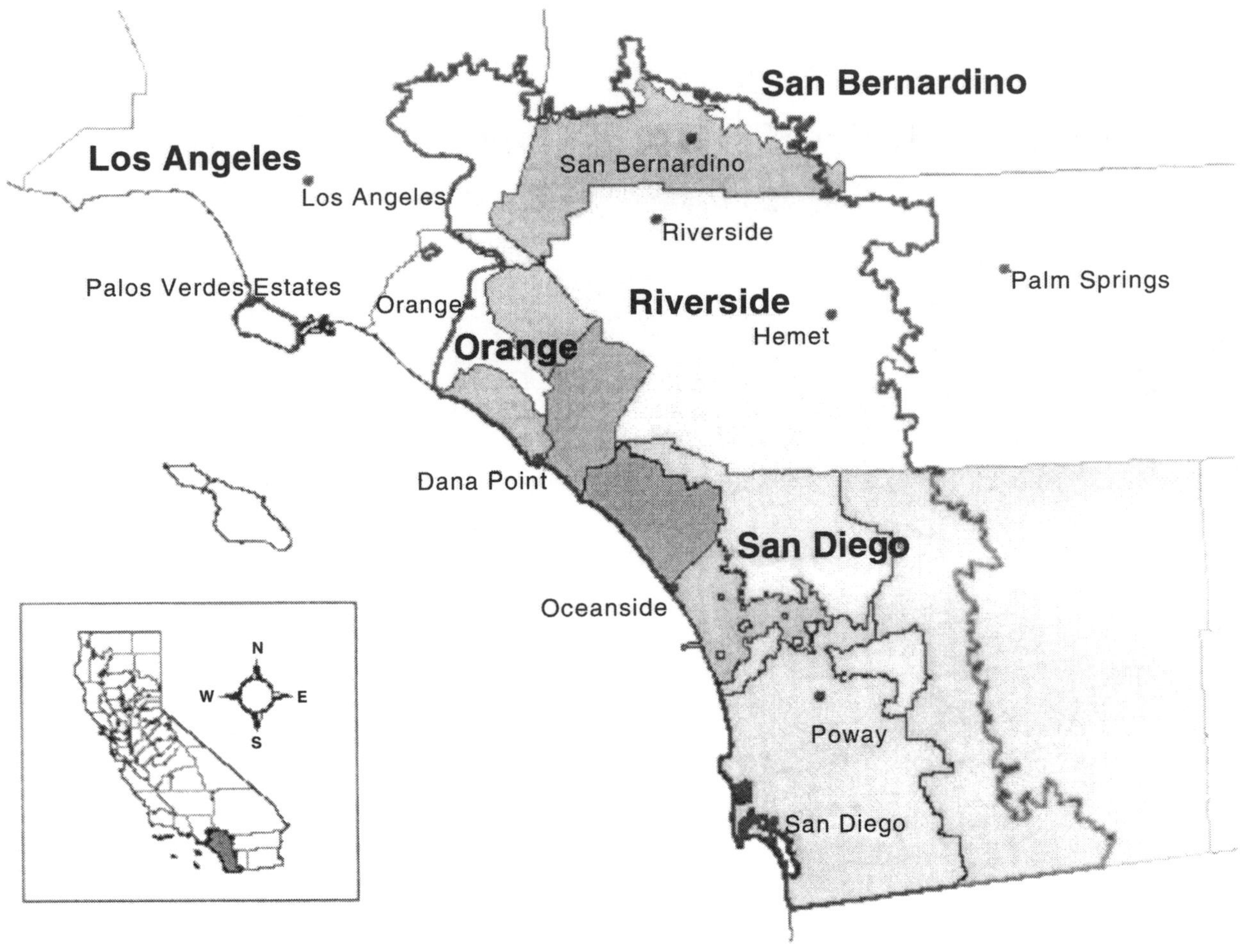

Figure A. *A map of the southern California Natural Community Conservation Planning region.*

the region was accomplished defensively and parcel by parcel. Using whatever means available—usually the threat of litigation under the Endangered Species Act—conservationists fought development on multiple fronts against scores of projects. Most successes were minor and hard-won. From a regional ecosystem perspective, the tidal wave of urban sprawl racing outward from San Diego, Riverside, and Los Angeles left mostly small fragments of habitat and overwhelmed conservationists. Clearly the strict protection of species provided by the Endangered Species Act (ESA) was not effective in creating broad conservation solutions necessary to resolve the conflicts on the ground between land use and conservation generated by southern California's growing population.

Land users and local governments regulated under the ESA were also less than satisfied with the outcome. In addition to incurring exorbitant costs and delays in the development process, they endured tremendous uncertainty about economic growth and use of private property. Political pressure against the ESA mounted rapidly and peaked when the California gnatcatcher was proposed for listing under both the state and the federal Endangered Species Acts. This widespread but rare songbird was the obvious tip of a regulatory iceberg that threatened to grind land use activity to a crawl in this populous, economically prosperous, and politically conservative region.

A common ground of dissatisfaction with the current land use system and a desire for certainty,

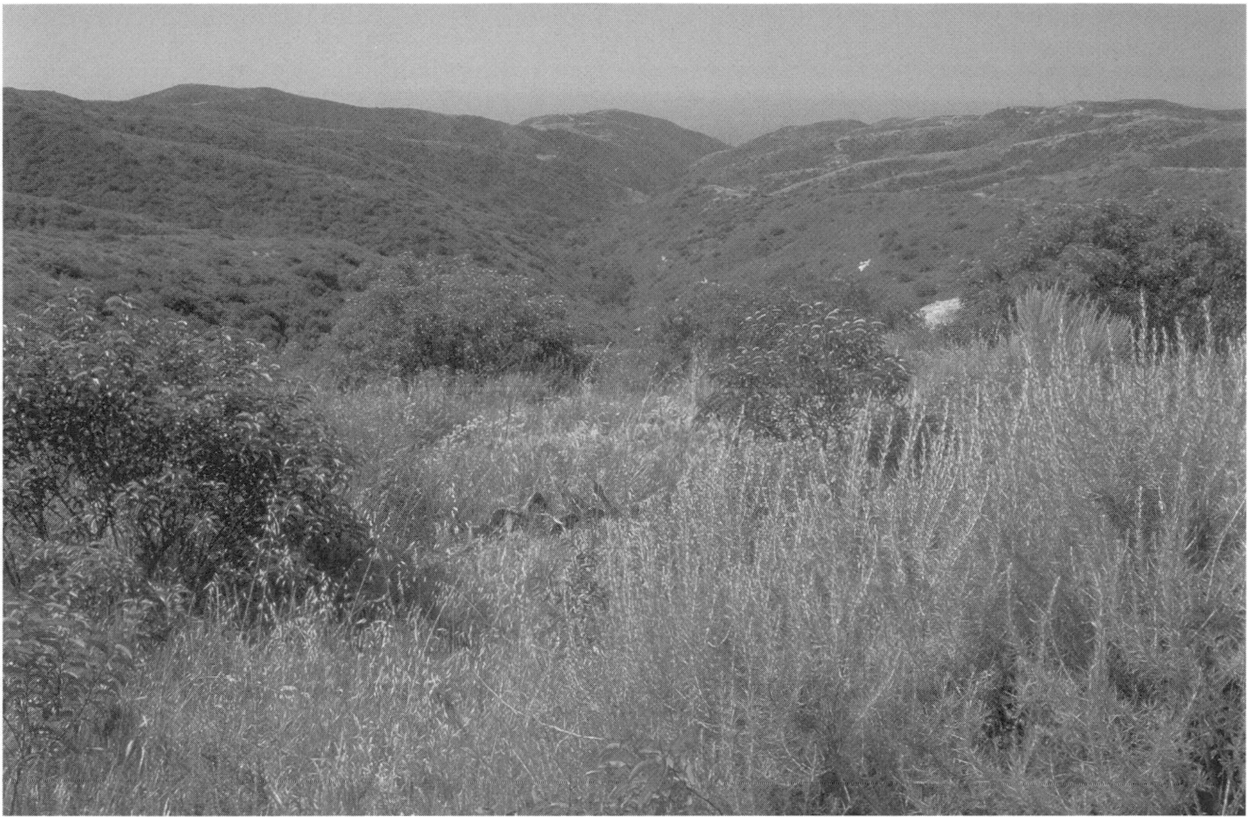

Figure B. *Coastal sage scrub habitat, which dominates the southern California NCCP region, supports numerous endemic species and is highly threatened by development.*

both for protection of the region's natural resources and for economic growth, drew together an unusual alliance of conservationists, government agencies, and land users. Among the key players were the California Resources Agency, the Irvine Company (a developer that creates large-scale developments and then sells lots to builders), The Nature Conservancy, and the Office of the U.S. Secretary of the Interior. They proposed a new process that would create large-scale conservation and development plans throughout the region and would seek to protect entire natural communities while streamlining the process for economic development in other areas. The Natural Community Conservation Program (NCCP) emphasized protecting, restoring, and managing large blocks of contiguous habitat—often in trade for developing highly fragmented but sometimes high-quality areas—and connecting these core reserves together through a system of secondary habitat reserves and less dense land uses than housing developments and shopping malls (such as golf courses and river parkways). At the same time, the program allowed the streamlining of the land-use process and strove to provide regulatory certainty essential to strong local economies.

Lessons Learned

When it began in 1991, the Natural Community Conservation Program was the largest experiment in conservation planning ever undertaken. It sought to bring community-based conservation to a vast and populous region with highly threatened and diverse habitats, while modifying the regulatory and land-use development process to bring certainty and efficiency. Ten years later, the program has a track record packed with strengths and weaknesses, all of which are valuable lessons for adaptive ecosystem management. The most notable successes and shortcomings are described here.

Successes. The most evident and visible success of the program is the collective focusing of financial resources and human energy on the conservation problem. In the past, the conservation community and regulatory agencies, such as the U.S. Fish and Wildlife Service, were the only participants working haphazardly toward habitat protection. But by making regional conservation commensurate with efficient regulation, former opponents to conservation have been persuaded to contribute greatly to its achievement and have built diverse coalitions. For example, local governments now actively implement land-use regulations that protect habitat, while contributing significant funding for additional land protection. San Diego County recently affirmed $6 million annually from its General Fund for land acquisition and management. Groups of environmental activists, landowners, developers, and local government representatives routinely lobby together for state and federal appropriations for land acquisition—something nearly unheard of previously.

In most urbanizing settings, habitat preserves are comprised of the leftovers of the development process: steep slopes, fragmented wetlands, and nonbuildable areas. The NCCP, however, permitted habitat protection to become a driver of local land-use planning. Although much of the protected land has come through private set-asides supplemented by public acquisition, habitat preserves result from a regional reserve design using the basic principles of conservation biology, not out-of-context, project-by-project negotiations between regulatory agencies and developers. Over 150,000 acres had been dedicated to conservation and management by 1999, with an ultimate goal of over 400,000 acres. The program is also responsible for establishing the San Diego National Wildlife Refuge and appropriating nearly $30 million for its acquisition.

Regional science provided sound footing for the Natural Community Conservation Program. An important early objective was to include independent scientific involvement in its design. To address this issue, the California Department of Fish and Game established a scientific review panel of distinguished academic scientists in 1991 to develop biological parameters for the southern California region. This team synthesized existing field data and other research and produced a set of broad conservation guidelines to protect and manage the focal ecosystems. These guidelines included reserve design principles, management direction, and a short-term research agenda. The science panel also provided interim regulatory guidance intended to avoid foreclosing conservation options, set limits for overall habitat loss, and divert urbanization away from key conservation areas. Each local plan developed under the regional program was compared to this biological framework before approval by state and federal regulatory agencies.

Shortcomings. The size, scope, and complexity of the conservation problem in the region have been the source of many of the difficulties and shortcomings of the southern California program. Among the most obvious deficiencies are issues that arguably are common to all regional community-based conservation efforts. These include a notable lack of resources and funding for the protection and management of habitat areas. Although the program clearly has resulted in more public and private conservation funding flowing into the region than would otherwise have been realized, it is still far short of the amount needed to acquire and steward the regional habitat reserves. Some estimate that achieving the biological goals of the program will require $1.5 billion in public funds.

A glaring and dangerous weakness in the program, and a crucial lesson for other community-based conservation efforts, has been the inability to make lasting changes in culture and attitude among program participants. Unfortunately, old habits die hard. Despite building uncommon and powerful alliances and coming to general agreement on regional land-use strategies, most of these successes have not survived the daily ups and downs of implementation without considerable facilitation. Recent years have seen many participants fall back to historic attitudes and methods.

What is clear is that although it takes tremendous effort and collaboration to reach agreement on NCCPs, it will take even more to successfully implement them. Some conservationists still fight local governments over every development project, including those approved under the regional plan that they originally supported, while landowners and lower-level agency staff members often square off over project-specific implementation of general policies in the regional plan. Ultimately, if the program fails during implementation, this will be a primary cause and a major lesson for other such efforts.

Application of the regional NCCP conservation guidelines developed by the Scientific Review Panel to individual local plans has also been inconsistent and somewhat controversial. Each local planning area was free to apply the regional guidelines through whatever process it chose, and the state wildlife agency ultimately determined whether the science in the local plan was consistent with the regional program. Some local planning areas, such as southern Orange County, employed an independent science team to translate regional science to local guidance, which heightened the public credibility of the result. Other planning areas used private consultants to conduct scientific analysis, design reserves, and create land management plans, which was more suspect within the conservation community. This lack of an accepted methodology for the application of scientific guidance has heightened controversy over the NCCP program.

Another and perhaps the most subtle flaw in the southern California program is that state and federal policy falls far short of supporting its regional conservation objectives. The federal ESA, despite its famous purpose clause—"to provide a means whereby the ecosystems upon which endangered species and threatened species depend may be conserved . . ." (16 U.S.C. § 1531b)—was not designed to protect ecosystems very well. Moreover, its regulations are based almost entirely on prohibiting impacts to individuals of listed species, not creating broad ecosystem conservation.

To enable the NCCP program to work, California passed its own state law in 1991. This legislation was general and rather brief, and despite several attempts over the last decade, it has yet to be updated and clarified using the lessons learned from the southern California pilot program. Federal participation was possible only through a special rule under ESA Section 4 when the California gnatcatcher was listed that creatively stretched the law around ecosystem conservation goals. These limitations mean that the full benefits of ecosystem conservation—such as protecting unlisted animals and plants, natural communities, and natural processes—will be difficult to realize until a comprehensive planning policy is developed.

Conclusions

The southern California example of the Natural Community Conservation Planning Program is an important and illustrative case study for community-based conservation. It tests the potential of regional ecosystem protection in one of the most severe contexts possible. The economic, biological, and political extremes in the region challenge the ability of even the most aggressive and creative conservationists to produce lasting, viable results. Nevertheless, the successes and setbacks of this effort have equal value in informing other attempts at regional conservation. Perhaps the most valuable lesson of all is that to succeed, any solution that advances regional conservation aims must also provide answers to the problems faced by landowners, local governments, and other stakeholders.

DISCUSSION QUESTIONS

1. The environmental community is sharply divided over whether the NCCP program is a positive benefit to species conservation or undermines the strict protections of the Endangered Species Act. Discuss the issues that arise for conservationists from the difficult choices that must be made in regional conservation planning. If NCCP is not the solution, is there a better alternative?
2. Examine the differences between conservation planning in urbanizing areas and rural/working landscapes. What kinds of issues are similar between the two situations? Design a regional conservation program that works in both rural and urban areas.
3. The Endangered Species Act is based on policies that prohibit detrimental activities, but it contains very few provisions to encourage beneficial activities for species. Moreover, the ESA's protections are only invoked when a species is nearing extinction. Can we effectively conserve biodiversity with such a policy alone? Discuss what other policies might be necessary to maintain biodiversity. Be both politically and biologically realistic in your answers.

References

California Department of Finance, Demographic Research Unit. 2000. Web site: Demographic research unit homepage. http://www.dof.ca.gov/html/Demograp/druhpar.htm.

Chaplin, S.J., R.A. Gerrard, H.M. Watson, L.L. Master, and S.R. Flack. 2000. The geography of imperilment: Targeting conservation toward critical biodiversity areas. Pp. 159–200 in B.A. Stein, L.S. Kutner, and J.S. Adams (eds.). *Precious Heritage: The Status of Biodiversity in the United States*. Oxford University Press, New York.

Noss, R.F., and R.L. Peters. 1995. *Endangered ecosystems: A status report on America's vanishing habitat and wildlife*. Defenders of Wildlife, Washington, D.C.

Ricketts, T.H., E. Dinerstein, D.M. Olson, C.J. Loucks, et al. 1999. *Terrestrial Ecoregions of North America: A Conservation Assessment*. Island Press, Washington, D.C.

CHAPTER 9

Managing Biodiversity Across Landscapes: A Manager's Dilemma

CLEARLY THE MOST EFFECTIVE WAY TO PROTECT biological diversity is to protect it where it occurs. This approach is not only more economically efficient compared to the costs of protecting it in zoos, botanical gardens, marine aquariums, arboretums, and insect pavilions, it has other advantages as well. For example, evolution in the wild can still occur normally, whereas domestic selection typically occurs in captivity, where the wants and needs of a species are provided with minimal risk of predation or competition. Most importantly, when biodiversity is protected in nature, it remains part of the natural landscape. This is where people still prefer to view it, and where it can function normally within ecosystems. In addition, if biodiversity can be nurtured on landscapes where some degree of multiple use and development can occur, then people and nature can renew their age-old bond of coexistence through observation, reverence, and utilitarian uses. Aldo Leopold (1938) had this in mind when he wrote:

> We end, I think, at what might be called the standard paradox of the twentieth century: our tools are better than we are, and grow better faster than we do. They suffice to crack the atom, to command the tides. But they do not suffice for the oldest task in human history: to live on a piece of land without spoiling it.

What Leopold meant by this is that when nature and people share common landscapes, people are more likely to develop a sense of responsibility for land health. The best environment for a sense of stewardship to develop is one in which people and nature are close to each other, not separated by fences and closely guarded borders.

Historically, we have protected biodiversity and ecosystems through a system of protected areas. These preserves are often carefully planned, as when The Nature Conservancy consults with Natural Heritage Programs to locate sites for conservation on the basis of representativeness. In this approach, a conservation organization acquires representative areas of all ecosystems, rather than continuing to acquire more and more areas of a few "popular" ecosystems, such as alpine meadows or forests.

Most of our protected areas, however, are national parks or wilderness areas, and few of these were selected for their diversity of species or ecosystems. Instead, these areas often were protected for their scenic beauty, recreational opportunities, or historical importance. Certainly parks and wilderness areas play critical roles in the persistence of biodiversity, particularly for species that may be area-, edge-, or dispersal-sensitive. Parks, however, tend to be overrepresentative of certain ecosystems, while not protecting many other ecosystems. In the United States, for example, coniferous forests and alpine tundra ecosystems dominate parks and wilderness areas, whereas swamps, prairies, and deciduous forests are rarer in our national role call of preserves.

Today, we will never be able to set aside enough land to ensure the continuation of the ecological processes and levels of biodiversity that were present when the human population was much smaller. The United States has slightly over 5% of its land mass strictly protected and an additional 5% in more relaxed types of protection. Considering our species' rapidly growing population, material consumption, and damaging technologies, it is a foregone conclusion that ecological processes and the diversity of life are condemned to only a small portion of the Earth's surface—unless we find alternative ways to allow humans and nature to coexist. This is not to deny the importance of protected areas, only to concede that we need to incorporate additional approaches into our toolbox for environmental protection.

In countries around the world, a change is occurring in how protected areas are viewed. In both developed and developing nations, we increasingly find conservation planners attempting to integrate conservation with sustainable development, such as ecotourism, livestock grazing, and a variety of forestry practices. Ecosystem management goes beyond the importance of protected areas, with their hard-and-fast geopolitical boundaries and detailed restrictions on what are appropriate uses. Ecosystem management works to ensure that entire landscapes are more friendly to biodiversity and ecological processes. This approach attempts to blend the needs of biodiversity and humans and, ultimately, integrate the two rather than keep them apart.

Ecosystem management is an innovative addition to the traditional approach of protecting biodiversity in preserves because it strives to promote human uses that are complementary to land health. In this chapter, we will examine a broad array of approaches to managing landscapes, from ecosystems where humans live to protected areas where humans only visit. Remember that there is not just one method for saving nature. Many ways exist that, to varying degrees, should be considered when striving to do what is best for human and natural communities.

EXERCISE 9.1

Talk About It!

For the scenario you are using, and for the species that have been identified in that scenario, prepare a list of areas where the species occur. Do they only occur in protected areas, or elsewhere, such as on private lands? Do the protected areas in your list receive equal levels of protection, or are different land uses allowed in some while prohibited in others? Is biodiversity secure in this scenario? What about where you live?

Ecosystems or Species? Coarse-Filter and Fine-Filter Approaches

Historically, natural resource managers focused on single species and their habitats. But there are too many species to manage them all, one species at a time. What to do? Consider a regional area, such as the northeastern pine forests and wetlands of the Untied States, that support a natural heritage that, even yet, has not been adequately inventoried. Your dilemma is that you must chose among two approaches that will protect the biodiversity of that region. Your two choices are these: You can protect the species, or you can protect the ecosystems. Which would you choose? If you choose the approach of protecting species, do you know all of the species that occur in the region, their popula-

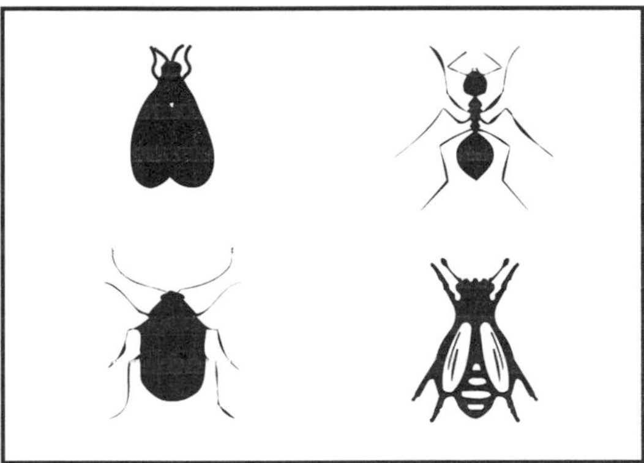

Figure 9.1. *Species associated with coniferous forests at different successional stages. The species of mammals are well known, whereas the species of insects are poorly known and sometimes not even catalogued. (Adapted from Hunter, 1996.)*

tion status, their distribution, and their complex interactions with other species? Might there be species of which you are unaware? If you choose to protect ecosystems, would you be, indirectly, offering protection to all species, including those not presently known?

Look at Figure 9.1. The center frame shows different successional stages of a northeastern coniferous forest. Above it are pictured some small to medium-sized vertebrates that might occur in that forest. These species are well known, and in some cases, population estimates exist because the species is of economic or conservation interest. In the frame below the forest, however, are a variety of species of insects, some of which may not even be known to science. Which of the two approaches outlined above has a greater likelihood of protecting the insects, managing for certain species or managing for ecosystems? If you believe that managing for ecosystems may be more likely to ensure the protection of all species, you are probably right. This approach is called the coarse-filter approach.

THE COARSE-FILTER APPROACH

The **coarse-filter approach** for managing for biodiversity emphasizes protecting ecosystems at appropriate landscape scales. This approach assumes that species richness—from plants to insects to vertebrates—will be maintained if the correct mix of ecological conditions is provided. The coarse-filter approach is popular with conservation planners because of the great likelihood that many species are not protected in approaches that emphasize single species. The persistence of species that are not economically important, do not have special legal protection, or may not even be known is often left in doubt when management efforts target single species. Species in these categories receive little consideration in traditional conservation planning. Because a coarse-filter approach emphasizes protecting ecosystems, managers feel more confident that species not subject to special consideration are offered some protection. There are several orders of magnitude more species than there are ecosystems, and there are certainly more species than can be ever be managed. Accordingly, emphasizing the management of ecosystems can be an effective method for protecting species not normally addressed by the single-species management approach, the so-called fine filter approach.

THE FINE-FILTER APPROACH

The **fine-filter approach** for protecting biodiversity focuses on providing suitable habitat conditions for individual species, guilds (species that exploit a similar resource a similar way, such as scavengers), or other groupings of species. This approach has several advantages. It complies with federal statutes, such as the Endangered Species Act and the National Forest Management Act (which requires the U.S. Forest Service to ensure viable populations of all native species). Even if land managers wanted to use the coarse-filter approach, federal legislation might very well insist they also employ the fine-filter approach for specifically targeted species. With few exceptions, the law protects listed species but does not yet address endangered or threatened ecosystems.

The fine-filter approach also appeals to stakeholders who may want to see species such as wolves and whooping cranes protected. Although people may not be particularly knowledgeable about some ecosystems, such as southeastern pocosins (swampy regions in pine savannas), they are often very much aware of a glamorous species, such as the bald eagle, whose image adorns their T-shirts! The ideas of PVA, metapopulations, and spatially explicit models discussed in Chapter 7 are appropriate for the fine-filter approach.

BLENDING COARSE-FILTER AND FINE-FILTER APPROACHES

Recall our reminder at the beginning of this chapter: Effective conservation programs use more than a single approach. It is not surprising, therefore, that resource managers have devised strategies that integrate both fine- and coarse-filter approaches to ensure native biodiversity and ecosystem health. A blend of both approaches has a number of advantages. By incorporating a coarse-filter approach, one strives to ensure that all ecosystems are represented, thereby avoiding the problem of managing for large numbers of species. Including the fine-filter approach ensures that the legal requirements of the Endangered Species Act and other laws are met. Incorporating species assessments for selected species provides useful feedback on whether the coarse-filter approach is working. In addition, a combined approach is more likely than either of the two used in isolation to include managing for essential ecological processes, which allows for ecosystem dynamics to be included in landscape planning.

EXERCISE 9.2

Talk About It!

In groups, for the scenario you are using, develop a strategy that blends both fine-filter and coarse-filter approaches for managing biodiversity. Describe how your approach would work, obstacles that would need to be addressed, and what it would look like on the ground. What are the strengths and weaknesses of using both methods in an overall approach to the maintenance of biodiversity?

Landscape-Level Considerations That Protect Biodiversity and Ecosystems

Natural resource managers often participate in the design of protected areas. In addition, many protected areas are already in place, and it might seem that little can be done to improve on their design. Not so! Even simple steps can vastly improve an existing protected area that was set aside decades ago. Consider this example. National forests in the United States are required by federal law to revise their forest management plans every 10 years. Most people would concede that the only truly protected areas for biodiversity in national forests are designated wilderness areas. That is true, but wilderness areas were seldom set aside for biodiversity values. Instead, they were the areas that were roadless and that did not contain valuable or readily accessible timber, water, and minerals. Wilderness areas are often what is left on national forests after human uses have been sufficiently limited to protect key wildlife and plant communities, and where important ecological processes are still allowed to occur naturally.

A natural resource planner who thought existing wilderness areas could not be improved for biodiversity protection, other than by making them larger, would be mistaken. During forest plan revisions, natural resource planners reexamine the existing road network in a forest. Forest plan revisions may suggest that some roads be closed and the land restored. A planner interested in enhancing the capacity of a wilderness area to protect sensitive species might propose that roads adjacent to the wilderness boundaries be closed. Closing roads near wilderness boundaries effectively increases the size of the wilderness area, thereby allowing larger populations of threatened species and even perhaps facilitating movement between protected areas isolated by the roads.

This example illustrates that certain "tools" may make landscapes more attractive to species sensitive to certain human uses, such as roads. An understanding of such concepts, and the ability to work with others across administrative boundaries, is often all that is necessary to promote biodiversity in certain areas.

AREA, SHAPE, AND ISOLATION

As a general rule, the greater the size of a protected area, the greater the likelihood that certain species may occur and persist in that area. Recall from Chapter 8 the concepts of area-sensitive species and the species-area relationship. Larger areas, all else being equal, contain more species. Larger areas also allow species with large-area requirements to occur, and larger areas are more likely to allow sufficiently large populations of these species to persist.

The proposal above of closing roads is an example of how protected areas can be made effectively larger. Other, more obvious, approaches would be to formally change the designation of land adjacent to protected areas, acquire these areas, or change the current land uses to ones more compatible with species sensitive to human intrusions.

Does the shape of protected areas affect biodiversity? Certainly. Again, recall how the shape of a patch affected the amount of edge of that patch. The more patch shape diverges from circular—the more indented and convoluted its boundaries—the more edge it has (see Figures 8.11 and 8.12). This becomes important when one considers edge-sensitive species and the biotic, abiotic, and human effects associated with edges, as well as the depth of these effects into a protected area. Therefore, the shapes of protected areas can influence what species are able to persist there. Natural resource managers can play an important role when designing new protected areas and in altering the shapes of existing ones. For example, if an existing area is more long than it is wide, then working to alter harmful land-use practices on adjacent lands might greatly enhance that protected area's capacity to serve an important role in attracting and maintaining species that are sensitive to edge effects.

Figure 9.2 illustrates this point. Due to historical human use adjacent to the Flat Tops Wilderness Area in the Routt National Forest of Colorado, two tongue-shaped areas of heavy human use associated with access to Trappers Lake and Stillwater Reservoir have greatly increased the amount of edge. Edge effects associated with the motorized use of these two areas undoubtedly affects the biodiversity otherwise protected by the wilderness designation. If these areas were incorporated into the wilderness boundaries, edge would decrease and species sensitive to motorized use would be better protected.

Because so little of the Earth is formally designated for the protection of biodiversity and ecological processes, it is not surprising that protected areas often are isolated from one another. Isolation results in protected areas being viewed as islands in a sea of humanity. Why is isolation not conducive to the persistence of certain species? As we discussed in Chapter 8, dispersal-sensitive species are those whose morphology, physiology, and behavior limits their ability to successfully cross alien landscapes, places where human uses have altered the natural ecosystem to such an extent that they are unable or unwilling to attempt dispersal.

Many species are not constrained from dispersing across human-dominated landscapes; however, they nonetheless prove to be unsuccessful because they experience elevated mortality rates from inadvertent encounters with humans and their activities. It is because of these factors that the persistence of many species is dependent upon

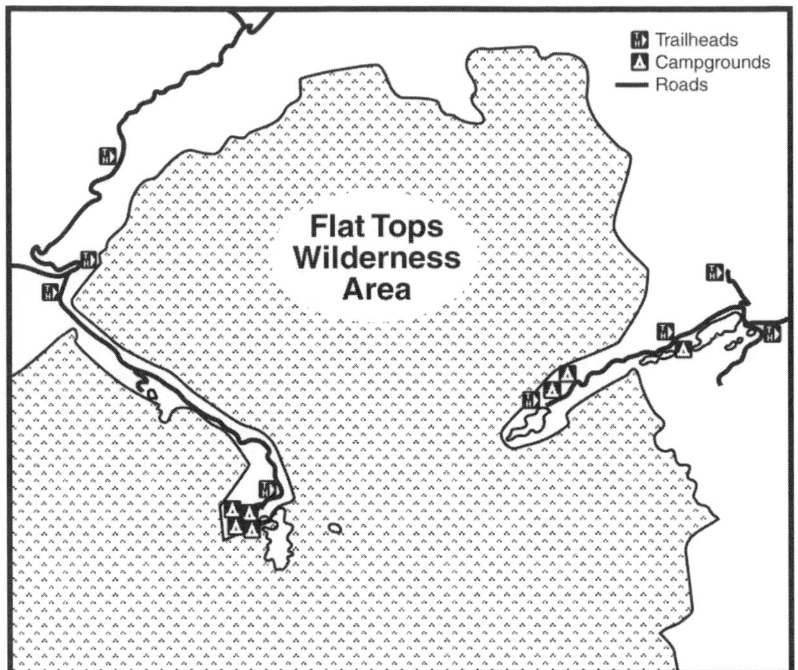

Figure 9.2. *A wilderness area in the Routt National Forest of Colorado. Areas of multiple use intrude into the wilderness area and increase the amount of edge along the wilderness boundary. Incorporating the two multiple-use areas into wilderness designation would alter the shape of the protected area, thereby decreasing the amount of edge as well as minimizing the human activities and the biotic and abiotic edge effects that may affect area-, edge-, and dispersal-sensitive species.*

being able to successfully move from one refuge to another. Furthermore, conservation genetics, metapopulation theory, and spatially explicit models all argue for the need for species to successfully disperse among areas. The acknowledgment that protected areas are increasingly isolated in landscapes, and the importance for species to be able to move successfully across landscapes, has spawned great interest in the idea of movement corridors.

EXERCISE 9.3

Collaborate on It!

In small groups and for the scenario you are using, develop a list of landscape-level actions that could be undertaken to enhance the capacity of the existing protected areas (refuge, preserve, state or national forests) to protect biodiversity. Be specific and practical.

MOVEMENT CORRIDORS

Movement corridors are linear strips of ecosystems intended to facilitate the movement of species between larger landscape patches. Movement corridors may facilitate the daily movement of individuals from one habitat patch to another, or they may be much longer strips that allow the successful long-distance dispersal of individuals.

There are two types of dispersal. **Natal dispersal** is largely an innate behavior that is expressed in young individuals that leave their site of birth and move in search of a mate and suitable, unoccupied breeding habitat. **Density-dependent dispersal** is movement from areas of high density and intraspecific competition to areas where density is lower and access to critical resources is more attainable. The Canadian lynx, for example, shows both types of movements (Figure 9.3). Juvenile lynx disperse outward from their natal sites in search of suitable but unoccupied areas as well as mates. Adult lynx periodically undergo density-dependent dispersal when their principle prey item, the snowshoe hare, undergoes population declines.

Movement corridors may also be used by species undergoing seasonal migrations, a two-way movement to and from breeding and nonbreeding areas. For example, several species of large ungulates, such as mule deer and elk, undertake elevational movements annually, traveling from their higher-elevation summering grounds to lower-elevation, and more protected, wintering sites.

Figure 9.3. *The Canadian lynx lives in boreal forests across much of North America. Juveniles disperse at a certain age, demonstrating natal dispersal. Adults and juveniles also disperse from northern latitudes of North America during population lows of their prey, demonstrating density-dependent dispersal.*

Types of Movement Corridors. Considering the diverse types of movements that plants and animals undertake, it is not surprising that there are various types of movement corridors. In reality, many movement corridors have tended to be along riparian habitats, which often are all that is left in forest ecosystems that have been systematically logged. Elsewhere, land-use regulations prohibit residential and commercial development in floodplains, thereby allowing them to remain in a somewhat natural condition. In addition, streams and rivers are linear, thereby serving as natural corridors connecting larger protected areas. Perhaps the strong association of movement corridors with riparian strips is due to the fact that riparian ecosystems contain the vital necessities for species survival: water, food, and shelter.

For some types of species, movement corridors need be nothing more than ecosystem remnants in a human-dominated landscape that connect one protected area to another. For example, lines of cliffs may serve as movement corridors if they facilitate the movement of species. The same is true for fencelines where shrubs and trees are allowed to persist. These linear strips allow a variety of species to successfully move across agricultural areas, traveling from one woodlot to another.

Sometimes movement corridors are human-engineered structures, such as the underpasses and overpasses that cross highways. These structures allow wildlife to navigate the landscapes that are fragmented by roads, and species mortality rates from motor vehicle accidents are thereby reduced. In Florida, motor vehicle collisions were responsible for nearly half the deaths of the endangered Florida panther, so researchers studied whether panthers used engineered underpasses beneath a heavily traveled interstate highway. In addition to panthers, animals that used the underpasses included bobcat, deer, raccoons, alligators, and black bears. In Canada, scientists found that black bears, wolves, lynx, and coyotes used underpasses and overpasses to successfully navigate across the Trans-Canada Highway (Figure 9.4).

Corridors should not be viewed exclusively as short connections between ecosystems; they can be thousands of miles long. For species whose dispersal ability is great, such as grizzly bears and wolverines, movement corridors may connect protected areas that cross national boundaries. Indeed, the Wildlands Project has a continental perspective that envisions movement corridors to allow wildlife movement from Canada across the United States to Mexico! Movement corridors would support species that are thinly spread and whose populations are not genetically or demographically viable within single regions, even those as large as a mountain range. Can jaguars survive if they are constrained to the Sierra Madre of Mexico? Possibly, but their chances of survival are enhanced if their range encompasses not only Mexican mountain ranges but also mountain ranges across the boundary with the United States that stretch through Texas, New Mexico, and Arizona.

Finally, another way of viewing movement corridors is to visualize them as spatially discrete habitat patches, or stepping stones where species that fly, such as migratory birds, bats, and insects, can stop to rest and feed during long-distance movements. More than half the species of birds breeding in the United States and Canada migrate to Latin America or the Caribbean Islands for the winter. If protected areas did not exist along their migration routes where they could feed and rest, the mortality rates of many of these species would be significantly higher. Partners in Flight is a coalition

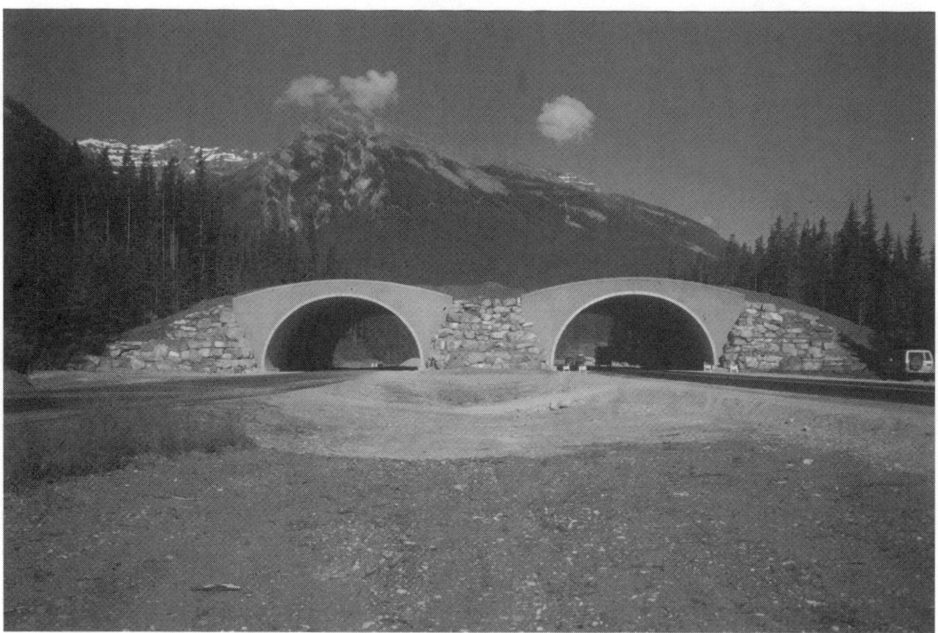

Figure 9.4. *An overpass crossing the Trans-Canada Highway in Alberta. This structure allows wolves and other large mammals to safely cross a landscape fragmented by highways. (Photo by Ben Alexander.)*

involving more than a dozen countries and hundreds of state agencies, nongovernmental organizations, and members of private industry from Canada to South America. It is dedicated to protecting these migratory species on the wintering and breeding grounds, but also along their migration pathways. Although it would be extremely difficult to provide continuous movement corridors over the length of their seasonal migration routes, it is more feasible to find and protect areas where migrating songbirds can stop to feed and rest before continuing their movement.

Much of the work that Canada, the United States, and Mexico does to protect wetland areas ensures that appropriate stopover areas exist along the migration routes of shorebirds and waterfowl. These birds cover thousands of miles in their movements twice each year. The North American Waterfowl Management Plan is a joint venture between Canada, the United States, and Mexico. It recognizes that waterfowl and shorebirds have needs transcending political boundaries and that land ownership of critical stopover areas is complex and requires international cooperation to protect. To date, they have protected and restored over a million acres of wetlands essential to waterfowl and shorebirds on their international travels.

EXERCISE 9.4

Collaborate on It!

For your scenario, identify the natural and human-made corridors. What do they connect? How heavily affected are they by human activities? What is or can be done to protect and enhance these areas? What government agencies and nongovernmental organizations are responsible for them? Do these groups recognize them as movement corridors?

Developing Movement Corridors. You might some day find yourself helping develop a movement corridor connecting protected areas—as an employee of a natural resource agency or private organization, or as a citizen on a land-use planning commission. How would you begin?

1. *Select a species of concern and evaluate its needs.* Although this appears to be single-species management, without picking a species and considering its requirements, it is hard to proceed with the other steps. Any design feature will have limitations for some types of species. If a species' life history needs are not considered, the corridor may not serve its purpose.

Figure 9.5. *Monitoring a movement corridor. (Above) An underpass for wildlife along the Kissimmee Highway in Florida, connecting habitat on either side and complete with a monitoring system. (Below) In this case, a bobcat is caught by camera passing through the underpass. (Photos by Melissa L. Foster.)*

2. *Identify the sites the movement corridor is intended to connect.* The sites will usually be protected areas, but what if one is a source and the other is a sink? This step entails understanding something about the populations of interest in the areas the corridor will connect.

3. *Map the corridor and evaluate its features.* This is a critical step because the sooner you confront the diverse jurisdictional and ownership boundaries that effectively fragment ecosystems, the sooner your planning will face reality. Movement corridors contain both human and natural features, including such things as rivers (are they passable by the species you have selected?), number and type of roads (can they be safely crossed or do they need to be engineered to facilitate movement?), ecosystem types (are these ecosystems where the species can find its critical resources while moving through them?), and surrounding land use (how natural or human-dominated is it?). Ownership and land use also affect whether movement corridors will work. If a corridor is meant to pass through an urban or suburban matrix, it will have to be wider than if it bisected a natural ecosystem.

4. *Design and implement a monitoring system.* What if you go to all the trouble and expense of creating a corridor that either is not used or does not ensure safe travel? Of all steps in the development of a movement corridor, this is the most neglected. By not monitoring whether corridors are used and whether the use is any different from adjacent areas without corridors, we lose opportunities to learn from our management actions (recall Chapter 4). To study whether the endangered Florida panthers used the underpasses beneath the interstate highway, the researchers used cameras to monitor the underpasses (Figure 9.5).

EFFECTIVE DESIGN FEATURES OF MOVEMENT CORRIDORS. Features that increase the effectiveness of movement corridors need to be studied more thoroughly. The adaptive approach of designing and monitoring corridors suggests that we will better understand what makes for effective movement corridors in the years to come. In the meantime, several general topics are helpful in considering effective design: gaps, width, and length.

Consider the gaps that will inevitably occur in movement corridors: the areas of poor habitat, roads, or other deterrents to movement. Gap width should usually be scaled to the movement ability of the species involved. Gaps can be relatively wider if the animal is very mobile (Figure 9.6a). If the animal is mobile and nocturnal the gap can be even wider because interference from or harassment by humans is less likely at night. For species that are more sedentary, and perhaps diurnal, gap width would need to be less.

The effective width of movement corridors depends on various factors. For example, consider the surrounding land use. The more similar the

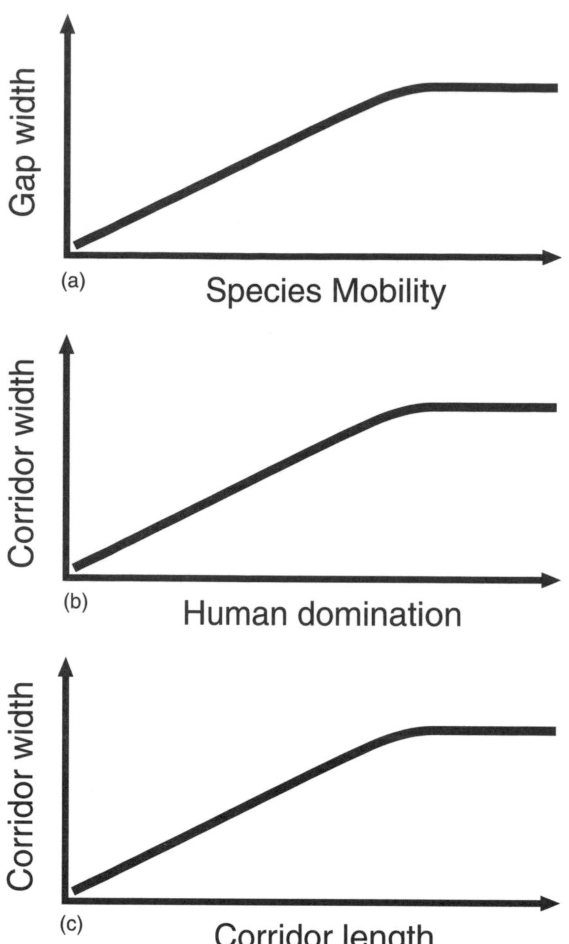

Figure 9.6. *Effective design features of movement corridors. (a) The width of gaps in corridors is scaled to the species' mobility; the more mobile the species the wider gaps can be. (b) The width of corridors depends on the surrounding land use; the more natural the matrix, the narrower the corridor can be. (c) The longer the corridor, the wider the corridor needs to be.*

land use and habitat adjacent to the corridor is to the corridor itself, the narrower the corridor can be, because the qualitiative difference between the corridor and the surrounding land is less (Figure 9.6b). A movement corridor leaving a national park and passing through a developed area should increase in width in order to protect wildlife from pressures emanating from the adjacent developed areas. This presents a real conservation challenge because the more developed a landscape is, the less likely there will be undeveloped areas through which a corridor can pass.

Corridor width also should be scaled with corridor length (Figure 9.6c). The longer the corridor is, the more time an animal is likely to spend in the corridor and the greater its needs may be. A short corridor that an animal can move through in a day will not require much in the way of food, water, and shelter. If the corridor is long and dispersing animals spend days or weeks in it, the corridor must have the capacity to sustain them.

It has been suggested that corridor width should be the average width of that species' home range. The idea behind this reasoning is as follows: If the corridor is as wide as the species' home range, then it should be large enough for the species to survive in while moving through. Because corridors are usually designed for large species, and because home range increases with animal size, a recommended corridor width might be quite large (Figure 9.7). A mediating factor in corridor width is the habitat quality of the corridor. Movement corridors that are parts of healthy ecosystems, where water, food, and cover are plentiful, can be narrower than those that are highly degraded.

Conservation planners may not always be able to design corridors that are as wide as needed. However, through ecological restoration, thoughtful planning, and design, they can make what they have more suitable for dispersing individuals.

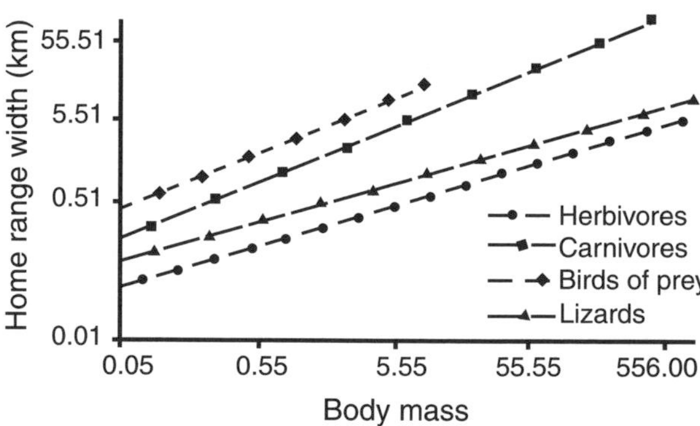

Figure 9.7. *The relationship between body mass and home range width for mammals (herbivores and carnivores), birds of prey, and lizards. (From Dobson et al., 2000.)*

EXERCISE 9.5

Collaborate on It!

Working in small groups and for your scenario, develop a movement corridor that connects at least two protected areas (refuge, forest, preserve). List the areas to be connected, the species chosen for your corridor design criteria, aspects of the species' life history that will determine design recommendations, and management and conservation actions that will be needed to ensure the corridor's utilization. Include suggestions for a monitoring program to see whether the corridor is used, and describe your plan in terms of adaptive management. Share your results with the class.

POTENTIAL DISADVANTAGES OF CORRIDORS. Sometimes there is a tendency in nature conservation to accept intuitively appealing ideas uncritically. This is true with movement corridors. Because ecosystems are increasingly human-dominated, we believe that linear strips uniting habitat remnants are essential for maintaining some level of connectivity for species that naturally disperse. What can be wrong with this idea? For most landscapes and in most situations, probably nothing. But what if movement corridors serve as conduits for the dispersal of invasive species, or of pathogens or parasites that might affect species in the remnants the corridors are connecting? Because corridors tend to be more long and narrow than wide and square they have a disproportionately high ratio of edge to area, thereby serving as habitat for edge-generalist species. Might these generalist species negatively affect other species that are moving through the corridors? In other words, could movement corridors serve as ecological traps?

At present we do not understand corridors well enough to evaluate these possible pitfalls. In the years to come, after more movement corridors have been designed, implemented, and monitored for success, we will better understand their potential advantages and disadvantages. In the meantime, we need to try and reconnect landscapes that were historically connected and that have become fragmented through human use.

Working Across Administrative Boundaries

Working across administrative boundaries is one of the pillars of ecosystem management. Because administrative boundaries seldom parallel ecosystem boundaries, and because legal boundaries frequently dictate quite different land uses, administrative boundaries often create sharp distinctions across ecosystems. Administrative boundaries almost always fragment a landscape, disrupting the ebb and flow of individuals and ecosystem processes (Figure 9.8). At the same time, boundaries often play important roles, such as marking the line protecting wilderness from mechanized vehicles or other human influences.

How did we get to where we are today, with so many different state, federal, and local agencies and private organizations—each with differing and sometimes conflicting mandates, policies, and regulations—all searching for ways to coexist in a common landscape? The reasons for today's fragmented management are many and we can categorize them as ecological or managerial.

Ecological boundaries were a necessary part of traditional vegetation descriptions developed by the pioneers of ecology in the early to mid-1900s. One of the tenets of this early ecology, embodied in the phrase "the balance of nature," was that ecosystems were internally regulated and in equilibrium with climate, inexorably moving toward a single climax, or stable, condition. These early concepts fostered the belief that ecosystem boundaries were tangible entities, rather than arbitrary constructs of our intellect and desire to understand a complex world.

Managerial boundaries were necessary for defining administrative jurisdictions and responsibilities; it was necessary for natural resource agencies to accept the notion of relatively fixed ecosystem boundaries. This combination of ecological and managerial factors led to a belief that lands managed by an agency were separate and independent from other lands—that what happened on one side of a border did not necessarily affect what happened on the other.

Ecosystem management is spurring a change in

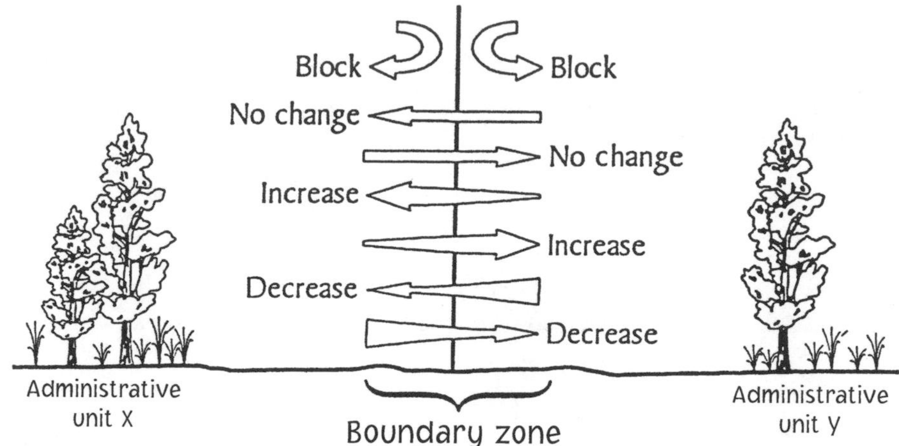

Figure 9.8. *Inflow and outflow across an administrative boundary. An administrative boundary can function like a semipermeable membrane that dramatically affects ecological flows across the line. Either side of the boundary may act as the "focal" side from which flow may increase, decrease, or not change; either side may therefore be a source or a sink for a resource, contaminant, influence, or organism. (From Landres et al., 1998.)*

this way of thinking. Today, managers increasingly recognize the importance of focusing beyond as well as within their boundaries, and ecologists recognize that the 1900s view of ecosystems as static does not reflect the actual spatial and temporal dynamism of ecosystems. Managers and scientists alike now see that administrative and ecosystem borders are arbitrarily defined and delineated. They are not closed but leaky, and they experience the inflow and outflow of things as diverse as water and pollutants, migrating species, and humans crossing borders to hunt, cut firewood, or picnic (Figure 9.9). With this shift from the belief in "the balance of nature" to a more realistic view of "the flux of nature," there is reason to believe that resource managers can be more responsive to the dynamic character of human-dominated landscapes. This new land perspective emphasizes that managers think beyond the boundaries for which they are responsible because what occurs beyond their borders affects what occurs within their borders, and vice versa.

The complex biological, socioeconomic, and institutional effects of boundaries are an important component of land-use decisions and land management practices today. Managers now face the difficult task of sustaining biological diversity while providing amenity and commodity uses from landscapes that have been delineated and affected by boundaries established in the past. These effects influence lands spanning a continuum of management goals, from designated wilderness to lands

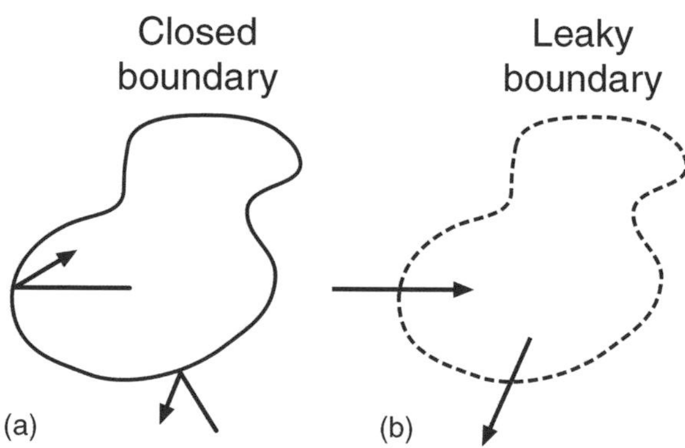

Figure 9.9. *Two views of administrative borders. (a) The traditional view was that boundaries were closed or impermeable; they did not allow things like pollutants, water, animals, or humans to cross over. (b) The revised view is that borders are highly permeable or leaky, allowing both biotic and abiotic elements, such as fire, to pass across dispersing wildlife, humans, and invasive weeds.*

devoted solely to commodity production. Boundary effects can be profound and require diligent cooperation when they lie between the ends of the management continuum, where ecosystem management strives to provide goods and services while maintaining native biodiversity and where managers strive to balance both amenity and commodity values. They are perhaps most difficult to deal with where public land boundaries abut private lands. Here land uses may be vastly different. Indeed, because this is where one most often finds the steepest contrast in land use, it is also where the greatest differences occur in ecosystems (Figure 9.10).

Consider, for example, the housing development adjacent to the Roosevelt National Forest shown in Figure 8.3. The gradients between land uses on either side of the administrative boundary are vivid. On one side there are nearly 100 homes with families and all their associated activities, while on the other side is public land with restricted uses and no permanent residences. Consider some of the effects. Water flows across the private-public land boundaries may be minor, with a low rate of change (Figure 9.10a), whereas the effects of predation by native predators and dogs and cats may show a low contrast but high rate of change (Figure 9.10b). The biological effects of non-native species the homeowners use for landscaping may reflect a sharp contrast and a low rate of change across the public-private land boundary (Figure 9.10c). Prescribed fire, on the other hand, might have a high contrast and have a considerable distance effect on either side of a border, because it will be largely unacceptable on the private lands while being prescribed on public lands (Figure 9.10d).

Considering the challenges of working across administrative boundaries, how does one proceed? The very fabric of our society, and our relationship with one another and with the natural world, have increasingly been under intense reappraisal. Land management agencies are now addressing the challenge of managing ecosystems in response to the realization that administrative boundaries are highly permeable. This has necessitated a shift from focusing on specific, small-scale units to an emphasis on the health of more broadly defined ecosystems.

With ecosystem management, it is not appropriate for agencies to operate as if their administrative boundaries form an impermeable wall, within which they have complete control and outside of which they are powerless. Because land management agencies commonly share borders with private lands, each must acknowledge their neighbors and find ways to cooperate. Cooperation is accomplished by creating interactive networks of

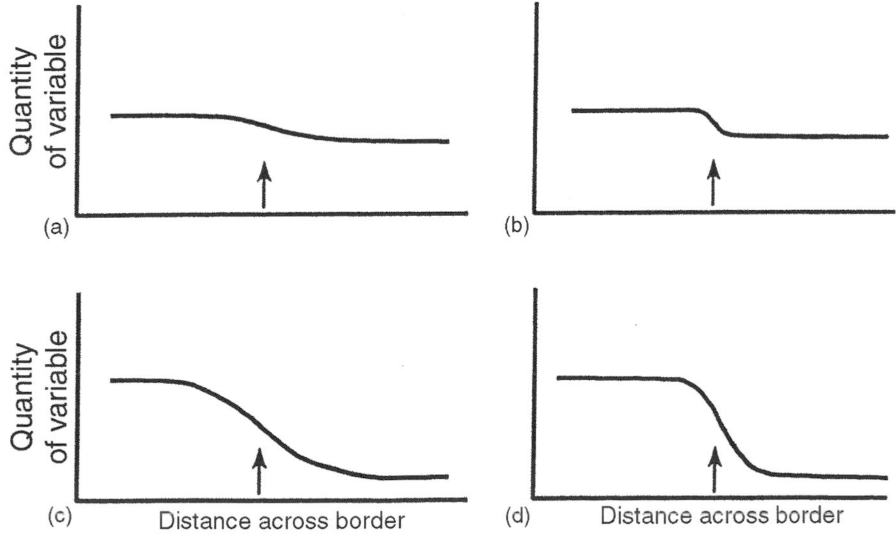

Figure 9.10. *Land-use changes on either side of an administrative boundary. These graphs show four possible combinations of the amount of contrast and depth of boundary effects: (a) low contrast and low rate of change; (b) low contrast and high rate of change; (c) high contrast and low rate of change; (d) high contrast and high rate of change. (From Landres et al., 1998.)*

participants, information exchange, and practices. These networks should represent the diverse and legitimate interests and capabilities of a pluralistic society. This approach also must address the problems arising from the highly fragmented distribution of information, resources, and power across geographic boundaries, social groups, organizations, agencies, and disciplines.

Boundaries between and within public and private lands require thoughtful discussion, because human activities on one side of the boundary can significantly affect what happens on the other side. In recent decades, urban and rural development near public lands has caused the alteration of biodiversity and the disruption of critical ecological processes. At the same time, home sites adjacent to undeveloped public lands have higher values and may sell at premium prices. It is important to address ecological, economic, and societal issues in a consistent, rational, and adaptive manner. By focusing on shared perspectives, and by tracing prior and possible future development and land-use trends, there is an increased likelihood of cooperation and collaboration.

General solutions to boundary dilemmas need to be applied site by site to be contextually relevant. Practical solutions require reliable information, planning, honest conversations, clear management recommendations, constructive dispute resolution, and ongoing policy appraisal. This approach may provide opportunities for sustaining the ecological, economic, and societal needs of diverse communities and organizations that share borders.

EXERCISE 9.6

Collaborate on It!

Working in small groups for your scenario, list the protected areas (from city, county, and state parks up to national forests and parks, refuges, and preserves) by ecosystem type and land use (recreation, logging, grazing, hunting, etc.). In a separate column, list the land uses across the protected areas' administrative boundaries (industrial, housing, business, agriculture, etc.). For each protected area, decide how similar and dissimilar the land uses are on either side of the borders. For each protected area, how secure is biodiversity within the borders? How intact are the natural ecological processes? Does this analysis stimulate ideas for additional management activities for these areas?

HCPs: Protecting Biodiversity While Promoting Cooperation

Once upon a time, nature conservation was viewed through a somewhat narrow lens. Where biodiversity was imperiled, the usual response was to stop the harmful action and protect the area. That approach usually involved purchasing the land and setting it aside, sometimes literally behind a fence. Although critical, such an approach will unfortunately not suffice in saving our planet's natural heritage. The human population is simply too large, and our activities are cumulatively affecting biodiversity on such a scale that other ways have to be developed. One of the benefits that has emerged from the idea of ecosystem management is that land managers are encouraged to look across their boundaries. By so doing, they are more likely to understand land uses that may threaten biodiversity—on both sides of the administrative boundary. In turn, this perspective increases the likelihood that solutions can be crafted to protect populations and encourage natural ecological processes.

A case in point is the response of the U.S. Fish and Wildlife Service (USFWS) to the onerous task of implementing the Endangered Species Act (ESA). The agency has been quite successful when dealing with federally listed species on federal lands, either national parks, forests, or refuges. But the story has been quite different when dealing with listed species on privately owned lands. Even though federally listed species that occur on private lands still receive protection under the ESA (although plants receive very little protection), trying to find ways to allow landowners to develop their land while protecting species has been vexing for all involved. The U.S. Department of the Interior (within which the USFWS is housed) has developed a series of creative ways to protect listed species while permitting limited landowner development.

One of the most imaginative approaches to emerge is called a **Habitat Conservation Plan (HCP)**. HCPs are agreements developed by private landowners in consultation with (and approved by) the federal government (usually the USFWS) allowing landowners to proceed with otherwise legal activities that might result in the inadvertent "take" of federally listed species on their property. (A "taking" is any activity that harasses, harms, pursues, traps, captures, or kills a listed species or subspecies, or significantly modifies its habitat.) Approval is granted if the flowing conditions are met:

1. The taking will be incidental to an otherwise lawful activity.
2. The impacts of the taking will be minimized and mitigated to the maximum extent possible.
3. There will be adequate funding (by the landowner) to conduct the HCP, and the landowner has established procedures for addressing some degree of uncertainty.
4. The taking will not appreciably reduce the likelihood of survival and recovery of the species in the wild.
5. The landowner agrees to include other measures that the government may require (Hood, 1998).

HCPs can guarantee to landowners that if unforeseen circumstances arise, the government will not require the commitment of additional resources, other than what was promised in the HCP. This clause, called "No Surprises," ensures the government will honor its agreements under an HCP, as long as the landowners are operating in good faith to carry out their responsibilities.

Another attempt to make HCPs more useful is called the "Safe Harbor" approach, which assures landowners who undertake voluntary conservation actions on their lands that the government will not further restrict their activities if their lands attract a listed species. It establishes a baseline of responsibilities for listed species below which the landowner may not go. If landowners exceed that baseline by improving habitat, they are not restricted in the future by having to stay at that new level. The purpose of the Safe Harbor approach is to reduce the disincentives that often cause landowners to avoid or prevent land-use practices that would otherwise benefit endangered species.

The HCP process was developed through the innovative efforts of agency personnel and landowners in the San Francisco Bay Area, where development interests collided with endangered species. What made this effort so unusual was that it attempted to resolve these conflicts of interest—development versus preservation—through cooperation rather than litigation. This approach was so successful that it was endorsed and codified by Congress when it incorporated the HCP process into the Endangered Species Act in 1982. As of 1998, over 200 HCPs were in various stages of development, covering almost 7 million acres. Although many of the early HCPs dealt with a single species and small areas (e.g., tens of acres), more recent HCPs deal with suites of species and may involve large areas—up to 1,000,000 acres.

There are clear dilemmas with HCPs. For example, the USFWS and landowners are required to monitor HCPs and apply adaptive management to the HCP implementation. The USFWS lacks the personnel and resources to ensure that HCPs are being carried out, and landowners seldom have

EXERCISE 9.7

Talk About It!

For a listed species of interest in the scenario you are using, discuss the steps necessary for the development of an HCP. What obstacles stand in the way? For example, who is responsible for initiating the HCP, and where will the resources come from to write it and do the appropriate mitigation and monitoring? Can private landowners be trusted? More information on HCPs can be obtained from the USFWS Web site, http://endangered.fw.gov/hcp/index.html.

Table 9.1. *Ecological Considerations for Protecting Biodiversity*

- I. Set objectives and priorities.
  - A. Prioritize species and communities based on local, regional, and global abundance.
  - B. Set goals for species populations and communities.
- II. Organize at the appropriate landscape or watershed scale.
  - A. Catalogue levels of genetic diversity, population size, and species richness.
    - 1. Heterozygosity, allelic diversity, N_e, genetic isolation, bottlenecks.
    - 2. Viable population sizes, metapopulations.
    - 3. Alpha, beta, and gamma richness.
  - B. Use landscape principles.
    - 1. Minimize fragmentation.
      - a. Cluster human activities.
      - b. Minimize barriers to movements (roads, power lines).
    - 2. Reduce edges.
      - a. Consider shape.
      - b. Soften "hard edges."
    - 3. Maintain connectivity.
      - a. Use existing corridors (riparian areas, cliff lines).
      - b. Construct underpasses, corridors for moving animals around and through human settlement.
      - c. Include corridors in reserve design and land-use planning.
- III. Communicate with landowners and agencies beyond your boundaries.
  - A. Public lands cannot be managed in isolation.
    - 1. Work to make adjacent land uses more compatible.
    - 2. Explore alternative approaches to compatible land uses.
      - a. Landowner notification.
      - b. Cooperative management agreements.
      - c. Land exchanges, leases, easements.
      - d. Work with local, regional, and national land conservation organizations.
- IV. Develop ecologically based management guidelines.
  - A. Set management guidelines that are compatible with objectives.
  - B. Implement ecological restoration.
    - 1. Restore fire.
    - 2. Remove exotics.
    - 3. Manage for appropriate herbivory.
    - 4. Simulate natural flooding regimes.
    - 5. Reestablish predator/prey dynamics.
    - 6. Restore natives.
  - C. Evaluate traditional and emerging commodity uses.
    - 1. Livestock grazing.
    - 2. Timber harvest.
    - 3. Recreation.
- V. Monitor and adapt management.
  - A. Establish monitoring programs at a variety of spatial and temporal scales.
  - B. Evaluate the effectiveness of management activities and compare with controls.
  - C. Evaluate goals.
  - D. Modify approaches.

the necessary expertise. On the positive side, this shared trust and responsibility fall within the spirit of ecosystem management, asking all of us to assume a greater responsibility in protecting our natural heritage. Clearly, conservation practices are never perfect. The important point is that they engage both institutions and stakeholders in efforts to ensure that nature and humans benefit; clearly one cannot thrive without the other for long.

Considering the diversity of ideas in this and other chapters, protecting biodiversity under the rubric of ecosystem management may seem like an overwhelming task. To help simplify, Table 9.1 presents an outline of steps that combine numerous points raised thus far. This outline should be viewed more as an ecological checklist than a rulebook. Innovation based on knowledge and conversations with others, and tempered with adaptive management, will be your best guide.

References and Suggested Readings

Baydack, R.K., H. Campa, III, and J. B. Haufler. 1999. *Practical Approaches to the Conservation of Biological Diversity.* Island Press, Washington, D.C.

Beier, P., and R.F. Noss. 1998. Do habitat corridors provide connectivity? Conservation Biology 12: 1241–1252.

Conroy, M.J., and B.R. Noon. 1996. Mapping of species richness for conservation of biological diversity: Conceptual and methodological issues. Ecological Applications 6:763–773.

DellaSala, D.A., D.M. Olson, S.E. Barth, S.L. Crane, and S.A. Primm. 1995. Forest health: Moving beyond rhetoric to restore healthy landscapes in the inland Northwest. Wildlife Society Bulletin 23:346–356.

DellaSala, D.A., N.L. Staus, J.R. Strittholt, A. Hackman, and A. Iacobelli. 2001. An updated protected areas database for the United States and Canada. Natural Areas Journal 21:124–135.

Dobson, A., et al. 2000. Corridors: Reconnecting fragmented landscapes. Pp. 129–170 in M.E. Soulé and J. Terborgh (eds.). *Continental Conservation: Scientific Foundations of Regional Reserve Networks.* Island Press, Washington, D.C.

Everett, R.L., and J.F. Lehmkuhl. 1996. An emphasis-use approach to conserving biodiversity. Wildlife Society Bulletin 24:192–199.

Foster, M.L., and S.R. Humphrey. 1995. Use of highway underpasses by Florida panthers and other wildlife. Wildlife Society Bulletin 23:95–100.

Hansen, A.J., S.L. Garman, B. Marks, and D.L. Urban. 1993. An approach for managing vertebrate diversity across multiple-use landscapes. Ecological Applications 3:481–496.

Hansen, A.J., J.J. Rotella, M.P.V. Kraska, and D. Brown. 1999. Dynamic habitat and population analysis: An approach to resolve the biodiversity manager's dilemma. Ecological Applications 9:1459–1476.

Hood, L. 1998. *Frayed Safety Nets: Conservation Planning Under the Endangered Species Act.* Defenders of Wildlife, Washington, D.C.

Hunter, M.L., Jr. 1990. *Wildlife, Forests, and Forestry: Principles of Managing Forests for Biological Diversity.* Prentice Hall, Englewood Cliffs, NJ.

Hunter, M.L., Jr. 1996. *Fundamentals of Conservation Biology.* Blackwell, Cambridge, MA.

Knight, R.L., and T.W. Clark. 1998. Boundaries between public and private lands: Defining obstacles, finding solutions. Pp. 175–191 in R.L. Knight and P.B. Landres (eds.). *Stewardship Across Boundaries.* Island Press, Washington, D.C.

Knight, R.L., and P.B. Landres (eds). 1998. *Stewardship Across Boundaries.* Island Press, Washington, D.C.

Landres, P.B., R.L. Knight, S.T.A. Pickett, and M.L. Cadenasso. 1998. Ecological effects of administrative boundaries. Pp. 39–64 in R.L. Knight and P.B. Landres (eds.). *Stewardship Across Boundaries.* Island Press, Washington, D.C.

Leopold, A. 1938. Engineering and conservation. Pp. 249–254 in S.L. Flader and J.B. Callicott (eds.). 1991. *The River of the Mother of God and Other Essays by Aldo Leopold.* University of Wisconsin Press, Madison.

Noss, R.F., and A.Y. Cooperrider. 1994. *Saving Nature's Legacy: Protecting and Restoring Biodiversity.* Island Press, Washington, D.C.

Noss, R.F., M.A. O'Connell, and D.D. Murphy. 1997. *The Science of Conservation Planning: Habitat Conservation Under the Endangered Species Act.* Island Press, Washington, D.C.

Pickett, S.T.A., and R.S. Ostfeld. 1995. The shifting paradigm in ecology. Pp. 261–278 in R.L. Knight and S.F. Bates (eds.). *A New Century for Natural Resources Management.* Island Press, Washington, D.C.

Poiani, K.A., B.D. Richter, M.G. Anderson, and H.E. Richter. 2000. Biodiversity conservation at multiple scales: Functional sites, landscapes, and networks. BioScience 50:133–146.

Prendergast, J.R., R.M. Quinn, and J.H. Lawton. 1999. The gaps between theory and practice in selecting nature reserves. Conservation Biology 13:484–492.

Scott, J.M., B. Csuti, J.D. Jacobi, and J.E. Estes. 1987. Species richness: A geographic approach to protecting future biological diversity. BioScience 37:782–788.

Stein, B.A., L.S. Kutner, and J.S. Adams (eds). 2000. *Precious Heritage: The Status of Biodiversity in the United States.* Oxford University Press, New York.

Soulé, M.E., and J. Terborgh (eds). 1999. *Continental Conservation: Scientific Foundations for Regional Reserve Networks.* Island Press, Washington, D.C.

Vogt, K., et al. 1997. *Ecosystems: Balancing Science with Management.* Springer-Verlag, New York.

Experiences in Ecosystem Management:

The Malpai Borderlands Group: Building the "Radical Center"

William McDonald

THE CONTROVERSY THAT HAS ARISEN OVER LIVESTOCK grazing in the American West has been characterized by extreme rhetoric and extreme actions. With government agencies nearly gridlocked and decisive legislation not forthcoming, activists increasingly turn to litigation and sometimes "monkey wrenching," or other forms of intimidation, in attempts to force their will upon a process that is often so mired in procedure that even the simplest management actions require reams of supporting paperwork.

The antagonists have been traditionally identified as "ranchers versus environmentalists" or "extractionists versus conservationists." Not liking the sound of those labels, some prefer "wise-use versus preservation." Those who graze livestock and their supporters have been expected to line up on one side of the issue, while the environmental community and their supporters line up on the other. Stories in the news media, together with the current spate of litigation over land use, have further solidified the grazing issue in the West as one that is black and white, us against them.

What is being lost in the rhetoric is the only thing that really matters—the eventual consequences for the land. I have purposely avoided the term "public land." In most of the West, the character of the public land depends in large part on what is taking place on the surrounding and intermingled private lands. Even in areas where the public acreage dwarfs the private, often the private land (the homesteaded land) may contain the only reliable water and/or the easiest ground (such as open meadow) for miles. It may be the piece that makes the area work ecologically for the wildlife inhabitants.

If the fate of the public lands depends to some extent on what happens to adjoining private land, it is even surer that the fate of much of the private land depends on the ability of the ranchers who own it to graze their cattle on adjoining public land. Denied that ability, many would no longer be able to maintain viable grazing livelihoods. The alternative source of livelihood, in many cases, has been to sell the land to developers, resulting in ranchette development and greatly increased human densities.

With these concerns in mind, in 1991 a small group of ranchers in southeastern Arizona and southwestern New Mexico, along the Mexican border, sat down with some folks from the environmental community to break from the traditional stereotypical positions and to try to find common ground, to begin to build, if you will, the "radical center."

At stake were nearly 800,000 acres of unfragmented landscape, the northern tip of the Madrean Archipelago, where Arizona and New Mexico join the Mexican states of Sonora and Chihuahua. As happened in many places in the West, the area had seen a major influx of people and livestock around the turn of the twentieth century, and the numbers proved to be unsustainable. Fire suppression, overgrazing, and other activities associated with nearly unrestricted settlement exaggerated the effects on the landscape of a climatic regime characterized by extremes. Harsh economic reality followed the ecological abuse, causing most to leave in search of other opportunities.

Today, about 30 families live on ranches within the huge area, possibly the fewest number of human residents in centuries. The concerns of those who gathered together in 1991 focused on two issues: (1) the continuing loss of grasslands to woody species, believed to be partially caused by century-long fire suppression; and (2) the anticipated threat of fragmentation of the area from a renewed influx of people. On three sides of the area, subdivision was accelerating. In looking for allies to address these concerns, the ranchers found them in, of all places, the environmental community.

Calling themselves the Malpai Group, the ranchers and their new-found allies met for discussions in ranch houses over a 2-year period. This discussion period had the effect, intended or not, of cultivating trust and friendships that became indispensable factors in the group's success when it turned later from discussion to action. An enormous advantage lay in the fact that the participants were far-sighted enough to address their concerns before they became crises.

The role of The Nature Conservancy (TNC) proved to be essential in helping move the group from being a forum for discussion into an action organization. TNC had been the area's largest landowner, having purchased the 320,000-acre Gray Ranch in 1990. They then confounded nearly everyone by selling the property to a local ranching family who purchased it with a conservation easement attached, which guaranteed that the Gray Ranch would never be developed. The relationship that developed between the family and TNC personnel led to their inclusion in Malpai Group discussion sessions. TNC brought organizational skills, fund-raising expertise, legal know-how, and additional contacts in the political world and in the scientific community. To some, however, there was a downside. Some ranchers feared the direct involvement of an international environmental group in a grass-roots organization, believing The Nature Conservancy would inevitably take over control. A few ranchers disengaged from the group, and some went so far as to begin a campaign of opposition.

One huge challenge for the group was trying to involve the Bureau of Land Management (BLM), the U.S. Forest Service (USFS), and two state land departments (collectively the owners of nearly 50% of the area's land) as true partners in an effort to realize an open-space future for the 800,000-acre landscape (Figure A). The Malpai Group addressed this issue by rallying the agencies around the idea of a regional fire management plan, which would include private landowner input. Agency personnel showed enthusiasm for the initiative and encouraged expansion of the idea to a whole ecosystem approach to management of the area's land. The timing was fortuitous. With a mandate for ecosystem management coming from Washington, and no one exactly sure what it meant, some of the progressive minds in the agencies saw this as an opportunity to define it "on the ground."

In 1994, the Malpai Borderlands Group (MBG) was born as a nonprofit organization, establishing official status in order to receive tax-deductible contributions and hold conservation easements. A board of directors was established, made up initially of the remaining participants from the Malpai discussion group. The USFS and the Natural Resources Conservation Service both assigned individuals to work with the fledgling organization. An additional boost came when a multiple-year grant was awarded to the research arm of the USFS to perform long-term fire and watershed studies in coordination with the group's efforts.

In addition to many tours and meetings with key officials and occasional trips to Washington, D.C., one of the things that has made the partnership with the agencies work has been the shared success in achieving stated goals. All parties (agencies, ranchers, scientists, and the environmental community) agreed that fire needed to be reintroduced into the landscape. The timing was right, as 1994 proved to be a big year for natural fires. Because of our working relationship with the agencies, over 100,000 acres were allowed to burn. Successful prescribed burns were carried out. The Baker Burn in 1995, the Maverick Burn in 1997, and the Miller Burn in 1998 all involved multi-

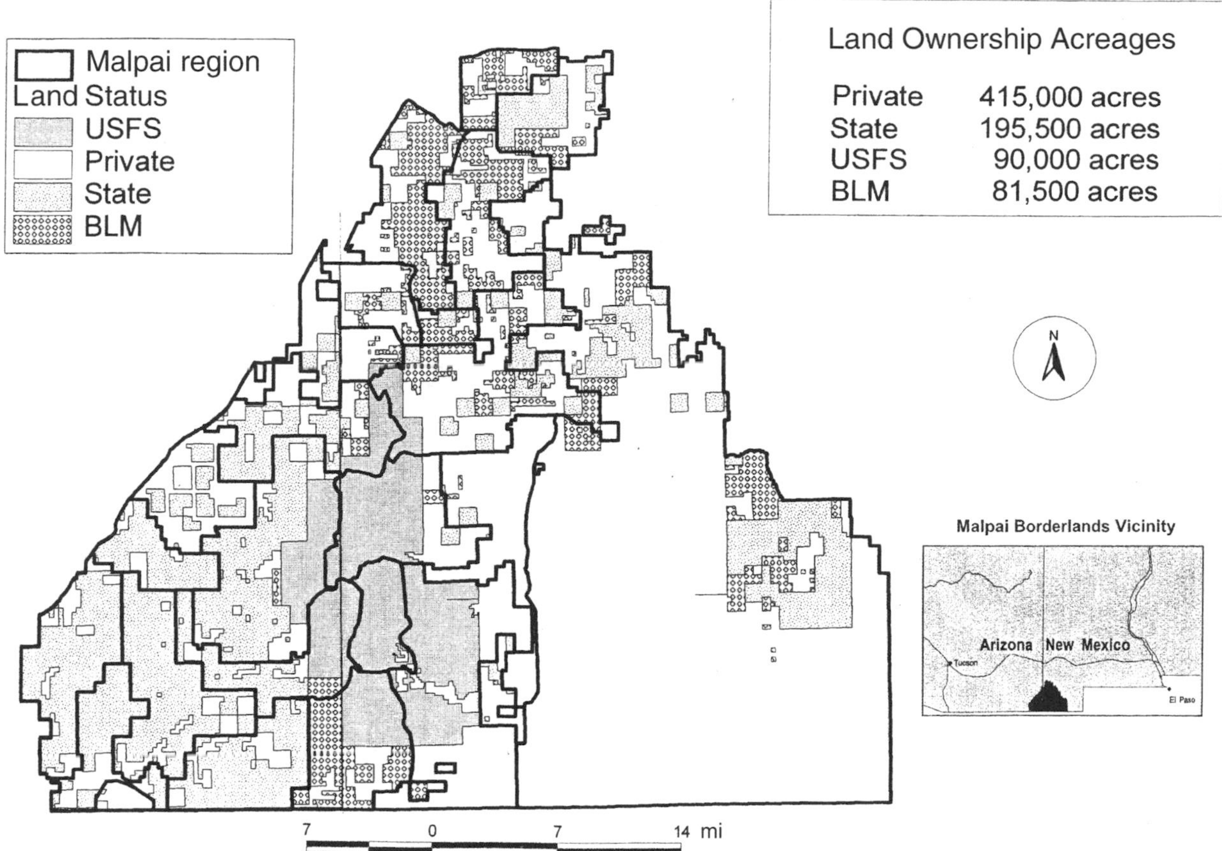

Figure A. *The Malpai Borderlands Group region, showing the complex land ownership patterns.*

agency and multi-landowner cooperative efforts. The prescribed burns allowed the use of "before and after" monitoring to document whether the results met expectations. Did the hoped-for impact on woody species occur? Was the grass invigorated? Is the anticipated increase in biodiversity taking place? Over 200 monitoring plots are now in place in the region, many measuring fire effects.

Different challenges presented themselves, depending on the land ownership involved in the burns. For prescribed burns on state and privately owned land, the biggest concern was being able to obtain the resources to actually implement a burn and ensure that the fire did not spread to places where it was not wanted. On federal land, abundant resources are available; however, planning costs and delays resulting from different opinions on the short-term effects of fire on endangered species present (or believed to be present) made for an excruciating process, leading right up to ignition. Currently, the MBG is immersed in a programmed approach to consultation on endangered species in the area. We hope this will result in a more efficient and predictable method of implementing prescribed burns in the future.

The Malpai Borderlands Group has been proactive in rare and endangered species issues. The group's work in helping an area ranching family with their efforts to save a threatened species of leopard frog (*Rana chiracahuaensis*) led to a cooperative effort that included the Arizona Game and Fish Department. This effort established a new water source on the ranch that benefits both the frogs and the family's livestock operation. Some of the tadpoles that hatch on the ranch are placed in ponds constructed at schools in nearby Douglas, Arizona, as part of an education and recovery project administered by herpetologists from the

University of Arizona. Eventually these frogs will be released back in the wild as appropriate habitats become available.

A chance encounter with a jaguar by another Malpai participant presented the group with an additional opportunity to be proactive. Instead of shooting the animal, the rancher took photographs, which were published in a booklet. The MBG helped initiate a conservation plan that became the template for the Jaguar Recovery Plan when the animal was listed as endangered in the United States. As a result of proceeds from sales of the booklet, the group maintains a fund to reimburse ranchers for any losses to livestock from a jaguar. The MBG actively funds and participates in research and monitoring efforts, most of which are conducted in Mexico.

These proactive efforts by the MBG have enhanced its credibility when it has been forced to react to court-ordered biological opinions involving federal grazing allotments in the area, which result from lawsuits being filed against the agencies. The group's ability to bring good science to bear on individual species issues has become respected in this arena, where the law requires answers to what is often unknown.

The most immediate threat to the Malpai Group's goal of securing a million acres of healthy, unfragmented landscape is the inexorable movement of people into the remaining open spaces of the West. In its attempt to keep development at bay, the group has been obtaining conservation easements on working cattle ranches in the area. Combined with the easement held by TNC on the Gray Ranch, approximately half of the land area is now permanently protected from development (Figure B). Conservation easements have been the single largest factor in the recruitment of participants in the group's activities. By being flexible in anticipating and meeting the needs of ranchers, MBG has been able to provide them with more than just protection from subdivision. In exchange for the first four easements MBG received, the landowners' cattle were given multiple-year access to forage on the Gray Ranch while their home ranches received needed rest from grazing following a severe drought. The Malpai Group paid for the forage by raising funds from individuals and grant-making institutions. The money is also used by the Malpai Group to share costs with the ranchers for the installation of watering facilities and fences. These will make the ranches more efficient and the ranchers better able to manage for droughts in the future. In two other instances, the MBG purchased the easements outright, and the ranchers used the money to purchase adjoining land that will make their operations more sustainable.

In attempting to find ways to improve the economic return and provide more security to the area's ranchers, the MBG has spent considerable time investigating the possibility of initiating an effort to market ranch beef directly to the consumer. The idea would be to establish a premium market for quality beef from cattle raised in a beautiful, unfragmented landscape by people who were committed to keeping it that way. As appealing as that concept sounds, the reality of putting a program together in this remote area, with a limited supply of cattle (approximately 5000 from all ranches combined), far from packing facilities, distribution centers, and urban consumers, has proven to be much more challenging than asking ranchers to work together toward conservation goals. The group hopes to take some steps cooperatively to position the ranchers' cattle to be part of a larger program, if a successful one emerges. It remains a challenge for American society to find ways to reward those who keep the land open and manage their livelihoods in an ecologically sound manner. The MBG has at least helped raise the visibility of the issue.

Although the Malpai Borderlands Group is being hailed as a success and a model for others after just 8 years in existence, it is clear to the group that its work is only beginning, and many challenges lay ahead. The novelty of ranchers and local environmentalists moving away from traditional adversarial positions and working together in the "radical center" has brought the group popularity and political strength outside the region. But it will take the group's staying power to bring the eventual acceptance of those who live in the region but have not yet participated in MBG's ef-

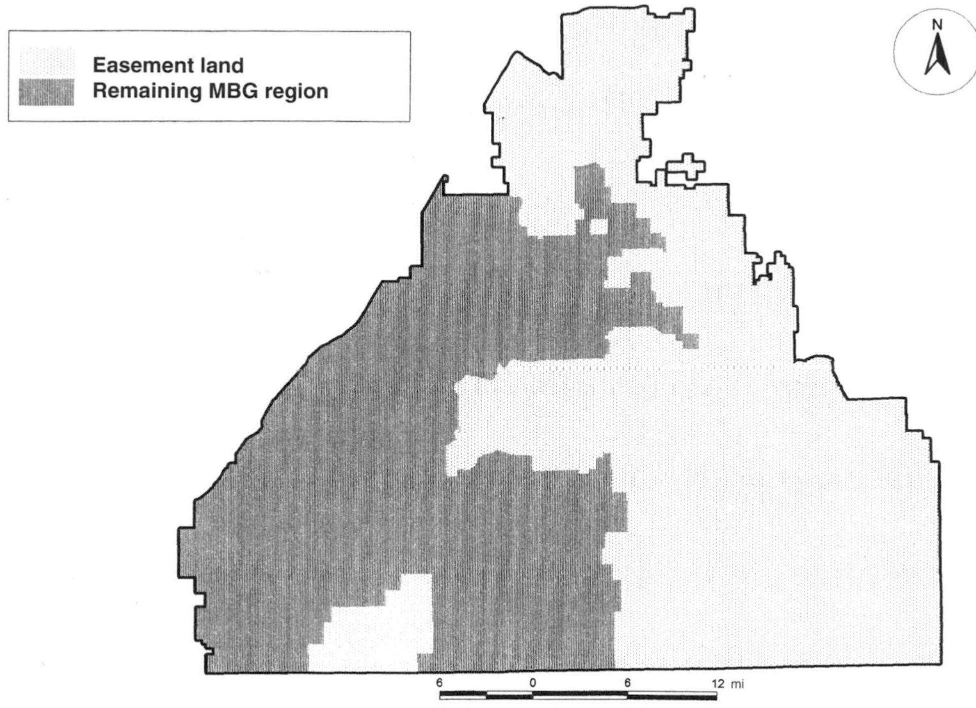

Figure B. *The distribution of permanent conservation easements on lands in the Malpai Borderlands Group region.*

forts. Nonetheless, it is apparent that the MBG has found a formula for success that has been elusive for many other similar efforts. Below are some observations from my experience after nearly a decade of involvement with the group's efforts:

- Have written goals, against which you gauge your actions and measure your success.
- Encourage and include. Do not try to force things on people but make opportunities available to them.
- Communicate, communicate, communicate.
- Provide everyone equal access to the tools of information and analysis.
- Teach and learn. There is ample opportunity to do both.
- Obtain and use the best science available.
- Don't start what you can't finish.
- Be aware that people work hardest when it is in their best interest to do so. They work hardest together when it is in their mutual best interest.

Discussion Questions

1. Many of the most difficult issues surrounding groups such as the MBG involve trade-offs between short-term costs and long-term benefits. For example, long-term landscape health in this area requires fire to eliminate woody vegetation and return to dominant grassland vegetation. However, burning also can kill individual ridge-nosed rattlesnakes, which are a federally protected species under the U.S. Endangered Species Act. Discuss how such a problem might be addressed.
2. Another trade-off in this region is the presence of cows and grazing on public lands versus elimination of the ranching lifestyle, with the likely de-

velopment of ranchettes and retirement communities on private lands. Neither land use might be perfect from an environmental perspective, and both have costs. Discuss the relative merits of such land uses, and explain how a long-term perspective might help determine how to proceed with a sustainable vision for the region.

3. Different groups involved in the Malpai region might have different interests and concerns, resulting in inherent conflicts. For example, ranchers are concerned with making a livelihood and carrying on a lifestyle, whereas scientists might be interested in collecting data and protecting endangered species. Discuss how these different interests might be balanced and addressed. Think in the context of short- and long-term interests, as well as flexibility.

PART III

The Human Dimensions

Working in Human Communities

MORE THAN A HALF-CENTURY AGO, ALDO LEOPOLD put people squarely into the conservation equation. He recognized that effective conservation requires understanding how people relate to one another and how they relate to the land. His land ethic is about how we—as individuals, communities, and organizations—take the long-term health of the land into consideration as we make decisions. The principles, concepts, and examples in this chapter build on Leopold's ideas by showing why working with the human community is essential for ecosystem management and how to do so effectively.

Many scientists and natural resource managers are uncomfortable working with people and consequently have had minimal influence in natural resource decisions. But we have defined and described ecosystem management throughout this book as an endeavor tied directly to an understanding and appreciation of human nature and value systems. Therefore, working effectively in ecosystem management requires the ability to develop durable relationships with people who may believe, think, and behave differently from us (Box 10.1).

As a society, we have also begun to realize that the task of resource conservation is too large and too important for government alone. The experiences of many successful planners and managers strongly indicate that even when agencies own and manage 70% of a watershed, they cannot manage those lands without the support and involvement of the local communities. In the eastern United States, where most land is privately owned, the successful management of public land requires the active support of adjacent private landowners. This is because of the permeability of land boundaries to ecological processes (see Chapter 9), as well as the intense interest of many people about what happens on their public lands.

The 1996 Keystone National Policy Dialogue on Ecosystem Management cites numerous examples in the United States where people are looking at the long-term effects of their local decisions on the land and on the human communities. In these and other examples, people are willing to set aside their differences and take action that goes beyond the obvious government tools of regulation, land purchase, and control (Box 10.2). To do so, we must develop

BOX 10.1

Ecosystem Management and People

Here are several examples of how people view their role in land management:

- "The federal agencies can't manage their public land—people manage the land." Jim Winder, New Mexico Rancher.
- "Departments of Natural Resources can't take care of the land, people take care of their lands. Agencies can only help local people do that. It is the person who works the land that is our greatest hope and challenge." Paul Johnson, Former Director, United States Natural Resources Conservation Service.
- "The role of stakeholders has begun to change from that of constituents to that of partners who may actually carry out the project as their own." Jim Addis, Former Natural Resources Administrator, Wisconsin Department of Natural Resources.
- "We are developing community-based solutions to coastal problems." Nina Garfield, Tiaunana Estuary Project.
- "It is people taking responsibility for their (human) community and the ecosystem that supports it. Community-based energy comes from the emotional, spiritual, and intellectual drives of people working together for a common good. Our goals are economic diversification and ecological sustainability." Riki Ott, Copper River Watershed, Alaska.

effective ways of listening, understanding, and interacting with human communities of stakeholders.

The Success Triangle

A concept called the success triangle is central to understanding how we make progress working with people. This concept asserts that successful collaboration shares three components: substance, process, and relationships (Figure 10.1). All three must be present for a successful interaction to occur and good decision to be made, but their relative importance varies with the issue and for each stakeholder involved.

Substance refers to the technical and factual content of a situation. This is the realm of the scientist and technician, for whom data collection and analysis are fulfilling and compelling. However, what is considered substantive may vary widely among different individuals and groups. Consider the debate surrounding old-growth forests in the Pacific Northwest. To a wildlife biologist, substance might be the content and probabilities associated with spotted owl populations such as MVP, minimum habitat requirements, and N_e. To the business investor, substance might be the allowable timber volume and its impact on profitability. To local loggers, substance might be pay rates, overtime availability, or health care costs. To community leaders, substance might be tax consequences to the county and municipalities, and job creation or loss. Most people want to be sure that the substance is correct, but that is not enough for a successful collaboration.

Process refers to the explicit and formal steps used in making a management decision. This is the realm of the administrator, lawyer, and special interest group. Process-oriented people watch how decisions are made to be sure that the process follows established steps and is open to all stakeholders. Almost all actions taken by groups today are governed by rules requiring public notification of meetings, open meetings, and ample time for

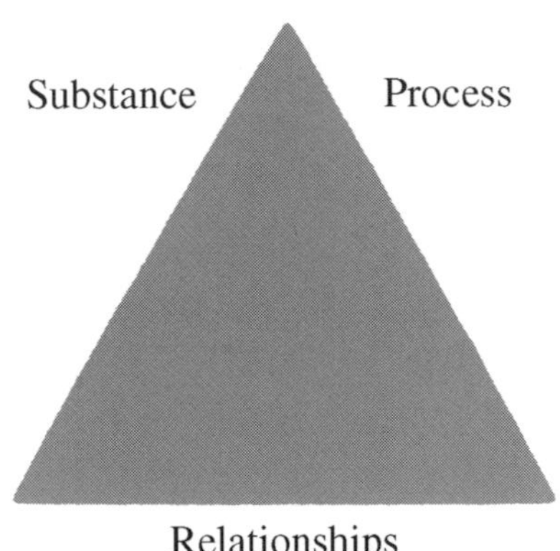

Figure 10.1. *The success triangle. Successful collaboration relies on a dynamic interplay among substantive facts, interactive processes, and interpersonal relationships. Approaches to ecosystem management that seek to blend all three tend to be more successful than other methods in working within the human community.*

BOX 10.2

The Kickapoo Valley Program

The Kickapoo River Valley of southwestern Wisconsin covers 500,000 acres of the unglaciated region just west of the Mississippi River. It is a region of steep hillsides, narrow valleys, and more than 400 miles of cold-water trout streams. Despite the terrain and its highly erodable soils, the region was heavily farmed and pastured. Soil erosion rates of more than 10 tons per acre per year and agricultural nutrient loading problems degraded streams, adjacent riparian areas, and eventually the economic basis of communities in the valley. In 1958, a Wisconsin Department of Natural Resources (DNR) report recommended that the state abandon the stream resources of the valley. But the people of the valley thought better of their region—and they took action. The result? In 1997, a national trout fishing magazine wrote: "Don't go to Montana to fish trout, go here." What happened?

A restored stream in the Kickapoo Valley, Wisconsin.

Economic hardship and ecological degradation combined to prompt local citizens to seek help from university sociologists, economists, the DNR, and several federal agencies. Their focus has been the implementation of compatible land uses and sustainable economic development while stressing environmental protection and ecological restoration. They have used community education and outreach, demonstration projects, and community-based restoration work to expand and improve a native trout fishery. They have considered their efforts to be driven not by agencies or private organizations but by "people coming together to look at their watershed from different perspectives." Participants have developed a common understanding of their objectives and believe that open, honest, and timely communication fosters mutual respect and trust. Stakeholders have exhibited a willingness to invest their time and money in shared work projects and to pursue work with an optimism and willingness to have fun. In 1998, Trout Unlimited chose this valley as its second national Home Rivers Initiative Project, supporting the continuing work through a full-time local project coordinator. Communication and community involvement have been maintained through an active newsletter, many public meetings, and local activities.

Their ongoing restoration priorities have been selected by a coordinating committee that channels federal, state, and private donations into specific work projects. Over a 1-year period (1998), they invested $335,000 from state and private donations to restore habitat in eight separate streams. Previous restoration work in one sub-watershed led to a restored trout fishery and an annual economic return estimated at more than $1,000,000 from anglers, according to a 2000 University of Wisconsin economic impact study.

After decades of poor farming practices, high soil erosion rates, and neglect of the natural landscape, people in the Kickapoo River Valley of southwestern Wisconsin chose to work together with public agencies to preserve and restore the rural character of their valley—both the natural and the human character of their landscape. Landowners, citizens, business owners, and government officials came together to share their differences and define their common purposes in relation to one another and to the natural landscape within which they live.

subsequent review and comment about proposed decisions. A fair process ensures that any decision about public resources is made or blocked openly through defined legal procedures or group processes (e.g., facilitation, negotiation, mediation, or arbitration). A well-designed public involvement process enables all interested stakeholders to hear and understand one another's concerns and needs, review facts, generate and evaluate alternatives, and then recommend a course of action. People interested in process generally believe that an open and inclusive process will ensure that the full substance of the issue is revealed and, therefore, that the best decision will emerge.

Relationships are the networks that develop among individuals with direct or indirect interest in or influence over a management decision. This is the realm of the politician, journalist, entrepreneur, and civic leader, who wish to know, trust, and have access to decision makers. Relationships are important to people who want to be understood and who demand confidence that their values and needs will be considered in decisions. Relationship building occurs outside the formal processes that collect official comments; it develops through frequent, informal, and nonspecific communication, and with a broad commitment to a community, whether localized or dispersed. People interested in relationships are good listeners who care about people, empathize, and can see the world through eyes of others. Relationships build the interpersonal trust and credibility necessary for mutual understanding and effectiveness, which leads to effective ecosystem management.

Ultimately, the success triangle is the very reason for thinking about the human dimensions of ecosystem management and being concerned with stakeholder involvement. If not for the importance of process and relationships to success, scientists and technicians could rightfully say, "We have the substance, so we are doing what needs to be done and everyone else can relax and just leave this to us." We all know the world does not work that way. Fair processes and personal relationships dictate what information is used and toward what ends; who is listened to and asked to contribute; and how seriously stakeholders, including scientists, are taken.

EXERCISE 10.1

Talk About It!

Suppose you are concerned about restoring the ecological integrity in a small sub-watershed. The sub-watershed has a mix of public forest and parklands, private agricultural lands, and developed urban/suburban areas. It suffers from poor water quality and degraded riparian habitat due to generally poor agricultural land-use practices, suburban runoff, and housing construction. A recent scientific study showed that the largest single problem, contributing 30% of the non-point-source pollution, came from one particular landowner.

Discuss the socioeconomic difficulties of ecosystem management in such a mixed-ownership watershed and the possible ways to overcome those difficulties. If the landowner contributing 30% of the pollution were a family-owned farm that was economically marginal and whose pastures and barnyard drained into the river, how would you approach the problem? Identify strategies for helping the farmer improve his or her farming practices and the people and organizations you would contact. How would your approach change if the problem were caused by storm water runoff from a wealthy golf course and condominium development?

Stakeholder Identification and Assessment

A critical aspect of successful ecosystem management is engaging and working with a broad range of stakeholders toward common goals. To do that we must be able to identify and assess the interests of these various participants.

WHO IS A STAKEHOLDER?

The most general answer is, "Anyone who wants to be!" Under an ecosystem management concept of expanded inclusiveness, a **stakeholder** is anyone who has an interest in the topic at hand and wishes to participate in decision making. Because interests vary, we suggest that stakeholders fit into one or more of five categories.

- *People who live, work, play, or worship in or*

near an ecosystem. People whose individual lives and well-being are directly connected to an ecosystem are the most obvious stakeholders. When jobs, neighborhoods, and sacred sites are at issue, people want to have a role in decisions. You could call this the "good neighbor" policy—involving local residents in an ecosystem plan is just good business. People near the site of a proposed road, timber harvest, housing development, or nature preserve will want to participate. Many local groups, from the Chamber of Commerce to the Ducks Unlimited chapter, are neighbors and stakeholders of this kind.

- *People interested in the resource, its users, its use, or its non-use.* Many people care deeply about the way natural resources are used, even if they do not live near the site under consideration. This is often called the community of interests. Some are interested in total protection of a resource for its intrinsic value or ecosystem function. They may advocate for or object to its use as a commodity (e.g., timber, grazing, irrigation water, hunting and fishing) or as an amenity (e.g., hiking, boating, photography). They also may perceive other values in the resources, such as therapeutic recreation, spiritual inspiration, or solitude. Many national and local NGOs, such as the Native Plants Society or the Farm Bureau, fit into this category.
- *People interested in the processes used to make decisions.* Though often overlooked, some stakeholders care deeply that all the legal requirements are met before a decision is made. As mentioned earlier, their interest is based on the belief that the right decisions will be made only if the right process is used—and they may use legal means to assure this. For example, an agency's failure to follow all the procedures for filing an environmental impact statement can result in an injunction to stop the work, regardless of its merit or public support for the effort. Groups such as Environmental Defense and Common Cause are likely to be stakeholders of this kind.
- *People who pay the bills.* Most people are concerned about how their money is used. For ecosystem management, this category of stakeholders includes taxpayers, some of whom will not believe that efforts spent on ecological projects are more important than those for public schools or improved roads. Traditional funders of natural resource agencies, such as hunting and fishing license buyers, have great interest in how their fees are being used and how the lands purchased with their earlier fees are being managed. Also, many private foundations, such as the Pew Foundation and the Rockefeller Fund, as well as NGOs, invest significantly in ecosystem management and, therefore, are key stakeholders.
- *People who represent citizens or are legally responsible for public resources.* These stakeholders include elected and appointed officials and agency staff members who have the legal authority to protect, preserve, and enhance natural resources. An important element of this legal responsibility involves Native American rights. Federal land management agencies, especially those in the U.S. Department of the Interior, have special responsibilities toward the lands, resources, and rights held by Native Americans. These are called **trust responsibilities**, meaning that the U.S. government holds lands and resources "in trust" for Native Americans and ensures that funds are maintained and expended to enhance the conservation of those lands (Box 10.3).

PRINCIPLES OF STAKEHOLDER INVOLVEMENT

Including all potential stakeholders in every decision or action, of course, is impossible. But it is important that all interested stakeholders or their chosen representatives (elected officials, designated spokespersons, or NGO officers) are invited and participate in ecosystem management. This is called the principle of inclusivity. Inclusivity may be troubling to some people because it means that stakeholders with opposing or conflíciting ideas are asked to participate. Others may object because subjective feelings, intuition, or traditional beliefs are as welcome as rigorous scientific studies or agency positions. The challenge of effective stakeholder involvement is to help people with different views recognize and understand their common interest in working together.

Another principle of stakeholder involvement is

BOX 10.3

Trust Responsibilities of Federal Agencies

Trust responsibilities arise from treaties, statutes, executive orders, judicial decisions, and other legal instruments that the U.S. government has entered into with Native American tribes or has otherwise issued. Such legal arrangements, which often date back to the mid-1800s, reserve lands and rights for Native Americans and commit the government to act on behalf of both tribes and their individual members. The legal documents also establish a government-to-government relationship which acknowledges that Native American tribes (555 tribes are recognized by the U.S.) are sovereign nations residing within the U.S. and limits the jurisdiction of the federal government in regard to tribes.

An example of the special trust responsibility is Secretarial Order #3206, issued by the Secretaries of the Interior and Commerce on June 5, 1997, defining the federal-tribal trust responsibilities with regard to the Endangered Species Act. The order commits the agencies to "carry out their responsibilities under the Act in a manner that harmonizes the Federal trust responsibility to tribes, tribal sovereignty, and statutory missions of the Departments, and that strives to ensure that Indian tribes do not bear a disproportionate burden for the conservation of listed species." The order identifies five principles to guide the actions of the departments:

Principle 1: The departments shall work directly with Indian tribes on a government-to-government basis to promote healthy ecosystems. This principle acknowledges that Indian tribes have inherent powers to make and enforce laws, administer justice, and manage and control their natural resources. It commits the departments to consult with and seek the participation of Native American tribes whenever an action might affect tribal resources, rights, or lands.

Principle 2: The departments shall recognize that Indian lands are not subject to the same controls as federal public lands. This principle implements the essential point that Native American lands—95 million acres in the U.S.—are not federal lands, but are held in trust for the tribes by the federal government. When the federal government implements law, policy, or regulation on tribal lands, it must consider both the purpose of the actions and the impact on other aspects of tribal life, including self-determination.

Principle 3: The departments shall assist Indian tribes in developing and expanding tribal programs so that healthy ecosystems are promoted and conservation restrictions are unnecessary. This principle commits the departments to helping tribes proactively by providing technical assistance and information, working cooperatively to address concerns for sensitive species and habitats, and working together to find ways to sustain the species without unduly restricting tribal activities.

Principle 4: The departments shall be sensitive to Indian culture, religion, and spirituality This principle requires the federal government to give wide latitude to Native American tribes in their noncommercial use of sacred plants and animals for medicinal treatments and cultural and religious purposes.

Principle 5: The departments shall make available to Indian tribes information related to tribal trust resources and Indian lands, and, to facilitate the mutual exchange of information, shall strive to protect sensitive tribal information from disclosure. This principle requires the departments to provide tribes with all the information they have about tribal lands and resources. At the same time, the departments must not disclose to others information they hold in trust on behalf of the tribes.

self-selection, which recognizes that stakeholders choose their own levels of involvement based on their interest and comfort (or discomfort) with how the issue is being addressed. This range of involvement can be visualized as **stakeholder orbits** (Figure 10.2). Stakeholders with a deep interest in an issue revolve in a low orbit, close to the center of the issue; they come up over the issue's horizon frequently, and they require constant attention. Those with less interest in the issue revolve in higher orbits, interacting with other stakeholders and decision makers less frequently and less intensely.

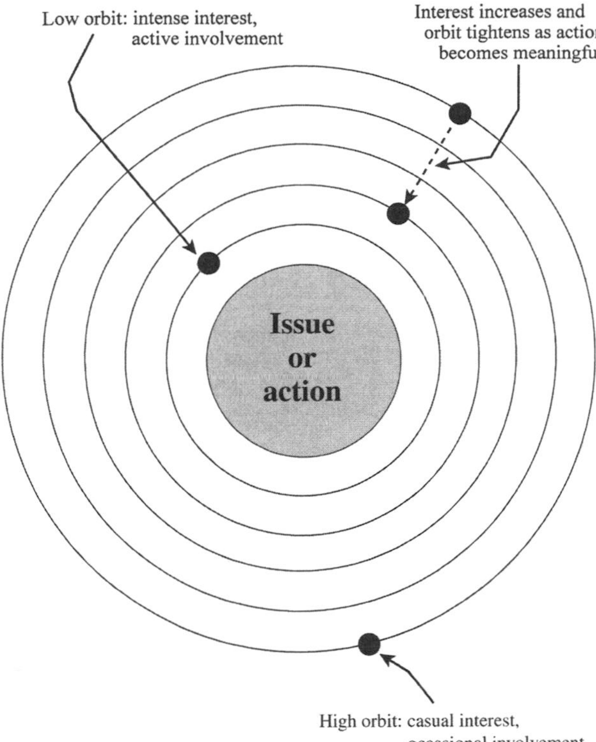

Figure 10.2. *Like satellites in orbit around Earth, stakeholders choose to place themselves at various distances from the center of an ecosystem issue. Low-orbit stakeholders express a deep interest in the issue and require frequent attention. Higher-orbit stakeholders are less interested and require less attention.*

Consistent with the principle of self-selectivity, stakeholders will change their orbits as their perception of the issue's impact on them changes. For example, a stakeholder may express little interest when an issue is first identified (e.g., loggerhead shrike populations are declining because of loss of grasslands), but may interact energetically when specific actions are proposed to address the issue (e.g., a proposal to convert harvested forestland to grassland). Other stakeholders may be very active in the early stages of identifying an issue, then move to a less active orbit when they perceive that they have been heard and their interest has been incorporated into the plans.

A third principle of public involvement is diversity of representation. The people involved should be a cross-sectional representation of the demography and interests of the community. Common sense tells us it is harder to work with strangers than with people we know and already trust. But this hard work will be necessary in ecosystem management, because stakeholders have many and different values (Box 10.4). A stakeholder involvement effort to examine forest management practices that includes only foresters is flawed; the participants should include recreationists, hunters, economists, local landowners, members of the forest products industry, preservationists, ecosystem scientists, and many others. Similarly, an effort to address community growth patterns and planning through a committee of eight representatives of land trusts and one developer—or the reverse—lacks balance and undoubtedly will backfire when recommendations are taken to the larger community.

People who are effective in community-based approaches owe part of their success to their willingness to get to know their local community. They work hard to understand the human ecology of their community, its stakeholders, and their knowledge, concerns, and needs. One particularly effective agency resource manager in Wisconsin invested 2500 hours getting to know and understand the people and resources of the watershed *before* he began to work on an ecological restoration issue.

STAKEHOLDER ANALYSIS

Once the major stakeholders for an ecosystem management activity have been identified, it may be useful to assess their probable relationship to the activity. We suggest gathering the following types of information from at least the low-orbit stakeholders:

1. *Primary information.* The name of the individual or group; affiliation, principal members/leaders, contact information.
2. *General characteristics.* The formal or informal mission and interests; related activities in other places or on other projects; size and scope of influence (e.g., membership, employment, land area); formal or informal authority over the activity.

BOX 10.4

Values of Americans Toward Natural Resources

Since the mid-1970s, Stephen Kellert of Yale University has been studying the ways that Americans respond to nature. His research began by looking at the ways Americans used and thought about animals, a subject now referred to as **biophilia**, or the biological dependence on and affinity that people have to the natural world. These nine values represent distinct responses, but most people probably hold more than one of these at the same time. In fact, most of us probably value nature in all these ways, but with differing importance.

- *Aesthetic: A value for the physical attraction and beauty of nature.* The aesthetic value reflects the inspiration and instruction provided by nature, such as the renewing aspect of watching a sunset or viewing a beautiful landscape. People who hang landscape paintings in their homes are expressing the aesthetic value.
- *Dominionistic: A value for the ability to master and control the natural world.* The dominionistic value sees enhanced physical and mental fitness through subduing nature, and it strengthens one's security and confidence. People who hang hunting or fishing trophies in their homes are expressing the dominionistic value.
- *Humanistic: A strong affection for and emotional attachment to the natural world.* The humanistic value expresses the human need to develop a sense of connection and kinship with nature and living objects. People who keep pets in their homes—especially more exotic pets, such as reptiles and birds—or who visit zoos are expressing the humanistic value.
- *Moralistic: A spiritual and moral affinity for the natural world.* The moralistic value assigns near equality to humans and to individual animals and plants, as well as to species and ecosystems. This value detests the destruction of nature and the mistreatment of animals and plants. People who are against cruelty and even the use of animals (e.g., for clothing and food) are expressing the moralistic value.
- *Naturalistic: A desire to experience the natural world directly.* The naturalistic value drives people to experience nature for deriving physical, mental, and emotional renewal. Time spent on the trail or on the water, especially in remote settings with few other people around, satisfies the naturalistic value. People who spend their leisure time—and perhaps their careers—in the outdoors are responding to the naturalistic value.
- *Negativistic: Fear, avoidance, and a disdain of nature.* The negativistic value regards nature as dangerous and uncertain, raising fear and anxiety. Although this may not seem to be a legitimate "value," natural resource managers must recognize that many stakeholders view nature this way. People who prefer to live and play in large cities or other highly controlled environments are expressing a negativistic value.
- *Scientific: An interest in understanding how nature works.* The scientific value recognizes that nature is a source of wonder and that seeking to understand nature will help humans understand themselves better. People who like to visit natural history museums or watch educational television programs about nature are expressing the scientific value.
- *Symbolic: Nature as a source of imagination and communication.* The symbolic value of nature is its ability to give us examples of how we might think and act. Fairy tales, children's stories, totems, legends, and religious parables all use nature to describe how the world works, to teach lessons about behavior, and to stimulate our higher purposes (e.g., the bald eagle symbolizes American national pride and patriotism).
- *Utilitarian: The material benefits of natural resources.* The utilitarian value expresses the physical comfort and security that we derive from using nature for food, clothing, shelter, medicine, and all other products and services. This, of course, is the most basic and pervasive value of nature.

Think for a moment about your own values with regard to nature. Are you a person who loves to "be in the woods"? Do you love animals and have always had pets? Do you enjoy the thrill of fishing and hunting? Now, consider whether or not your view of nature and the values you hold most dearly are representative of the larger population where you live. Probably not. Students of natural resources often have values that are much different than the rest of society. Leaders in ecosystem management need to be prepared to acknowledge, accept, and respect the values and views of others—a difficult task, especially for people who are deeply committed to nature.

3. *Interests.* The future outcome likely to be desired by the stakeholder, in both the short term and long term.
4. *Probable levels of involvement.* The ways in which the stakeholder may wish to participate in the activity.
5. *Stakeholder needs.* What the stakeholder will require in order to participate, ranging from basic information to an invitation to take a leadership role.

EXERCISE 10.2

Collaborate on It!

Consider the following proposals for development (one for each scenario):

- ROLE Model: Creation of the Round-About Bikeway.
- SnowPACT: Buffalo ranching by the Semak Nation.
- PDQ Revival: Golf course development near the Muir Wildlife Refuge.

Think about the many different stakeholders in the scenario, their different value systems, and their relative power and influence within the community. Using the concept of stakeholder orbits, select and place ten different stakeholders in high, medium, or low orbits, based on their interest in the issue. Select one or more stakeholders who are in low or medium orbits, and perform a stakeholder assessment. Discuss the assessment with others, and add their ideas to your own.

Imagine you are the manager of the wildlife refuge in the scenario and that you have assembled your staff into a work group to assess how you might help bring each stakeholder group into active and positive participation in the ecosystem work of the community.

1. What other interests does each stakeholder have with which you could help them?
2. What common interests or values does your staff share with the stakeholders that could be the basis of a long-term relationship?
3. What other stakeholders do they interact with, and how could those relationships be used to help address other ecosystem issues in the community?

LEVELS OF INVOLVEMENT

The amount of stakeholder involvement appropriate for any particular activity will depend on the nature of the activity and the interests of the stakeholders. Although there are no definitive rules for selecting level of involvement, experience shows that more involvement is needed for activities that are:

- *Special,* rather than routine; for example, routine road maintenance on a wildlife preserve needs little involvement, but plans to close an existing road to local traffic need extensive involvement.
- *Major,* rather than incremental; for example, changing from paper to electronic bids for timber harvests needs little involvement, but changing the requirements for erosion and sedimentation plans needs extensive involvement.
- *Required,* rather than voluntary; for example, asking visitors at a wildlife viewing platform to turn off their cellular telephones needs little involvement, but prohibiting users from riding mountain bikes on hiking trails needs extensive involvement.
- *Controversial,* rather than unanimous; for example, controlling mosquitoes needs little involvement, but plans for controlling wolves need extensive involvement.

Levels of involvement can be ranked according to their extent, intensity, and required commitment into five major types.

NO INVOLVEMENT. No stakeholder involvement is appropriate in some cases. During emergencies, such as a forest fire, responsible individuals need to act promptly to protect life and property, without consultation about other concerns that stakeholders might have. Similarly, regulatory actions that are firmly grounded in case law that has withstood legal challenge do not require stakeholder involvement (e.g., stopping someone who is spraying illegal toxic chemicals does not need discussion). Recognize, however, that our ability to proceed in these cases has probably been established through a rule-setting or procedure-setting process that included extensive stakeholder involvement in the past.

NOTIFICATION. Notification is a form of "good neighbor" policy in which an individual, organization, or agency, although having the legal authority to act without consulting stakeholders, chooses to inform them of planned activities. For example, a decision by a land trust to conduct timber harvest on a portion of its holdings would probably be received better by neighbors, members, and the press if they were informed about the rationale, scope, and legality of the harvest before it began. Such communication fosters good community relationships and builds trust.

REVIEW AND COMMENT. Opportunities for **review and comment** are used to seek stakeholder reaction to a proposed activity. By exposing their ideas to open review, decision makers signal that, although they may have narrowed the possible options somewhat, they are still open to improving their plans. If there is little negative reaction to the proposal, it is likely to proceed. If the reaction is strongly negative, however, more work is needed. Stakeholders may just need more information before they feel comfortable with the proposal, or they may be deeply opposed to it.

Review and comment are also often used to gauge stakeholder preference among several options for action. This is the required practice, for example, for U.S. government environmental impact statements, and it is good practice for a decision in which the options are fairly limited but their consequences are quite different. For example, if a community group were considering where to conduct a stream restoration project, the available sites might be well defined (headwater area, middle of town, below a mill pond dam), but the consequences might differ widely in ecological, sociological, and institutional ways. Most techniques in public involvement have been created to conduct reviews and comments, including public meetings, workshops, surveys, press reports, media stories, and on-site visits.

CONSULTATION. Consultation is the process of requesting substantive input from stakeholders at the early stages of thinking about proposed actions. Leaders may take a crude proposal to small groups of stakeholders and ask for their help in developing it more fully. At the extreme, leaders may ask stakeholders themselves to identify the issues, generate alternatives, and evaluate them before any decisions are made (Box 10.5). For example, when deciding what projects to undertake to combat the spread of non-native invasive plants, a community group might consult with stakeholders to determine their priorities among many available strategies, from educational programs to direct eradication.

Decisions about how a limited resource will be allocated among several competing interests can benefit from this type of consultation. Active consultation can be a starting point from which to build community support for long-term commitments to ecosystem restoration and management. Consultation may also lead diverse stakeholders to commit to specific actions in support of ecosystem management activities.

LIMITED PARTNERSHIPS. A limited partnership is an agreement among stakeholders to pursue mutual goals with shared assets of time, money, equipment, or authority. Also called *participation* in many management texts, in a limited partnership

EXERCISE 10.3

Collaborate on It!

Consider the following issues, one for each scenario:

- ROLE Model: Managing the resources to protect and enhance the lily bush.
- SnowPACT: Managing the lands and waters of Cigueña Marsh.
- PDQ Revival: Managing the Swamp Fox Wildlife Management Area.

For your scenario, select an action, proposal, or decision that might be made with each of the levels of stakeholder involvement described in the text (no involvement, notification, review and comment, consultation, and limited partnership). For each level and action, describe the rationale for your decision, which stakeholders you would ask to participate initially, and how you would go about the process.

BOX 10.5

Nominal Group Technique

Producing the desired product from a group meeting can be a tricky process. People like to talk, discussions get off-track, an initial idea is pursued for too long—and then it is time to adjourn! The **nominal group technique** can help keep a meeting focused on its task. The group follows a series of carefully defined and monitored steps to ensure that the meeting ends with a product. The general process is as follows.

1. *Posing a trigger question*. The facilitator stimulates and directs thinking by stating a previously agreed question or statement to which the participants respond. The trigger question is often in the form of a sentence completion.
   - We will know that our ecosystem is sustainable when we see . . .
   - A desirable riparian area is characterized by . . .
   - Our community's biggest needs for environmental education are . . .
2. *Silent generation*. Each participant is asked to write a series of responses to the trigger question, individually and silently. Each response should have only one answer, rather than several. Thus, in response to the last trigger question above, a person might write these responses:
   - . . . increasing teachers' knowledge of ecological principles.
   - . . . teaching the business community about recycling.
   - . . . adding 4-H programs about forestry.
3. *Round-robin listing*. The facilitator then asks each person to read his or her most important response—just one. The facilitator writes the response on a flip chart. No discussion is allowed, except to clarify what the response actually means. The facilitator continues around the room until everyone's ideas are recorded. It is very important to get only one idea at a time from each participant, so everyone has a chance to contribute. A list of 30–50 items usually results.
4. *Clarification and combination*. The facilitator reads each response in order, ensuring that everyone understands it before going on. Still, no discussion of the merits of the response is permitted. As the facilitator proceeds down the list, he or she asks if each item might be combined with an earlier response (only an *earlier* item; if participants start looking ahead, chaos breaks out!). If the originators of responses agree, the responses are combined. This process ensures that minor items are incorporated into major items and that very similar responses are covered only once. It reduces the list to 20–30 items. The facilitator may need to stop the group from reducing the list to just a few items that are too general to be useful.
5. *Voting*. Each participant is then given a small number of votes, usually 3–5, to assign to the items he or she thinks are most important. Participants place stickers next to the items they think are most important. Only 1 vote is allowed per item. The facilitator tallies the votes and may rewrite the items in order, from most to fewest votes.
6. *Group discussion of outcome*. The group discusses how they wish to consider the outcome. Most of the votes will usually be concentrated on a few responses, indicating the group's highest priorities or interests. Participants might also wish to combine some of the other responses that received only a few votes because they were too specific. This list can then be the basis of subsequent decisions about the group's priorities and interests.

stakeholders are given the formal authority to actually decide what will be done. Most of the exam ples used in this book are a form of limited partnership in which NGOs, citizens, and government agencies voluntarily share responsibility for protecting or restoring an ecosystem (Box 10.6). This might mean that an agency and NGO, for example, each assigns one employee to a task force whose work is directed by a community committee; in a true partnership, the contributing organizations agree not to interfere with how the employees spend their time. Typically, however, the govern-

BOX 10.6

The Applegate Partnership

During the 1990s, a 500,000-acre valley nestled between the Coastal and Cascade Ranges of southern Oregon was the scene of an intense emotional and economic controversy over the future of old-growth forest. The Applegate Valley is heavily forested and contains a significant population of the endangered northern spotted owl, in addition to several other endangered or threatened species. The area is ecologically diverse and includes habitats at elevations ranging from 700 to 7000 feet. More than 70% of the area is federally owned, 10% of the land is managed as commercial forest, and the remaining 20% is owned by private citizens, 12,000 of whom live in the valley. Logging, limited ranching, and small farming comprise the primary economic base of the area.

The Applegate Partnership was begun in 1992 when a logger and an environmentalist decided to resolve the divisive issue of timber management through community involvement rather than through the courts. The community had split into pro-preservation (anti-logging) and pro-logging (anti-environmental) factions. The listing of the spotted owl as an endangered species exacerbated the issue because of the threat to halt all logging. From the beginning of the partnership, the participants sought input from diverse groups within the community, and they used a professional facilitator to structure meetings and dialog sessions for the participants. Their meetings were conducted using a set of ground rules that included a formal agenda, rules of conduct for participants (e.g., common courtesy and respectful listening), a defined process for making group decisions, and agreement about how participants would interact with the news media. These rules helped to balance power differences among the participants and ensured that each had an equal and fair opportunity to speak and have their concerns heard. The formal meeting structure helped build mutual respect and a sense of trust among the participants.

The sixty original participants elected a nine-member board to resolve differences and recommend actions. Community-wide involvement, a regular newsletter to all residents, and open meetings characterize how they operate. Instead of having a single leader within the board, they make decisions through general agreement. Their membership includes loggers, ranchers, environmentalists; representatives from the Sierra Club, the Audubon Society, the Bureau of Land Management, the Farm Bureau, the U.S. Forest Service, the USFWS, and community leaders and private citizens. Their motto is: "Practice trust— them is us." Their goals include building community-wide agreement about responsible resource extraction, protecting the long-term ecological health of the forest, and providing for a stable local economy.

Several early agreements included no clearcutting of timber, no routine use of pesticides in the forest, and a desire to manage for a diverse and healthy forest. Other actions included restoration of riparian areas, tree planting, selective thinning and logging, controlled burning, maintaining and recreating old-growth forest, and the application of natural resource science to guide their efforts. Their work, and the sense of community they have built in the valley, have drawn national attention. Their criteria for selecting projects to pursue include ecological soundness, social acceptability, and economic viability.

---

ment members retain their authority to regulate and usually assume the bulk of the liability for the partnership (thus the word "limited" in describing the partnership). This represents the highest degree of stakeholder involvement and will require the largest investments of time, money, and energy by each partner. Partnerships also require a more substantive commitment—the willingness to abide by the collective decision of the partners.

Techniques for Stakeholder Involvement

In its simplest form, stakeholder involvement is a systematic approach to working with people that emphasizes openness and inclusivity. All interested parties can share their concerns and ideas and can help develop common solutions to community issues. The techniques available for stakeholder involvement range from simple conversations to highly formal public meetings. A description of some of these techniques, modified from James Creighton's classic 1981 public involvement manual, is included here, arranged in order from simple to complex.

- *Interviews with key informants.* A quick assessment of public sentiment can be completed by informally interviewing key stakeholders in a community. Key informants should be chosen to represent a wide range of the community, rather than a subset who would tend to respond similarly. Interviews with key informants can provide the political context for a project, illustrate how various stakeholders might want to participate, identify other groups or individuals that need to be interviewed, and begin building relationships in the community. However, poor interviewing skills, or the appearance of advocacy for a particular outcome, can derail a project at the very beginning.
- *Establishing a local office.* Opening a local office in a community is a mechanism for encouraging regular and informal communication flow with citizens and community leaders. Local offices are particularly useful where residents and other interested people may be present for only limited times (such as resort communities) and would therefore be unable to participate in one-time public meetings or other events. Local offices should be highly visible (storefronts in downtown areas or shopping malls), providing an easy way for all citizens to learn about the project at their convenience. Such offices also demonstrate that the organizers are serious about the project, and provide a mechanism for participants to learn much more about the community.
- *Electronic communication.* Establishing an anonymous method for stakeholders to learn about a project and offer their own views can be extremely effective. Traditional methods are telephone hot lines, including 800 numbers, but Internet-based communication strategies are equally valuable today. A Web site for a project, including basic information, personnel contacts, updates, opportunities for viewers to respond, chat rooms, and live interview times with project leaders and specialists, should be included. Electronic communication has many advantages, but it should be used as a supplement to the face-to-face communication that many stakeholders desire.
- *Displays and exhibits at local events.* Stakeholder involvement often can occur at other community events. County fairs, community "days," special events at shopping malls, and similar occasions provide opportunities for project representatives to show, tell, and listen. Such events bring a wider variety of individuals than those who might attend a project-specific event, providing a method for working with the broadest possible segment of the community.
- *Informal meetings with community groups.* One of the best methods for building understanding and gaining input about a project is to visit with community groups. Civic organizations (e.g., Rotary, League of Women Voters); neighborhood groups (e.g., homeowner associations, newcomers groups); special interest groups (e.g., county ranchers association, birding clubs); and private groups (e.g., business clubs, golf course boards of directors) provide the opportunity to explain a project and receive feedback regarding views of particular stakeholders. Informal meetings must be presented as just that—chances to converse with interested individuals—rather than as decision-making meetings or substitutes for broader and open stakeholder involvement.
- *Focus groups.* A **focus group** is a highly directed meeting of selected participants who are homogeneous demographically or in other important characteristics. A typical focus group consists of 5–20 individuals who meet once for a few hours to discuss their feelings about particular projects or ideas. The group is led by a facilitator who triggers the discussion with questions; the conversation is recorded, and the general views of the group are assessed through analysis of the conversation. Focus groups are designed to be homogeneous, so that participants can feel comfortable about airing their views in the company of people who are likely to think similarly. Consequently, a series of focus groups would be needed to represent a community fully.
- *Workshops.* A **workshop** is a small meeting (i.e., 15–25 participants) that usually lasts several

hours and is intended to generate a specific output. Outputs are usually sets of priorities for programs, criteria that should be used to judge projects, or a listing of the likely consequences of a project. Workshop participants are chosen to represent a cross section of stakeholders, and they are specifically invited to attend, so that the size of the group can be controlled. To be effective, workshops must use facilitators, established processes (e.g., the nominal group technique), and clear ground rules.

Workshops generally have four major elements:

1. An *orientation*, in which the purpose and background are reviewed.
2. A *group activity,* in which the participants collaborate to develop their thoughts.
3. A *reporting* of the outcomes of the group activities.
4. An overall *discussion and evaluation* of the outcome, to be sure that it is representative of the group.

• *Charrettes.* The most intensive kind of workshop is the **charrette**, a form of retreat in which important and deeply interested stakeholders spend one or more days in focused deliberation on a topic. The desired outcome of a charrette is a formal plan or position on which all participants can agree. Charrettes can be very effective at resolving conflicts among stakeholders, but they are also highly charged events that are likely to work only when major stakeholders agree that such a format is their best, and perhaps only, chance to emerge with a common solution.

• *Town meetings and other large-meeting formats.* Large meetings provide the opportunity for all interested stakeholders to listen to experts, share their own views, and listen to the views of others. The **town meeting** is one example of a large-meeting format in which highly active participation by the attendees is expected. Other formats, including panel discussions and briefings by project representatives, provide for more organized sessions aimed at sharing information. Large meetings also can be operated to resemble small meetings by beginning with a general session and then breaking into smaller discussion groups led by facilitators.

• *Public meetings.* **Public meetings** are the most formal and complex kind of stakeholder involvement. A public meeting is typically used, when required by law or regulation, by a government agency or quasi-government group (e.g., a group operating with a government grant) to make formal decisions. The meetings involve a hearing officer, requirements for public notices well before the meetings, and a formal public record of the proceedings. Although public meetings are the most common form of stakeholder involvement used by government agencies, they are now considered to be less useful than other forms because they tend to become either ritual (i.e., agencies go through the motions, but offer little real opportunity for citizens to participate) or confrontational (i.e., factions take the opportunity to voice their views in extreme and emotionally charged ways).

A few words of caution are needed about these techniques for stakeholder involvement. First, working with stakeholders is difficult, and it requires training and experience in order to be effective. This is particularly true for ecosystem management, where stakeholders are often considering new ideas in new contexts and settings. Therefore, we recommend extensive training before conducting stakeholder events.

Second, effective stakeholder involvement almost always requires a professional facilitator. Facilitators are experts at "process"; their job is to make the event proceed as planned and to deliver the desired product. It is equally important that the facilitator be external to the group or project, rather than a trained person who is also a stakeholder. Facilitators must be able to remain neutral about the content of the topic, and they must have the freedom to control the stakeholders. For example, a stakeholder acting as a facilitator might have difficulty telling a disorderly participant to obey the ground rules, whereas an external facilitator could do so without concern for later repercussions.

Third, stakeholder involvement works best when it is highly flexible. Although a technique may have worked in a similar situation earlier, it might not work the same way next time. And although a technique may have a standard format, any particular situation may require instant adaptation—another reason for using an experienced facilitator. Accomplishing what the stakeholders want is more important than remaining true to any particular technique or format.

EXERCISE 10.4

Collaborate on It!

Consider the following issues, one for each scenario:

- ROLE Model: Removal of the Northeast Power Company Dam.
- SnowPACT: Zoning the Red Cliff escarpment into areas open and closed to climbing.
- PDQ Revival: Relocation of the Camp Fraser waste dump.

For your scenario, describe 5–10 groups of stakeholders that you would put into focus groups, remembering that each focus group should be homogeneous with regard to some important characteristic of the issue. For each focus group, list several questions you would pose to get their discussion started.

Keys to Successful Collaboration

How can stakeholders interact in a productive way that builds a long-term relationship of openness, mutual respect, and trust? And how can they work together so that they achieve the real purpose of ecosystem management—a sustainable environment, economy, and community? Collaboration is not easy (Table 10.1), largely because it requires a commitment to change. Unless people decide they cannot achieve a better outcome without both change and collaboration, there will be no reason to work together. Collaboration often begins when some threat—a natural disaster, the loss of a major employer, a legal challenge—appears in a local community. Once such a defining moment has occurred, however, the success or failure of the subsequent response will depend on how leaders and citizens handle the situation. We offer the following ideas for ensuring success.

- *Seek to understand others.* Successful collaboration depends on the willingness of participants to understand a situation from all perspectives. This is one of Stephen Covey's habits of effective people—they "seek first to understand, and then to be understood." The key is to imagine how you would think and feel if you were in another person's position. This helps each participant appreciate and accept the needs and views of others. A logger and a preservationist openly talking and listening to each other for the first time may seem like a small thing, yet these small steps are significant (Box 10.7).
- *Listen empathically.* To understand how others see a situation, you must be an empathic listener. This means that you listen from the other person's frame of reference (what he or she is feeling), rather than from your own. We all tend to think we are good listeners, but our individual behavior often is not very effective. For example, when someone else is talking, do you listen carefully to what they are saying, or are you concentrating on what you will say in response, formulating your arguments and waiting to pounce at the first available break in the conversation? Effective listening is an important part of communication, and therefore of collaboration.

The elements of active listening are (1) a focus on the other person, rather than on yourself; (2) absence of judgment about what the other person is communicating; (3) absence of personal distractions, like doodling or excessive note taking; (4) concentrating on the meaning the speaker is trying to convey, rather than picking apart each point, word choice, or speaking ability; and (5) patience while the other person presents his or her message (Box 10.8).

- *Use many methods for communication.* Effective communicators—and, therefore, effective collaborators—do not expect their messages to be heard the first time, using just one technique. They repeat their messages many times and in different ways. They convey messages through words, pictures, stories, analogies, personal testimonials, demonstrations, field trips, one-on-one meetings in

Table 10.1. *Challenges and Approaches to Building Collaboration*

Common Challenges	Example	Common Approaches
Complacency or a lack of a visible crisis or opportunity.	An ecologist notes an acute decline in the population of endangered salamanders but fails to create community awareness until a development proposal is stopped by a lawsuit brought under the Endangered Species Act.	Collect and share information broadly. Foster responsibility and ownership of the problem and commitment to solutions among the stakeholders. Regulatory threats may compel people to act before legal action begins.
Lack of a compelling vision for the community or failure to effectively or adequately communicate the vision to a spectrum of interests.	A canoe club member talks to Trout Unlimited about why TU should support a proposal to remove a dam to improve canoeing. He fails to link his issue to TU's interest in habitat improvement.	Build on common ground and existing interests within the community. Think broadly, and identify mutual goals.
Lack of a sufficiently powerful or knowledgeable guiding coalition.	Three landowners develop their own proposal to restore a wetland complex on private and public land and then demand that the local community use tax dollars to pay for it. The local commission turns them down.	Create opportunities for interaction among diverse interests. Approach the problem from a holistic perspective that integrates interests and aligns knowledge, assets, legal authority, and actions. Build alliances. Recognize and reward openness and honesty.
Allowing obstacles (challenges) to become absolute barriers.	During a public meeting to address land use, agricultural non-point pollution, and municipal storm water runoff, an influential mayor says, "Your idea will never work, and I won't support it!" Because it is the mayor, others quickly nod their heads in agreement.	Use neutral facilitators, well-run meetings, and a set of participant ground rules to balance power differences among stakeholders and work through difficult or divisive issues. Focus on understanding problems and generating multiple solutions.
Failure to create short-term successes.	A coalition of local interests sets a goal "to make our community sustainable by 2050" but fails to generate much attention or action.	Produce tangible results early in the process. Set realistic objectives, and celebrate small successes. Share the credit, and recognize all the partners.
Declaring success too soon.	Biologists measure a 35% increase in the number of salmon returning to spawn this year, and the headline reads, "Salmon recovery assured, restoration a success."	Recognize and communicate uncertainty during planning, set realistic expectations about stochasitic variability, and monitor results jointly with stakeholders.
Failing to institutionalize the change.	After a successful 3-month effort to restore 10 acres of natural prairie, a community coalition dissolves without a plan for managing the prairie.	Build community and agency capacity to organize, initiate, and sustain a collaborative effort. Establish and commit to long-term and short-term agendas, often with support from government.

Source: Common challenges adapted from Kotter, 1996. Common approaches adapted from Fisher and Ury, 1981; Keystone Dialog, 1996; Senge et al., 1994; Wondolleck and Yaffee, 2000.

BOX 10.7

Logging Versus Spotted Owls?

An ecologist who argues against logging and for the protection of spotted owls, and a landowner whose husband works for the local timber mill, find themselves on opposite sides of the timber management issue. The ecologist attempts to persuade the woman with facts about the intrinsic value of the spotted owl, its value as a component of biodiversity, and the importance of its old-growth forest habitat to the hydrology of the valley. His case is brilliantly presented and illustrated with slides, data, and GIS displays.

The woman replies, "You don't care about us. All you care about are owls! If you want to save them, put them in a zoo!" Her response is delivered with conviction, deep concern in her voice, and fire in her eyes.

The moment is ripe for argument. Or is it? Under these circumstances, more data and facts about ecosystem protection are likely to fall on deaf ears. Defensive behavior by the ecologist is also likely to be unproductive. But what if the ecologist responded with, "Obviously this is very important to you. Can you help me understand what your concerns are?" If spoken with genuine empathy and honesty, this approach may evoke a response that reveals an underlying and emotional concern.

"You want to put the mill out of business. Without that job and the extra money it brings to my family, I'm concerned that we'll loose our farm, and it's been in my family for five generations" replied the woman.

In this exchange, the woman's concerns are about the loss of her family farm. Logging is one way for her family to maintain an outside income, which they need to keep their farm. Their problem is financial—not spotted owls or forest protection *per se*. The ecologist probably does not want this family to loose their farm either; their economic failure is not the ecologist's goal. For him, the owl may be part of a bigger issue of protecting large contiguous blocks of old-growth forest. Discussions that focus solely on the owl will probably fail to address either person's underlying concerns. Each party can benefit by better understanding the "truths" and real concerns of the other. Alternatives that fail to address the farm family's economic and personal concerns (and those of other such individuals) will prove unfeasible from their standpoint. Alternatives that fail to provide for the long-term survival of the owl and the protection of old-growth forest will be equally unacceptable to the ecologist and endangered species law. In this example, alternatives that provide other forms of security for the farm and protect old-growth forest offer the potential of a collaborative win-win solution for this pair of stakeholders.

people's homes, newsletters, press reports, radio and TV, Web sites, and any other way that will work. Collaboration needs the same commitment to multiple messages, multiple styles, multiple sites. Trying to collaborate by just using public meetings or just sending out surveys will probably end in failure. Instead, a series of techniques will be needed to ensure that everyone has an opportunity to participate.

• *Become engaged together.* Throughout this book, we have urged the need to "walk the land." Walking the land is a technique for getting stakeholders to become engaged in the place about which they all care and to focus on the real target of their concerns. By becoming familiar with the nature of their ecosystem together, stakeholders learn more about where they live and work, and the new knowledge becomes shared knowledge (Figure 10.3). By learning together, the participants also contribute their expertise to the common good, building their ownership in the place and the process, and also creating a bond between the teachers and learners—they become a team. Such a process also builds confidence among the people for whom *land* has ultimate meaning. One Colorado rancher put it this way:

> Our ranching culture values hard work. Don't tell us what to do or show us data. Invite us to your ranch. If we see that you've worked hard, that your calves are fatter than ours in the spring, and that your riparian areas are green and look good, then we'll ask you to show us how you did it.

BOX 10.8

The Art of Listening

As ecologists, we like to communicate in the explicit language of science. We use hypotheses, principles, and heuristic models to communicate with each other. Our conclusions are often based at the $p < 0.05$ level, and we most readily accept information that is similarly generated. However, could nonscientific information, anecdotes, and traditional knowledge also be sources of valid information? To many people, the answer is yes. One challenge that scientists face in working with people is recognizing that, although scientists should become skilled at translating scientific information into common terms, they must also become skilled in listening to stakeholders who use their own languages to convey implicit knowledge. Information about long-term trends or cycles in an ecosystem is more often found as implicit knowledge in anecdotes or stories than in explicit long-term data sets. Using that knowledge requires that we translate the imprecise and often metaphorical language of stakeholders into the explicit concepts of science.

Consider the following. A Native American tribal elder tells a simple story about his native homeland, which also happens to be the ecosystem of interest to you as a resource manager:

"My people have always lived in the shadow of the Great Mountain. The earth, our mother, was good to us. She fed us and cared for her children. When we hungered in the early spring, she sent the great shaggy buffalo, our brothers, to feed and clothe us. In the summer we feasted on berries and deer. When the leaves turned the yellow color of the sun, the red salmon came to fill our cooking pots. And with the first snows, our mother gave us a sign to prepare for the great buffalo hunt. Many buffalo cows were sent so our children would not hunger in the time of the deep snows. Our earth mother made it so."

The story can be interpreted in different ways. At one level, we could simply accept it as a quaint traditional story. But if we accept the story as having some basis in fact (and perhaps utilitarian value for the tribe), then we might look for its deeper meaning. The elder speaks in the past tense. Does this mean something has changed? If so, what is different today? What could the story tell us about the natural state of the ecosystem and the perturbations to which it has been subjected? Were salmon once indigenous and reproducing with a fall run? What species could have been called red salmon by the tribe? Are salmon or trout still found in this ecosystem today? If not, why not? Could or should they be reintroduced? Where were the tribe's traditional fishing streams?

The story also suggests that buffalo used this area on a seasonal basis. The elder spoke of a semiannual buffalo migration in which the bulls ("our brothers") arrive first and are hunted in the spring. The fall hunt is based on cows and may imply that hunting pressure was differentially applied by the tribe (i.e., hunt bulls in the spring, but hunt the cows only after they have weaned this year's calves) or that cows migrated through this area only in the fall and took a different route in the spring. The "great hunt" in the fall suggests that buffalo were available for a short time in the fall and not available during the "deep snows" of winter. If true, what are the implications for the reintroduction of a resident buffalo herd? What scientific or other evidence might support or refute the inferences you draw from the story?

Most ecosystems have an associated indigenous culture with oral traditions, which may extend back millennia (as in this example) or perhaps only a few decades. Either way they may be valid for understanding not only the local ecology but also the local culture. What might a third-generation commercial fisherman know about long-term population cycles or historic spawning reefs in the waters he fishes? What might a fourth-generation farm family members know about long-term weather patterns, recurring droughts, floods, or prairie fires? How might they express what they implicitly know through stories handed down from generation to generation?

Rarely do we have scientific data sets that extend back 100 years, or even 30 years, but we often can access stories and anecdotes that extend back farther than that—if we're willing to listen carefully and translate from the implicit knowledge of oral tradition or personal anecdote to the explicit concepts and data of science. Wisdom and knowledge can be found in many places.

Figure 10.3. *Members of the Rock River Coalition meet to discuss land-use issues within their watershed in a process that builds mutual understanding. The large round buttons worn by several members read "I live, work, and play in the Rock River Watershed," a sentiment that bonds community members in working toward common goals and understandings. (Photo by Dennis A. Schenborn.)*

Because local knowledge and connections are so valuable, government agencies and national NGOs often hire people from the local community to coordinate ecosystem demonstration projects, rather than using experts from the state or national headquarters.

• *Focus on interests rather than positions.* Once stakeholders begin to develop as a team, collaboration can then move to the stage of developing workable solutions to ecosystem issues. Fisher and Ury, in their 1981 book *Getting to Yes,* pioneered the idea that successful negotiation depends on thinking about what really matters to participants, rather than stating and then defending predetermined positions. When negotiators—which is just another name for stakeholders—take positions (e.g., no wetland conversion versus full development of all land), they become trapped in a contest where they either win, lose, or compromise. Successful negotiation, however, rests on paying attention to the interests that stakeholders bring (e.g., maintenance of species that live in wetlands and a chance to secure a comfortable retirement). Other options may be available that can satisfy all the interests, creating a win-win solution (e.g., a development plan that protects high-value wetlands on some land and actually makes the developed land more valuable).

• *Seek what is held in common.* When stakeholders begin to concentrate on their interests rather than their positions, they can develop the most powerful stimulant to collaboration—the recognition of common interests. Most people who live, work, play, and worship in an ecosystem share most of the same values and interests. People living and working in an ecosystem not only want to make a good living, they also want a clean and beautiful place to live. People who play and worship in an ecosystem not only want the benefits of leisure pursuits and spirituality, they also want good schools and stable communities. The growth of community-based conservation owes much of its popularity to this power of common interests. Daniel Kemmis, the former mayor of Missoula, Montana, and an expert on place-based management, refers to this as the *res publica*—the public thing—that unites the people who live together. When stakeholders can focus on what they all care about together, they can begin to work on the few things that may divide them.

• *Start small.* It has been said that community-based conservation efforts move only slightly faster than glaciers. Although not quite true, this expression underscores the reality that building relationships within a community takes time—and therefore patience. The journey is enhanced if small successes can be achieved along the way, encouraging people to remain engaged and optimistic. By concentrating on a few ideas that are most central and least threatening to stakeholders, a track record of success can develop (e.g., working together on a local park or conducting field days about local history for school groups).

• *But think big.* Henry Ford once observed that you can say that you can do something, or that you can't do something, and you will probably be right either way! Most of the time, our ability to succeed depends not on having the best idea, but on having the vision, courage, patience, and perseverence to succeed. Gerald Nadler and Shozo Hibino call this "breakthrough thinking," the mind-set to consider an idea positively at the outset rather than requiring it to be perfect before even trying it. By concentrating on possibilities rather than deficiencies, a person or a group can make big changes happen. The successes that we discuss in this book and that have become the classics of ecosystem management, such as the Applegate Partnership and the Malpai Borderlands, occurred because people refused to be dissuaded by those who said it was impossible. Anything is possible, if we are willing to work and think together for our common good.

Three Little Words

The essentials of stakeholder involvement, and the promise of its effect on ecosystem management, can be summed up in three little words. The first word is *we*, perhaps the most important word in the English language. It should replace the word *I* as often as possible in ecosystem management. *We* conveys interest in our common purposes and needs, and assigns value to our common contributions and accomplishments. When we speak, write, think, and act about natural resources, we should be thinking *ours*, not mine. It is not my refuge, my park, my plan, or my species—all of these belong to all of us. I don't manage them, *we* do. Experience has shown over and over that stakeholders accept responsibility—and act responsibly—in proportion to how much authority they have. When I becomes *we*, then mine becomes *ours*, and conflict can become *collaboration*.

The second little word is *and*; it should replace *or* whenever possible. *Or* can be a nasty word, because it generates conflict and restricts our imagination. Thinking in terms of *and* allows us to create possibilities that we had not imagined before. *Or* makes us choose, but *and* allows us to reframe issues to expand the options and maybe find solutions that are better for everyone than the choices with which we started. This is critically important in ecosystem management. Almost always, someone tries to quickly narrow a situation down to an either-or proposition. Jobs or environment. Forests or deer. Recreation or timber harvest. Our challenge, and our greatest potential contribution, is to explore new ideas, listening first and judging later—we do that by thinking *and*.

The third little word is a preposition: *with*. Prepositions are words that show relationships between other words. The important preposition for ecosystem management is *with*, and we need to use it in place of *to* and *for* whenever possible. *To* implies a conception of the public interest in which institutions or groups are protectors and regulators. *For* is a little less dictatorial, but still implies that institutions exist to provide the technical expertise to handle complex problems for people. But *with* is the desirable word for linking stakeholders in ecosystem management. Partnerships, in which we share information, beliefs, authority, and responsibility, can access the goodwill and hard work needed to accomplish ecosystem goals. Without the help and participation of everyone, we will never have enough time, money, and talent to address our challenges. But *with* them, we can do almost anything.

References and Suggested Readings

Addis, J.T., and B.J. Less. 1996. Expanding perspectives on gaining public support for management. American Fisheries Society Symposium 16:42–46.

Bryk, A.S. (ed.) 1983. *Stakeholder-Based Evaluation. New Direction for Program Evaluation:* 17. Jossey-Bass, San Francisco, CA.

Cortner, H.J., and M.A. Moote. 2000. *The Politics of Ecosystem Management*. Island Press, Washington, D.C.

Covey, S. 1990. *The Seven Habits of Highly Effective People*. Fireside Books, New York.

Creighton, J.L. 1981. *The Public Involvement Manual*. Abt Books, Cambridge, MA.

Crowfoot, J.E., and J.M. Wondolleck. 1990. *Environmental Disputes: Community Involvement in Conflict Resolution*. Island Press, Washington, D.C.

Decker, D.J., T.L. Brown, and W.F. Wiemer (eds). 2001. *Human Dimensions of Wildlife Management in North America*. The Wildlife Society, Bethesda, MD.

Decker, D.J., C.C. Krueger, R.A. Baer, Jr., B.A. Knuth, and

M.E. Richmond. 1996. From clients to stakeholders: A philosophical shift for fish and wildlife managment Human Dimensions of Wildlife, Spring: 70–82.

Dennis, S. 2001. *Natural Resources and the Informed Citizen.* Sagamore Publications, Champaign, IL.

Fisher, R., and W. Ury. 1981. *Getting to Yes.* Houghton Mifflin, Boston, MA.

Hardin, G. 1968. *The Tragedy of the Commons: Managing the Commons.* W.H. Freeman, San Francisco, CA.

Kellert, S.R. 1996. *The Value of Life: Biological Diversity and Human Society.* Island Press, Washington, D.C.

Kemmis, D. 1990. *Community and the Politics of Place.* University of Oklahoma Press, Norman.

Keystone Dialogue. 1996. *The Keystone National Policy Dialogue on Ecosystem Management.* Keystone Center, Keystone, CO.

Kotter, J.P. 1996. *Leading Change.* Harvard Business School Press, Boston, MA.

Leopold, A. 1938. Engineering and conservation. Pp. 249–254 in S.L. Flader and J. Baird Callicott (eds.). 1991. *The River of the Mother of God and Other Essays by Aldo Leopold.* University of Wisconsin Press, Madison.

Leopold, A. 1947. The ecological conscience. Pp. 338–346 in S.L. Flader and J. Baird Callicott (eds.). 1991. *The River of the Mother of God and Other Essays by Aldo Leopold.* University of Wisconsin Press, Madison.

Main, M.B., F.M. Roka, and R.F. Noss. 1999. Evaluating costs of conservation. Conservation Biology 13: 1262–1272.

Nadler, G., and S. Hibino. 1990. *Breakthrough Thinking.* Prima Publishing & Communications, Rocklin, CA.

Nielsen, L., B.A. Knuth, C.P. Ferreri, S.L. McMullin, R. Bruch, C.E. Glotfelty, W.W. Taylor, and D.A. Schenborn. 1997. The stakeholder satisfaction triangle: A model for successful management. Pp. 183–189 in D.A. Hancock, D.C. Smith, A. Grant, and J.P. Beumer (eds.). *Developing and Sustaining World Fisheries Resources: The State of Science and Management.* CSIRO Publishing, Collingwood, Australia.

Ostrom, K., J. Burger, C.B. Field, R.B. Norgaard, and D. Policansky. 1999. Revisiting the commons: Local lessons and global challenges. Science: 278–282.

Schenborn, D.A. 1985. Environmental scanning: The difference between strategic success and failure. Transactions of the 50th North American Wildlife and Natural Resources Conference: 304–312.

Schenborn, D.A. 1989. Public involvement as a management tool. Alces 25:175–177.

Schwarz, R. M. 1994. *The Skilled Facilitator: Practical Wisdom for Developing Effective Groups.* Jossey-Bass, San Francisco, CA.

Senge, P., A. Kleiner, C. Roberts, R. Ross, and B.J. Smith. 1994. *The Fifth Discipline Fieldbook: Strategies and Tools for Building a Learning Organization.* Currency Books, New York.

Experiences in Ecosystem Management:

Collaborative Stewardship: Views from Both Sides Now

Mark W. Brunson

MOST UNDERGRADUATES CARRY IMAGES IN THEIR heads to remind them what they're working toward. Students in biology and natural resources imagine themselves climbing tall firs to band nestling spotted owls; riding horseback through lush, sage-studded meadows to check the success of a rangeland restoration project; silently paddling across a mist-shrouded lake to conduct a loon census; sitting around a table with backpackers and ranchers who are forcefully declaring that they can't *possibly* consider sharing a particular mountain valley.

OK, they probably don't focus on that last image. But maybe they should, because ecosystem management entails more than personal contact with natural landscapes. It also means countless hours spent talking and listening to those who want to help shape the future of contested landscapes. Collaboration—that is, cooperative effort among multiple landowners and stakeholders to achieve goals that are thought to be infeasible using traditional means—is a cornerstone of cross-boundary stewardship. Single agencies or owners often lack the personnel, legal authority, and/or financial resources for effective conservation at landscape or larger scales. But by collaborating with other agencies or owners, they can share skills and resources while accommodating diverse and sometimes conflicting values. Moreover, collaborative solutions are designed to fit particular places, avoiding the "one size fits all" mentality that has plagued natural resource management in the past. And collaboration may be the best way to avoid the delays and gridlock that accompany actions by the courts and Congress.

Yet not everyone sees collaboration in a positive light. Some people worry that it fosters exclusiveness. Those who form collaborative groups tend to seek out people they believe they can work with, sometimes ignoring important constituencies. Moreover, because collaboration usually requires frequent on-site participation, it can favor local concerns over equally important national interests. Other critics are alarmed for exactly the opposite reason: Collaboration tends to be more purely democratic than traditional natural resource management, weakening the "clout" of interests that have learned how to get Congress or the courts to do their bidding. And for natural resource professionals themselves, collaboration is daunting because it is hard to do. It can add considerably to workloads, and it requires skills that are quite different from the analytical methods of science that dominate natural resource education.

I will offer some advice to help natural resource professionals become effective leaders and participants in collaborative stewardship efforts. My recommendations are based on a decade's worth of research on ways to bring people together more effectively in working on ecosystem management projects. They are also based on hands-on experience during the 27 months I spent as facilitator of a collaborative conflict-management group in northern Utah.

Why Study Collaboration?

In the early days of ecosystem management, collaborative stewardship was uncommon. Despite a few well-known examples of model partnerships—Oregon's Applegate Partnership, the Malpai Borderlands Group in the Southwest, South Carolina's ACE Basin Bioreserve Project, to name a few—it took a while for community-based groups to appear across the U.S. Some agencies and interests were slow to embrace ecosystem management; others enthusiastically focused on the new land management tools, such as landscape-scale modeling or alternative silvicultural practices, but were slower to adopt the new social ideas in ecosystem management.

Also, there were clear barriers that kept federal land managers from initiating or joining collaborative partnerships. Some barriers were institutional: rules of public involvement set by judicial fiat rather than common sense, laws against backroom collusion that make it hard for agencies to seek advice, agency "cultures" that discouraged managers from giving up any of their decision authority. But there were personal barriers, too. Most public land managers—and many of the constituents they serve—have educational backgrounds that are strong in understanding the land but weak in understanding people. Often they are reluctant to discard established ways of making and influencing decisions—NEPA processes, courts, legislative lobbying—in favor of the new collaborative approaches.

Still, there was slow but steady growth in the ranks of people who chose the collaborative route. If this truly represented the future of land management, someone needed to find out what was working—and not working—about collaborative natural resource decision making and make those findings available to land managers and their constituents.

In 1994, I began studying what makes collaborative group participants believe their efforts are fair and/or effective. I hoped to identify the elements that should be designed into collaborative processes for partners to feel their time has been well spent. I used the traditional tools of social science: a thorough review of the relevant literature in both the natural resource journals and the writing of conflict management and negotiation experts, a series of interviews of potential collaborative partners, and a survey of existing groups designed to test and refine that theory. With the aid of a capable graduate student, Kimberly Richardson Barker, I set out to develop guidelines that could help managers and public land stakeholders feel a bit less queasy about collaborative stewardship efforts.

Exploratory Steps. Much has been written in the past few years about collaboration, conflict, and related topics. Although many lessons can be learned from this work, it was not as helpful as we initially hoped. Some authors seemed more interested in selling a particular formula for collaboration than in giving basic advice. Others offered case studies of a particular locale but could not say if the same process would work anywhere else. Studies often were not comparable because each researcher defined "collaboration" and "success" a bit differently. Nonetheless, by synthesizing what we read, we gathered ideas to explore in the next phase of our research.

That next phase was a series of interviews of natural resource stakeholders, mostly ranchers, in six rural Utah counties. We focused on that group of potential collaborators for several reasons. First, we felt it was important to understand what potential participants saw as prerequisites to their involvement in collaborative stewardship, since no collaboration can be fair or effective if it cannot attract the relevant partners. Second, prior attempts at collaboration in Utah had been largely unsuccessful, and we wanted to know if that would make future efforts more difficult to achieve. Finally, since rural Utahns were wary about ecosystem management in general, we wanted to understand their beliefs and concerns about it.

There was quite a bit of overlap between what the ranchers told us and what the experts had written. We were able to synthesize those two sources of information to develop a "theory" about collaborative processes that could entice Utah's rangeland stakeholders to join with federal land managers in a stewardship effort. Both groups said goals and objectives of the partnership had to be realistic and that decisions should be made via consensus rather than a simple majority vote. They also agreed that all participants in a process should be knowledgeable about the specific landscape affected by the partnership and that there must be an atmosphere of mutual respect—especially on the part of the agency decision makers, who had to be truly committed to implementing the outcomes of the collaborative effort.

In addition, the experts emphasized "voice and control"—the perception by partners that they would be listened to and could influence the outcome. Many experts said success would be best accomplished with the aid of a professional facilitator. Meanwhile, the ranchers also told us that groups needed to be relatively small, should consist solely of local community citizens, and should efficiently use the time spent in meetings.

Testing Theory Through Surveys

Our next step was to test our theory. We did this by seeing whether existing collaborative efforts included the elements that the experts and potential participants said were necessary and also whether the presence or absence of those aspects mattered to the success of the collaborative effort. We surveyed members of eight range stewardship groups that had both private landowner and public agency members. The groups operated in four states (Idaho, Nevada, Utah, and Wyoming), and they varied in their specific land management goals, histories, and procedures. For example, four groups used Coordinated Resource Management, a process specifically recommended for managing range conflict. The others used rival approaches or hybrid processes of their own design. Some had water quality goals, others wanted to improve wildlife habitat without eliminating grazing, and one sought to make ranching and recreation more compatible.

We received completed surveys from 101 respondents who answered questions about how their partnerships worked and about whether the efforts were considered fair and/or effective. We measured perceived fairness with questions derived from "procedural justice theory," which states that people are more likely to accept decisions—even those that are personally unfavorable—if they believe the process leading to the decision was fair. Perceived effectiveness was measured by asking respondents if their efforts had made the land, the group as a whole, or themselves personally better off. Effectiveness was defined in terms of *improvement*, rather than *achievement* of goals, because some groups had not yet had time to reach those goals and also because we believe conflicts among competing interests are more effectively managed if one seeks to "improve a situation" rather than "solve a problem." By examining statistical correlations among responses to the various questions, we were able to identify the aspects of collaboration that were most important to perceived fairness and effectiveness.

Participants in our eight collaborative groups generally felt the groups were a good way to get their own views heard (72% agreed; 36% strongly agreed) and would lead to fairer decisions than could be reached otherwise (82% agreed; 48% strongly agreed). People were more likely to feel a process was fair if its goals were feasible, decisions were made via consensus, time was spent efficiently, all partners were committed to the outcome, mutual respect was shown among group members, and partners had a chance to be heard and help shape the outcomes. We did *not* find that local involvement or knowledge were critical to perceptions of fairness, and there was no difference in perceived fairness between groups that did or did not use an outside facilitator.

Most partners also felt their efforts had been effective. Some 70% felt they personally were better off as a result of the partnership (25% much better off), and 78% felt the land was better off (38% much better off). Collaborators were more likely to believe the effort had been effective if goals were feasible, decisions were reached via consensus, time was spent efficiently, all partners were committed to the outcome, all participants were knowledgeable about the area and issues, mutual respect was shown among group members, and they were given adequate voice during the process. We did not find that effectiveness required compromise among competing interests, involvement solely by local citizens, or a feeling that respondents personally had a strong influence on partnership decisions.

Thus, we concluded that public rangeland agencies and stakeholders *can* collaborate successfully, but that their efforts are most likely to be fair and effective if they rely on open dialogue and consensus-based processes to reach decisions that all participants—especially the land management agencies—are committed to implementing. Membership should be open to any knowledgeable person, local resident or not, who is willing to work hard and treat others with respect. Trained facilitators may not be necessary, but leaders must maintain some impartiality and be able to keep members focused on the task.

Testing Theory Through Participation

As a researcher, my job was finished. But once I began telling people they should collaborate because it works, someone called my bluff. I was asked to facilitate a collaborative group myself. At that time, there was considerable conflict among winter recreationists in the national forest near my home, and the local district ranger felt a conflict management process might help ease the situation. I was nervous about agreeing to the request, but saw no way around it: it was time for me to walk the walk I'd been talking for years.

In January 1996, we assembled a group of snowmobilers, backcountry skiers, recreation business owners, government officials, and interested citizens to work collaboratively toward improving how winter recreation was managed in our area. That first year we developed a list of issues, shared ideals (e.g., wildlife protection, back country etiquette), and procedures based largely on my research findings. We met biweekly through May, took 6 months off, then spent the second year on the nuts and bolts of a plan. Working our way across the ranger district watershed by watershed, we agreed on detailed, place-specific suggestions for improving access and reducing conflicts among winter recreation users. Along the way we devised other improvements (better maps, new etiquette signs, temporary boundary markers for non-motorized use areas) that the U.S. Forest Service could use immediately without a NEPA process. We chose to use a consensus-like process whereby we would vote on contentious issues, but a single "no" vote could block a recommendation as long as the voter offered an alternative recommendation that was an honest attempt at compromise. Although our group included both government officials and private citizens, the government folks' role was advisory; only the private citizens could vote. Although the initial partners had joined by invitation, we left the door open to new volunteers, and about half our membership changed over the course of the process.

Finally, in March 1998, we held a public meeting to describe our ideas to the other people who would be affected by them. At first the meeting was tense because our "constituents" were not sure we truly represented their interests, but by the evening's end, our ideas were well received. Moreover, we ourselves felt good about the final product. We built relationships that helped make conflict more manageable. Because the Forest Service immediately acted on suggestions that did not require public comment, we could point to tangible improvements as a result of our efforts. It was tremendously affirming to see my research truly put into useful practice.

What Research Alone Could Not Tell Me

Some people in the community-based collaboration "movement" argue that research on collaborative stewardship is useless; each situation is so different that you cannot learn much from anyone else's experience. I disagree. But I've also learned that while research is valuable, there are some things you can only learn by doing.

One thing I learned was that some of the most important rules are ones that are never actually invoked. No one in our group ever actually used his or her veto power to block a proposed recommendation. But people seemed to look harder for compromise because they knew they could not move on to the next issue otherwise. We spent a lot of time developing our initial list of shared ideals and goals, because most experts feel the search for "common ground" is vital to the success of any collaboration. But I learned that too much emphasis on such tasks is self-defeating. People want to get on to the task at hand, or else they feel their time is not being well spent.

I also learned a lot about collaborative dynamics. I found that two or three participants often set the tone for everyone else. These participant-leaders may have more influence than others in the group, but they also tend to ensure that those others feel they are being heard. I learned that knowledge outweighs locality—that is, people from outside the area *can* be valuable participants, but only if they contribute new knowledge that rings true with local understanding. I found that busy people could make time during business hours if they felt something would be achieved. But I also found a need to occasionally stop and revisit decisions—even if it felt like we were spinning our wheels—to meet the natural human need to reaffirm important choices.

Perhaps most importantly, I learned that collaboration does not take place in a vacuum. As we continued to meet, our feelings about "the other side" began to change. Our members began to feel loyal not only to the people we represented, but also to our group itself. At the same time, the situation we were trying to improve was changing, too. It's critical to be an adaptive and watchful collaborator, shifting gears as needed to meet new challenges as they arrive.

Collaboration is not easy. It is time-consuming, often frustrating, and takes us away from the natural places we love best. But after viewing the collaborative process from both sides, as an observer and as a participant, I have to conclude that the potential alternatives—gridlock, court cases, failure to accomplish even the simplest management objectives—are at least as frustrating and a lot less rewarding.

Discussion Questions

1. Are there decisions in biological conservation or natural resource management that *should not* be made via a collaborative stewardship process involving multiple stakeholders? If so, discuss what they are and why collaborative solutions are unlikely to be useful.
2. Some interest groups will not take part in collaborative stewardship as a matter of policy, generally because they feel there is no room for compromise on key issues. What should be done if one or more of these groups has an important stake in the outcome of a collaborative process? Ignore them? Abandon the collaboration? Something else?
3. How can the views and rights of nonlocal stakeholders—for example, urban residents who have a strong love for a larger western national park—be accounted for in a collaborative stewardship process?
4. Why do people who participate in collaborative processes generally feel those efforts are fairer than traditional natural resource decision processes?
5. Does the program of study in your academic major give you coursework that will help you be an effective participant in collaborative stewardship? If not, what course(s) might you want to take to be better prepared for this aspect of ecosystem management?

Strategic Approaches to Ecosystem Management

A ROCKFISH DECIDED TO MAKE HIS WAY IN THE world, so he began his journey. He soon discovered, however, that he had no idea in which direction to travel. He met a grouper and asked what direction he should take. The grouper was puzzled by the question, but advised that slow and steady swimming took him wherever he wished. Next, the rockfish met a flounder and asked what direction he should take. The flounder was also uncertain of the best direction, but advised the rockfish to keep his head down and stick to the beaten path. The rockfish met a tuna and asked for the same advice. The tuna demurred on the question of direction, but asserted that speed was what mattered—swim fast and you will get where you want to go. Finally, the rockfish met a shark and asked his question again. The shark asked what the rockfish's goal was. "I have no goal," the rockfish answered, "I just want to make my way in the world." The shark opened his mouth wide, pointed down his throat, and said, "Then this is surely the direction you want to go." The rockfish followed the shark's direction and reached the end of his journey quickly. The moral is that, no matter how you travel, if you do not know where you want to go, you will probably end up somewhere else (Figure 11.1).

The moral is applicable in all aspects of life, including ecosystem management. Earlier chapters have described the complex basis of an ecosystem approach from ecological, socioeconomic, and institutional perspectives, including the need to embrace uncertainty, work collaboratively with others, and learn continuously from management actions. Add to those the complexities of expanding scales of time, space, and inclusion, and we have outlined a situation that requires careful thought, open and honest communication, and a commitment to do what matters over the long term.

Imagine a landscape in which groups were independently working toward what they thought was best for an ecosystem and its human residents. Managers of a wildlife refuge might make elaborate plans for restoring native ecosystems on specific parts of the landscape, including herbicide treatment to kill non-native invasive plants. Adjacent landowners might be developing an organic

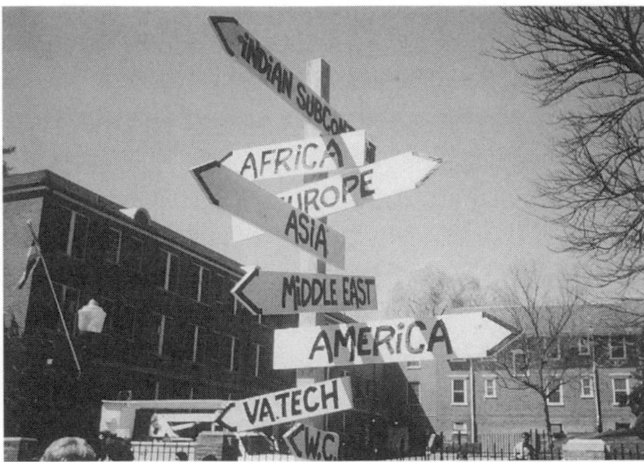

Figure 11.1. *Every crossroad presents a staggering array of choices in ecosystem management. Strategic management provides the map for where a community wants its ecosystem to go. (Photo by Larry A. Nielsen.)*

BOX 11.1

The Evolution of Strategic Management

Is strategic management a passing fad? Not likely. Various concepts of strategic management have come and gone, as described below, but the basic idea has been amazingly resilient.

MBO (Management by Objectives)
The current generation's initial exposure to strategic management was MBO, in which many objectives were written to govern all aspects of an organization's work. Popular in the 1960s, MBO had a parallel in education; teachers were expected to write learning objectives that defined and guided all their classroom activities and time.

ZBB (Zero-Based Budgeting)
In the late 1970s, President Carter's administration decided that budgets appropriated to executive agencies should be based on performance. They created ZBB, with the idea that an agency would justify its budget request in terms of strategic management by listing their plans in detail. In practice, only the last 15% of an agency's budget was subject to this program.

PPBS (Planning, Programming, Budgeting System)
Strategic management became systematic in the 1980s, with formal systems to link planning, budgeting, and operations. The idea was that the strategic intent of an organization, as expressed in its plans, needed to be linked directly to its budget allocations.

GPRA (Government Performance and Results Act)
In the 1990s, Congress passed a law requiring that all federal agencies must justify expenditures (and future budgets) in terms of strategic "outputs and outcomes."

farming cooperative that would allow no chemicals. At the same time, local officials might be planning to buy land for a new landfill they anticipate needing in the next decade. They would naturally be looking at lands that are distant from residents and currently undeveloped, exactly the same location that a land trust might have targeted for purchase or easements as part of a wildlife corridor. Local landowners might be considering offers to sell their land to a manufacturing company being courted by tax breaks and low-interest loans from the governor's rural revitalization program. And the state transportation department might have already drawn engineering designs for a new limited-access road that has been placed here to save wetlands in the adjacent valley. Despite the good intentions of all these people and groups, the final outcome might leave behind a damaged ecosystem and a damaged community. Working successfully at the ecosystem level requires a mechanism to keep everyone talking, thinking, and planning together—the exact conditions for a strategic approach to management.

Strategic management has evolved continuously as a management concept (Box 11.1). Recent manifestations began with Management by Objectives (MBOs), which developed the foundation for strategic management: We should set objectives for what we wish to accomplish. After many other forms since MBO, U.S. government agencies are now implementing the Government Performance and Results Act (GPRA), which requires all agencies to identify and measure "outputs" and "outcomes." Soon, some other terminology will replace this scheme. Despite the transience of specific approaches, however, the principle has remained: By selecting a clear and purposeful destination and then taking the journey by the most appropriate and economical route, we have less chance of being eaten by metaphorical sharks.

EXERCISE 11.1

Think About It!

Consider a group in which you have used strategic management—a student group, community group, sports team, school committee, agency team. The activity might not have been called strategic management, but it involved people discussing what the group was going to do. How did you feel during the process? How did others react to the process? What frustrated you? What energized you? After you finished planning, how did others react? Did you implement the plan? If not, why not? List some conditions that you think led to success or failure.

Characteristics of Strategic Management

Throughout this book, we use the term "strategic planning" sparingly. We do this on purpose, because for many people, planning has become synonymous with long, boring meetings in which committees struggle to reach a consensus about where a group should go, write an elaborate (or elegant) document to explain their thoughts, and then see it ignored. Whether the group is a student club, a community volunteer organization, or a public agency, such a process can sap the energy, creativity, enthusiasm, and commitment of the participants, and many may leave before any real work happens.

We prefer the term **strategic management**, which means "acting with a purpose." Although strategic management is often associated most closely with businesses or public agencies, it is relevant to any human endeavor. Popular self-improvement books, such as Stephen Covey's *The Seven Habits of Highly Effective People*, generally promote a purposeful, strategic approach to life. Regardless of whether a strategic approach is used by individuals, families, communities, clubs, or governments, certain characteristics are universal (Box 11.2).

• *Strategic management is explicit.* Strategic approaches define what an organization, group, or individual intends to do, how and why those actions were selected, and eventually whether or not they have been accomplished. Explicitness is desirable for several reasons. First, everyone knows what the group is doing, from the newest member or volunteer to a newspaper reporter, a potential funder, a legislative oversight committee, or the general public. Second, hidden agendas are exposed. If a community group states that it is

BOX 11.2

The Top Ten Attributes of Strategic Management

Skeptics often say that strategic management is where the rubber meets the . . . sky! But when practiced well, strategic management will help ensure that when the rubber meets the road, the vehicle is heading where we want it to go. Therefore, we should practice strategic management because it is:

1. Explicit—we say what we intend to do and what we will not do.
2. Accountable—we can measure whether or not we have done what we said we would do.
3. Directional—we all aim for the same endpoints.
4. Flexible—once we are aimed in the right direction, we can use all available tools and routes to get there.
5. Empowering—once we know our direction, we can act without constant supervision.
6. Focused on top priorities—we always know that we are doing important work.
7. Collaborative—we are part of a diverse team that sets direction.
8. Widely communicated—we all know what we are about and why.
9. Simple—we can understand the process and the products.
10. Evolutionary—we seek continuous improvement, not perfection.

committed to restoring native biodiversity and then plants exotic species in restoration projects, the mismatch will be obvious. Third, accountability is possible. If an agency commits to implementing recovery plans for five endangered species and has accomplished only two, the agency can adjust to do better in the future (if it has accomplished seven, it can enjoy the praise!). This, of course, is one reason some people dislike strategic management; because knowledge is power, sharing knowledge shifts power from the leadership to the members and stakeholders. Because ecosystem management depends on broad community support, spreading knowledge and power through strategic management is essential.

• *Strategic management sets direction, but allows flexibility.* Although it may sound oxymoronic, strategic management actually blends planning and flexibility. It begins with setting strategic direction through planning, focusing people on the agreed intentions of the group. This is followed by an operational environment that gives individuals freedom to use their innovation and creativity to pursue the group's intentions. An ecosystem group may set the protection of a riparian corridor as its primary aspiration for the next decade; individuals within the group can then pursue that goal through their own skills and capabilities, whether in land management, education, fund-raising, or new opportunities. By focusing on the longer-term outcomes, people are free to use the best means to advance their common goals.

• *Strategic management promotes action.* A group that has made its purpose explicit can empower its members to act on that purpose—and effective action is what we seek. For example, a community group might decide that it needs to help local children learn about the history of their lands and resources; it might then create resources to help that task, such as a historical comic book and a biodiversity coloring book, and give a supply to each member for their use. Armed with a purpose and tools, the members are empowered to tell the story, perhaps in school classrooms, church groups, or youth clubs, and to distribute the resources as they wish. The members will also know that they should focus on local children, rather than on adults or politicians or national groups. When decisions are made and actions taken in alignment with the framework of strategic management, they fit together into something larger than the sum of the parts—a desired outcome of ecosystem management.

• *Strategic management starts at the top.* Does this mean that the board of directors or team leader makes all the decisions? Absolutely not. Starting at the top means "at the top of the *purpose* of the organization or group." As later sections will describe, strategic management moves through a series of steps that begins with setting the fundamental purpose for the group. Subsequent decisions then fit within the context of this broader effort or purpose. This ensures that an individual or a group is doing what it ought to be doing, rather than using its resources on actions that are not central to stated goals or aspirations. If an ecosystem team is focusing on stream habitats and someone proposes a worthwhile lake project, the team will be faced with a difficult decision; the higher purpose will be there to guide them. Effective work is about choices, and strategic management helps make good choices.

• *Strategic management involves all stakeholders.* Who sets that top-of-the-group purpose? It isn't a single person or small team, sitting at the head of the table. Especially in ecosystem management, the best way to set direction is to involve a cross section of members, staff, executives, decision makers, opinion leaders, clients, and interested citizens—representatives of all the stakeholders identified in Chapter 10 (Figure 11.2). Of course, legally designated representatives may eventually need to "sign off" on behalf of the team, but that formality should never be confused with acceptance by the larger group.

• *Strategic management requires good communication.* An important corollary to full stakeholder involvement is good communication, especially in ecosystem management. Communication about the process and progress of the team's work must be regular and meaningful. This also means providing continuous and accessible feedback opportunities for those involved, so improvement will continue. Strong two-way communication is especially im-

Figure 11.2. *Strategic management works best when the team is composed of diverse people. Along with all the legal aspects of diversity, teams should include many levels within a group or organization and many types of stakeholders, especially those with a major stake and a willingness to participate. (Photo by Larry A. Nielsen.)*

portant in ecosystem conservation, because of the expanded time, space, and inclusion of the work. Actions will be occurring in many places, implemented by different individuals and groups, over many years. Good communication will sustain the energy and commitment of participants, as well as ensuring that actions remain aligned with the team's purpose and that the purpose remains relevant to the ecosystem.

- *Strategic management seeks improvement, not perfection.* One of the downfalls of much strategic planning is that the planning team works to exhaustion, usually over several years, to produce a perfect plan. In the meantime, life goes on, enthusiasm dries up, the group evolves, and the ecosystem changes. The supposedly perfect plan may be obsolete before it ever gets finished! It is better to create an unfinished plan that implements a few clearly desired actions and that can be improved as the work proceeds. Moreover, an ecosystem plan that requires massive changes in resource use, economic structures, community conditions, or operational budgets will seldom get very far. People who live and work in an ecosystem have too much invested to change dramatically. As in most aspects of life, evolution works much better than revolution.

A Simple Strategic Management Model

Given the characteristics of successful strategic management described above, we propose a simple approach for setting and implementing strategic intention for ecosystem conservation (Figure 11.3). The model includes the four steps of *inventory, strategic thinking, implementation,* and *evaluation,* depicted as a progression. However, a fundamental aspect of the process is that it is continuous. All four steps are in operation continuously—another way to say that strategic management is a "way of doing business" rather than a one-time burden for a strategic planning team.

THE INVENTORY

The first step in strategic management is making an **inventory** of the system under consideration, answering the question "Where are we?" Before

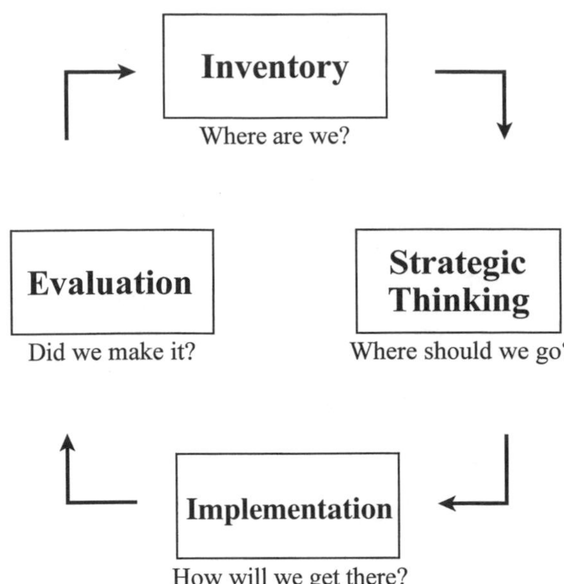

Figure 11.3. *The four-step model illustrated here is a simple approach to strategic management. Other models may be much more complex, but they all contain these four steps in one form or another.*

deciding the future course for a group or an ecosystem, the current conditions must be known. For ecosystem management, this involves assessing ecological, socioeconomic, and institutional conditions (Figure 11.4).

Many organizations, such as government agencies, NGOs, and universities, have established programs for collecting inventory data, including surveys of plant and animal diversity, abundance of major species, harvest of species, land uses, soil types, habitat quality conditions, use by humans of various resources and for various purposes, economic values, agency budgets and allocations, laws and regulations, and staff characteristics (the technical content of Chapters 5–10). Community groups and informal ecosystem teams, however, may find this step particularly challenging. Therefore, a partnership with established agencies and organizations may be needed to acquire and catalogue useful data. Because so much of a group's time and energy are used in inventory, decisions

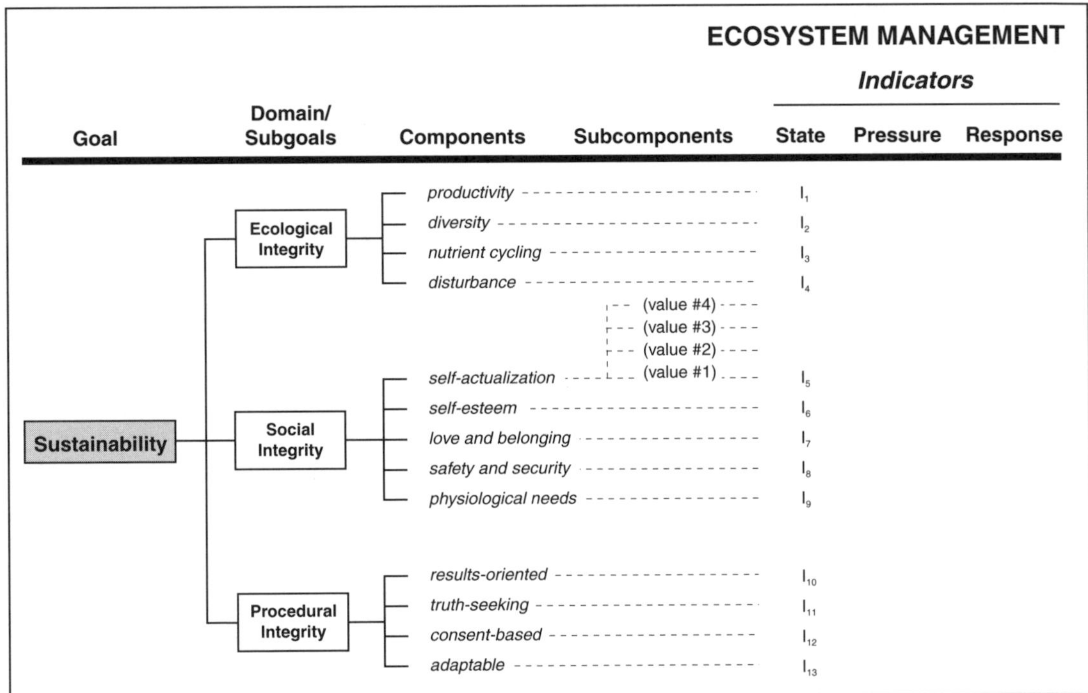

Figure 11.4. *Creating an inventory system to support ecosystem management is a major step. The scheme shown here was developed by Paul Pajak of the U.S. Fish and Wildlife Service to ensure that all aspects of ecosystem concern—ecological, socioeconomic, and institutional—were covered in meaningful ways. Full elaboration of the model would require explicit and case-by-case selection of the components, subcomponents, and indicators (I_1, I_2, etc.). Then, a data collection and reporting system would be needed to determine the state, pressure, and response of each indicator. (From Pajak, 2000.)*

about what data to collect and then how to collect, store and communicate the data are crucial.

A word of caution about inventory: The desire to have more information can overwhelm individuals and groups, a pitfall called "paralysis by analysis." Ecosystem management is highly vulnerable to this problem, partially because so much of ecosystem management seems new. Earlier chapters described new ideas to consider, from genes to landscape processes, any of which might take a lifetime to document. Remember, though, that the goal of strategic management is action, not just information gathering. Adaptive management (see Chapter 4) is one way to ensure that knowledge of the system increases simultaneously with actions—use it whenever possible!

EXERCISE 11.2

Collaborate on It!

What is the difference between need and want? We all suffer from the desire to have lots of information with which to make a decision, and in today's world of instant access to electronic information, that trend will continue. For ecosystem management, however, we will often find that the information we desire is not available. Think about the difference between the information one *needs* to make a decision versus the information one *wants* to help make the decision. Make a list of the criteria you would use to judge whether information was truly needed for a decision versus that which was just "nice to know."

Use those criteria to list what information is "needed" and what is "wanted" in this situation (one for each scenario):

- ROLE Model: Writing a plan for the conservation of the minnow community in Little Lake.
- SnowPACT: Assessing the hydrologic effects of removing the Pine Lake dam from KARMA.
- PDQ Revival: Developing a conservation plan for the coastal fox in the Swamp Fox Wildlife Management Area.

STRATEGIC THINKING

The next step in strategic management is **strategic thinking**, in which the group sets its direction. The question here is, "Where should we go?" This step sets the goals and objectives for a group and then works down through the decision-making process to guide day-to-day activities. Strategic thinking is vital to ecosystem management because the scope of possible action is so large. Should an ecosystem team focus on the composition, structure, or function of the land and water resources; on education, land purchase, technical assistance, or regulation; on hot-spots of biodiversity or the overall landscape? Without establishing its priorities, a team's efforts may end up unfocused or even internally contradictory. We explore strategic thinking in detail a little later.

Many groups and individuals undertake this step, but then fail to achieve the promise of strategic management because they stop there. The fruits of strategic management, however, ripen in the next two steps.

IMPLEMENTATION

The third step in strategic management is **implementation**, which asks the question, "How will we get there?" and then answers with, "Let's try this." In this step, actual projects are designed, with allocations of time, funds, and other resources, and then implemented. This stage of strategic management is crucially important and often difficult, because this is when the group's activities are judged against its goals and objectives. If a group has been operating for some time, its members have skills, tools, routines, and relationships that may encourage them to continue doing old activities and rejecting new ones. To be truly effective, however, a group's strategic intent must be transformed into strategic actions in the implementation stage.

EVALUATION

The fourth step in strategic management is **evaluation**. This step asks the question, "Did we make it?" It also goes further to ask the more thoughtful questions, "If we did not make it, why not?" "If we did make it, are we satisfied?" Evaluation, therefore, is the feedback loop in the four-step model. Evaluation is tricky because it involves both objective and subjective information, and because the specific questions that need to be asked about a goal or objective cannot all be anticipated in the previous strategic thinking step. We cover the principles and practices of evaluation in detail in Chapter 12.

These four steps together compose a cycle, where the information from one step moves the process along to the next. This is a feedback process at a large scale. Feedback, however, should also occur continually within each of the steps (Figure 11.5). Inventory can improve internally through better technical and statistical procedures and a sharper focus on what data are essential; inventory topics must be added when new goals, objectives, or projects require new information, and old topics must be dropped. Goals and objectives (the strategic thinking step) can be changed, added, or abandoned if short-term feedback suggests unanticipated problems or new opportunities. And implementation should respond continually to new techniques, possibilities, and problems. Remember that successful strategic management is *flexible*, allowing learning and adjustment to make actions more effective and efficient.

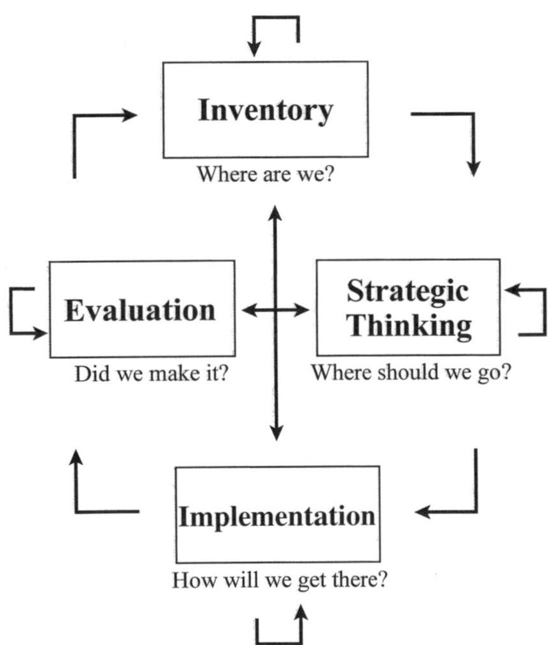

Figure 11.5. *This version of the four-step model shows that feedback must be a part of all stages in strategic management. Links between steps, as well as internal feedback within each step, are fundamental to good performance.*

In its entirety, strategic management should look familiar: It is a form of passive adaptive management (see Chapter 4). Decision makers use inventory data to make plans for action, they implement activities, and they then collect data to assess the outcomes. Consequently, they make changes in the next round of activities. In this sense, strategic management is also a learning process, making each iteration of management better. For this reason also—to ensure that learning can occur—strategic management is the ideal vehicle to carry ecosystem management.

The Strategic-Thinking Step

Strategic thinking directs actions through a step-down process, from the highest purposes of the team to the actual implementation stage. In other words, strategic thinking provides the "thought path" for converting intention into action. The thought path must be logical, from the broadest statements of intention to the most local and immediate actions. And to communicate that pathway clearly, a team needs to adopt names for the steps they follow. We recommend the step-down terminology commonly used in natural resource management (Box 11.3). But remember that it is not important if one team's terminology differs from another's; all that matters is that at any time within a team, everyone knows what the words mean.

MISSION AND MANDATE

The highest level of stated purpose for a group, organization, or agency is the **mission** or **mandate** (often, though not always, used synonymously). This is typically established in the law or charter that created the group. (For an informal group, such as a set of adjacent landowners, the mission may not be explicit, but the individuals will have a strong sense of why they are working together.) Most federal agencies, for example, have an organic act that defines the legal purpose and scope of the agency; organic acts are passed by Congress and signed into law by the president.

BOX 11.3

The Language of Strategic Management

Examining a stack of strategic plans or strategic-planning guides will reveal a nasty reality—they do not all speak the same language! Many of the terms are used interchangeably within one plan, and many are used differently among plans. No matter—we have to live with it. For the sake of clarity, we have adopted these terms and definitions, presented in the step-down order we will discuss them.

- Mandate. The legal authority for a group, defining its official purpose, usually in the broadest possible terms; also called a charter.
- Mission. The highest-level statement of purpose, covering the entire group, usually written to be inspirational and usually brief.
- Strategy. The general approach or approaches used by the group to address its mission. Groups often skip designating a strategy, because it is embedded in their goals.
- Goal. A general description of what the group seeks to accomplish and for whom. Groups typically write a few goals (3–10), each of which covers one aspect of the mission.
- Objective. A specific statement of what the group intends to accomplish, stated in ways that can be measured and monitored. Several objectives may be written to address each goal.
- Problem. An explicit obstacle that stands in the path of accomplishing an objective. Groups often skip these statements, because they are assumed as part of the objective and project.
- Tactic. An operational approach chosen to overcome a stated problem. Groups often skip these statements, because they are assumed as part of the objective and project.
- Project. A particular item of work, and its associated costs, that will be undertaken to accomplish an objective. The project should follow from the problem and tactic, if stated explicitly.

BOX 11.4

The Mission Statement

Mission statements vary in detail, but they all share general characteristics of being comprehensive, visionary, general, and brief. Here are three mission statements from conservation organizations.

- The Nature Conservancy: The mission of The Nature Conservancy is to preserve plants, animals, and natural communities that represent the diversity of life on Earth by protecting the lands and waters they need to survive.
- U.S. Fish and Wildlife Service: The U.S. Fish and Wildlife Service's mission is working with others to conserve, protect, and enhance fish, wildlife, and plants and their habitats for the continuing benefit of the American people.
- Pennsylvania Bureau of Forestry: The mission of the Bureau of Forestry is to ensure the long-term health, viability, and productivity of the Commonwealth's forests and to conserve native wild plants.

A community-based ecosystem management group might produce a mission statement like this:

> We commit ourselves to conserve, protect, and enhance all living creatures in our watershed as renewable and sustainable resources for the benefit of all our citizens, those alive today and those who will be born in the future. To accomplish this mission, we work in collaboration with our neighbors, elected and appointed officials, business leaders, and others to make our watershed a good place to live—for people and for the other creatures with which we share the Earth.

What specific guidance does this statement contain? In what ways is the statement clear and not clear? How does it make you feel? Of what use is it?

Generally, a group's mandate is much broader than the work that it actually does, which must be restricted because of other instructions that it receives, the limits of its budget, and the priorities it sets. For example, the hypothetical mission statement for a community-based ecosystem group described in Box 11.4 pledges the group to protect and enhance all the natural resources of the community for all the people; the actual work will probably focus on a much smaller subset, such as restoring a specific tract of degraded municipal land. Mandates and missions often serve as inspiration for staff and stakeholders, but much more definition is needed to direct exactly what the group will be doing.

> EXERCISE 11.3
>
> **Collaborate On It!**
>
> In a small group, consider the situation in your scenario—the resources, setting, stakeholders, issues, and opportunities. Based on these, and other information in the scenario, draft a mission statement for the team of people who would lead an ecosystem management effort. Ask others to evaluate the mission statement in terms of its ability to capture the spirit and aspirations of the ecology, people, and institutions of the scenario.

STRATEGY

The next level in the step-down process is called **strategy**, a general approach for accomplishing a mission or mandate. For example, in attempting to have a winning football season, a team could choose a strategy based on a strong defense, a strong offense, or even a passing versus a running offense. These represent different ways to go about achieving the same mission, but they would require much different resources, skills, preparation, and operational decisions. Thus, an explicit strategy significantly influences later decisions.

Traditionally, governments have used four basic strategies for getting their work done:

- Direct management—for example, owning the resource, such as a wildlife refuge or park.
- Regulation—for example, requiring permits for filling or draining wetlands.
- Financial incentives—for example, paying farmers to replace crops with conservation plantings.
- Technical assistance and education—for example, employing extension agents or service foresters to develop management plans that landowners can adopt if they wish.

Similar strategies have been and could be adopted by community groups, NGOs, or informal associations of landowners. For example, The Nature Conservancy's early strategy for accomplishing its mission was to acquire lands and later deed them to public agencies as parks and refuges. More recently, organizations have added other strategies, including partnerships (e.g., community-based management) and co-management (sharing decision-making authority among organizations).

Selecting a strategy is important because it focuses the group on a narrower range of subsequent choices and actions. For example, a group whose strategy was regulation (or legal challenges based on regulations) would employ mostly lawyers and law enforcement officers, whereas a group that chose technical assistance would employ mostly educators. An NGO whose strategy was direct management would raise funds to purchase land, whereas an NGO choosing partnerships might direct the same funding to pay landowners for improving their lands. Often, however, groups do not explicitly state strategies; instead, they embed strategies in their goals.

> EXERCISE 11.4
>
> **Collaborate on It!**
>
> Typical strategies include direct management, regulation, incentives, technical assistance, and partnerships. Describe how a community-based team might use each of these strategies to accomplish the following task (one for each scenario):
>
> - ROLE Model: Protecting the native minnow community in Little Lake.
> - SnowPACT: Conserving the Snow River cutthroat trout in the Snow River watershed.
> - PDQ Revival: Ensuring that the wetlands of the PDQ watershed do not get converted to other land uses.

GOALS

The next stage in strategic thinking is selecting a series of goals for the group. A **goal** is a general description of what the members seek to accomplish. Based on the mission or mandate and directed by the chosen strategy or strategies, the group selects a small number of primary interests, expressed in about 3–10 goals. Groups usually enjoy this process because goal setting feels good, makes great press, and generates little controversy. Goals are visionary, general, and qualitative. They are not designed to be achieved, but to express intention; they point the direction, but do not define the route. Goals represent a team's aspirations, the vision they have for what their efforts can bring. Ecosystem management often expresses this concept as **desired future condition**, a description of the intention a community has for its lands and waters (Box 11.5).

The real value of goals, however, is to begin establishing what the group will emphasize. Mission and strategy still leave a great range of choice. For example, a community-based ecosystem group might have a mission to "maintain and restore the ecosystem" and might choose a direct management strategy based on the single-species approach. However, the group still needs to decide which lands to protect, which species to use as the basis for management, and what techniques to apply. These choices would be defined first as goals. Hard choices begin to be made at the goal-setting step because an organization can only do so much.

BOX 11.5

Desired Future Condition: A Goal for Ecosystem Management

According to the Committee of Scientists, a team of experts assembled in 1998 by the U.S. Secretary of Agriculture to help reform management planning for national forests and grasslands, desired future condition has the following characteristics:

- Links assessments of current conditions to decisions about what to do.
- Requires extended public dialog because it is a social choice affecting current and future generations.
- Seeks to protect a broad range of choices for the future.
- Must be dynamic, allowing for learning through monitoring, review, and evaluation.
- Seeks to sustain ecological integrity over the long term, under the premise that preserving natural disturbance patterns and processes will ensure ecological resilience.
- Seeks to sustain the social capacity for future generations to support cultural patterns of life and adapt to evolving societal and ecological conditions.

As part of the Northwest Forest Plan, the Little River Adaptive Management Area created a shared vision of desired future condition through an assessment process and a collaborative planning process among agencies, local entities, and the public. They wrote their desired future condition as follows:

Biophysical: Terrestrial

- Highly productive timber management areas.
- Landscape more resilient to wildfire, disease, and insects.
- A network of late-successional forest, with an emphasis on riparian areas.
- Legacy habitat components left or developing in all stands.
- Diversity of native plant and wildlife habitats and populations.

Biophysical: Aquatic

- Riparian areas dominated by late-successional condition with increased levels of large in-stream wood.
- Water quality that meets Oregon Department of Environmental Quality standards.
- Sediment and flow regimes that result in high-quality aquatic habitat.
- Increased populations of healthy, native fishes and other aquatic organisms.

Socioeconomic

- Public knowledgeable about ecosystem management.
- Sustainable and dependable harvest level.
- Private landowners and local public participating in collaborative land management processes.
- Improved and increased recreational opportunities.
- Funding continuity through interagency cooperation.

EXERCISE 11.5

Collaborate on It!

Goal setting continues the hard decision making of strategic thinking. Within a very broad mission and a given strategic approach, the next step is selecting a few goals for a group. Under the mission statement in Box 11.4 and the strategy of partnering, goals might include the following:

1. Protect and enhance undisturbed habitat patches of the watershed, on all ownerships, and, through partnerships, across ownerships.
2. Restore at-risk native plants and animals to their former abundance and distribution, using both public and private lands.
3. Teach our children about the plants, animals, and special places in our watershed.

Consider these questions:

- Do these statements provide more specific guidance than the mission statement?
- Are these goals clearer and less ambiguous than the mission?
- If these represented all the goals for the community group, have important parts of the mission been ignored?
- What other goals would you add?
- How would you know when you had enough, or the right, goals?

After considering the overall description of a scenario, and the mission that your team wrote in Exercise 11.4, convene a small team of "community leaders" and write a set of goals for the scenario. Remember, these define what the ecosystem team will address, and they will not address aspects that are not listed.

OBJECTIVES

The difference between goals and objectives, and our willingness to shift logically between them, is the distinguishing characteristic of strategic management. Here is where the magic happens, where a group steps from its general purpose to specific accomplishments and time frames.

Moving from goal to objective is a particularly daunting step in strategic thinking. Dedicated members of an ecosystem group, whether natural resource professionals or community members, hesitate to make such choices. Recognizing that we cannot do all things is hard enough, but actually listing those things that we will do and, by their absence, those things that we will not do is a tough intellectual and emotional task. When we fail to make explicit choices, however, we run the risk of wasting our capacity by spreading ourselves too thinly—never doing anything well enough or thoroughly enough to make a difference. But when we focus our capacity on a smaller number of actions, we can significantly advance conservation.

For each goal, this step requires the creation of one or more specific objectives that will be accomplished. **Objectives**, therefore, are very different than goals, because objectives are quantitative, specific, and designed to be achieved. Furthermore, from a large universe of possible objectives, only a few can be selected.

How does a group choose its objectives? Although there is no formula, certain perspectives and practices can help. First, objective setting depends on the knowledge and commitment of the people in and around a group. It depends on a sense of what is important, as well as what is technically possible, socially and politically acceptable, and financially responsible. Choosing objectives requires the best and most selfless thinking that the members of a community or an ecosystem partnership can muster.

Second, the best objectives generally emerge from a team of folks who will be responsible for the work that will occur. Therefore, the objective-writing team should include a diversity of people, especially those who will implement the work to achieve the objectives and those who will pay for the effort.

Third, the best objectives are linked to outputs and benefits, rather than inputs and processes. These terms refer to the way systems-oriented specialists look at any activity (Figure 11.6). An activity begins by amassing the needed *inputs*, which are then used in a series of *processes* to produce *outputs* that provide *benefits* for the recipient (customer, client, neighbor, community). In terms of land conservation, an ecosystem group may ac-

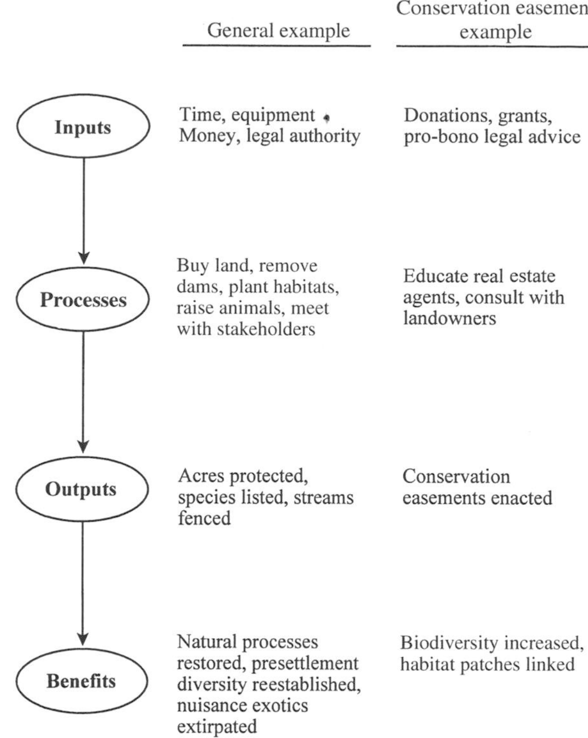

Figure 11.6. *The systems approach includes the steps of inputs, processes, outputs, and benefits.*

quire inputs (funds, expertise) that are used in processes with landowners (informational meetings, individual consultations) so that the group can provide outputs (conservation easements, areas protected, acres of habitat restored) that result in benefits (biodiversity increases, greater ecosystem resilience).

Fourth, good objectives are information-rich. They contain details that relate to their achievement and mileposts along the way that also should be watched. Information-rich objectives make the work obvious and allow better evaluation, both at the end and along the way, so that better decisions can be made about continuing, modifying, or abandoning an objective (see Chapter 12).

Objectives must be put into a form that allows strategic thinking to continue on to become strategic action. The technical characteristics of good objectives can be remembered through a simple acronym: Good objectives are SMART (Box 11.6).

- *S* is for *Specific*. A good objective defines a positive change that can be made in the condition of the ecosystem. For a goal of "restoring native biodiversity," an objective will address the status of a particular ecological community, species, population, habitat, or process. An objective to "increase the abundance of rare species" is not specific enough; an objective to "double the number of suitable home ranges for the American marten" is specific.
- *M* is for *Measurable*. A good objective is quantitative, providing a way to measure whether or not the objective has been achieved. Quantitative objectives can take many forms—numbers (e.g., 20,000 acres covered by conservation easements); comparisons (e.g., reduce vehicle-caused tortoise

BOX 11.6

SMART Objectives for Chesapeake Bay

The Chesapeake Bay Program is a collaborative effort of federal and state agencies, communities, and NGOs to restore and sustain the Chesapeake Bay ecosystem. To direct its strategic intent into specific actions and accomplishments, the Chesapeake Bay Program has stated the following objectives:

- Increase riparian buffers on 2010 miles of streams and shorelines by the year 2010.
- By 2003, open 1356 miles of stream for fish passage to restore spawning habitat for migratory fish.
- Increase the recovery of native sea-grass beds in the Chesapeake Bay to 114,000 acres by 2005 and 225,000 acres by 2010.
- By 2010, reduce total nitrogen and phosphorus loading to the Chesapeake Bay, including that from atmospheric deposition, by 50% from 1985 levels.
- By 2010, reduce the amount of toxic substances entering the Chesapeake Bay from all sources by 50% from 1995 levels.
- The annual loss of existing streamside buffers should be reduced by 75%, and 5000 additional miles should be restored by 2010, using 1995 as a baseline.

mortality each year by half); or ranks (e.g., achieve the lowest rate of legal challenges to the plan among all national forests). An objective to "reduce water pollution in the watershed" is not measurable; an objective to "reduce phosphorus loading in the New River by 50%" is measurable.

• *A* is for *Accountable.* Accountability means that the group has accepted responsibility for addressing the objective—and, more importantly, that someone has agreed to tackle the work. The best objectives have deep commitment by the group and its members, compelling them to dig into the projects that will follow.

• *R* is for *Realistic.* Realistic objectives have a reasonable possibility of happening. This can mean the technical capacity is adequate, the sociopolitical climate is accepting, the land or water resources can support the intended organisms or uses, or the organizational resources are available. Realistic also means that the objective is within the group's sphere of responsibility or influence. An objective to "restore native prairie ecosystems to 50% of Illinois" is not realistic; although it is technically feasible, few people would trade the economy of Illinois (or our food supply) to accomplish such an objective. However, an objective to "restore native prairie ecosystems on 50% of abandoned railway rights-of-way" might be realistic.

• *T* is for *Time-Fixed.* Every good objective states when it will be done. Along with a final deadline, good objectives also include intermediate deadlines, or milestones. An objective to "quadruple the number of conservation easements in four years" might not give enough guidance; a better version would be to "double the number of conservation easements in 2 years, and double again in the next 2 years." The time frame also needs to be specific, to avoid confusion. Thus, an objective to "break ground on the environmental education center in 2005" could mean January 1 to some people and December 31 to others; "break ground by March 31, 2005" leaves little room for confusion.

EXERCISE 11.6

Collaborate on It!

Analyze the following objectives in terms of how SMART they are, and revise them to make them as SMART as possible.

1. Improve habitat for red wolves by 10% by 2003 in Harnett County, North Carolina.
2. Harvest the excess production of elk from Yellowstone National Park each year.
3. Collect all needed data on threatened species in the Everglades by 2005.
4. By 2004, implement a youth education program for all second graders in Fort Worth.
5. Bring all private, undeveloped lands in the Smith River watershed under some form of conservation easement by 2010.
6. Answer 90% of inquires by reporters within 2 days of initial contact, starting immediately and continuing indefinitely.
7. Write a recovery plan for every endangered species within 18 months of listing.
8. Reduce the stocking of non-native rainbow trout in Lake Champlain by 25% per year for the next four years.

For the goals that you created for a scenario in Exercise 11.5, prepare several objectives. Have the objectives reviewed internally by your team and externally by members of other teams.

PROBLEMS AND TACTICS

A logical assessment is needed that carefully defines the specific **problems** that stand in the way of achieving a stated objective. The problems may be ecological, socioeconomic, or institutional, and they must be stated explicitly. (Remember the old saying that recognizing a problem is half the solution?) As the examples in Box 11.7 illustrate, an organization cannot protect critical habitat if it does not know what an organism needs.

Having recognized what problems stand in the way of achieving an objective, one or more **tactics** can be chosen to overcome them. If the problem were lack of a habitat model for a species, a tactic could be chosen from options such as conducting a research project, adapting a known model from a similar species, convening an expert panel to write

BOX 11.7

From Goals to Projects

After setting goals, strategic thinking becomes even more difficult. From now on, the decisions become specific and measurable—no more platitudes! The example below picks up on the mission statement from Box 11.4 and one goal from Exercise 11.5. Many possible projects could have been developed to satisfy the objective and use the tactics identified; a careful process would be needed to conclude that the two projects listed were the ones with the highest chance of success.

Goal 2
Restore at-risk native plants and animals to their former abundance and distribution, using both public and private lands.

Objective 2.1.
Reestablish five breeding populations of blue grouse in the ecosystem in the next 5 years.

Problem 2.1.1.
The critical habitat for blue grouse has not been defined, prohibiting managers from identifying and characterizing the best possible locations for reestablishing populations.

Tactic 2.1.1.
Consultation with the Ruffed Grouse Society indicates that many states have grouse restoration projects under way and that several university researchers have developed preliminary models for explaining blue grouse distribution. Therefore, the tactic chosen to overcome this problem will be convening an expert panel to develop a critical habitat description.

Project 2.1.1.
With help from the state DNR, the Ruffed Grouse Society, the Wildlife Society, and the extension service of the land-grant university, conduct a nominal group technique to construct a quantitative model of blue grouse habitat requirements. (Fifteen top experts will be gathered at the next annual meeting of the Wildlife Society for a 2-day work session, followed by computer analysis and subsequent review.)

Problem 2.1.2.
Blue grouse do not fly long distances, making it unlikely that they will disperse effectively across the landscape rapidly enough to colonize new habitats within a few years.

Tactic 2.1.2.
Blue grouse from existing populations within the ecosystem will be captured and moved to suitable habitats (as identified in Project 2.1.1).

Project 2.1.2.
With the help of grouse hunters and students from the university, 100 fledgling blue grouse will be located, captured, banded, and released, 20 each in the five most promising unoccupied habitat patches.

a new model, or instituting an active adaptive management experiment. After some assessment of the available tactics, one or more must be selected; in Box 11.7, we have chosen to convene an expert panel to create a model.

PROJECTS

Strategic thinking is a prelude to the actual accomplishments that happen in the third step of strategic management. The selection of projects links the strategic-thinking step to the implementation step—the real work of ecosystem management.

Implementation converts objectives, authority, funding, and people's time, energy, and hopes into accomplishments through tangible work projects. In its simplest definition, a **project** is an interrelated group of activities needed to achieve a planned objective. The emphasis in the definition is on two terms: interrelated and planned. Interrelated activities are those components essential to completing a project from start to finish (including measuring and reporting the results of the work to stakeholders). A planned objective is one found in the strategic plan, generated from the higher-level mission and goals, and further described by

BOX 11.8

Selecting Projects for the San Francisco Bay Ecosystem

A good example of how strategic thinking is converted into projects is the San Francisco Bay ecosystem restoration program, commonly known as CALFED. CALFED funds a broad range of projects that include habitat restoration, improvements to a community's drinking water, improved irrigation practices, the installation of Best Management Practices (BMPs) on farmland, and projects that emphasize ecological farming and enhance the economic stability of farm communities. Some of these are large in scope; one 4-year project on Battle Creek was begun in 1999 with the objective of restoring habitat on 42 miles of stream at an estimated cost of $51 million. Other projects are smaller but equally well focused on overall results that support the program's strategic direction.

CALFED, like many ecosystem efforts, invites its stakeholders to submit projects for consideration each year. It receives hundreds of proposals from state agencies, federal agencies, local governments, for-profit and nonprofit private organizations, universities, and tribal governments. Decisions by its committees of experts about what to fund are posted on the Web site, along with a description of each project, its relationship to the overall Bay-Delta plan, the project's specific objectives, estimated costs, expected outcomes, and partners.

The projects submitted compete in a two-step review process. First, a committee of experts and stakeholders looks at the technical merits of each proposal. They ask questions such as: Is the project focused on objectives that support the overall CALFED goals? Does it use sound science? What ecological benefits will likely result from the project? Are the engineering figures correct? Is the project cost-effective? To what degree are local partners committed to the project? How much money, material, or effort are they contributing to its success? What are the measures of success for the project? Have they built adaptive management into their approach? Are the applicants qualified to conduct the project?

The recommendations of the technical reviewers are sent to a second committee, where the projects are reviewed from the broader standpoint of their integration with other CALFED work: How do the projects fit together with other CALFED work? Is there a balance between research and implementation projects? If funded, how will each project fit with potential future actions? How does the project fit with past work?

Funding is competitive; in a typical year, CALFED receives many more proposals than they can afford to fund. As with many smaller community-based approaches, the high degree of openness and emphasis on timely communications build trust, show accountability, and actively engage stakeholders in the effort of accomplishing actual work.

problem analysis and tactic selection. In this way, a project is part of a greater whole; those involved in preparing, conducting, and evaluating projects can see why their project matters (Box 11.8).

As with objectives, the project is explicit. It is a written declaration that answers at least the following questions:

1. What specific work will be done, and where?
2. Why is it necessary to do this work?
3. What work will be accomplished?
4. How will the work be done (i.e., the technical steps)?
5. Who will do the work, and how are these people qualified?
6. What will the work cost, in money, time, equipment, and land and water resources?
7. What is the schedule for the work, including the firm completion date?

Of these seven questions, the first three link the project to the overall goal, objective, problem, and tactic that the project addresses. The remaining four questions provide details that enable decision makers to select among possible projects as they allocate the time, expertise, funding, equipment, and land and water resources. The answers to these questions also provide the data for the formative evaluation that will help make the decision to undertake the project or not (see Chapter

BOX 11.9

Writing an Ecosystem Project

Projects are the ways that strategic intention is turned into actual work. The project below responds to the strategic elements described in Box 11.7, the restoration of blue grouse populations. Specifically, the project addresses the capture, relocation, and release of blue grouse.

Tactic: Introduce fledgling blue grouse into suitable habitat patches.

Project: Fledgling-grouse capture and relocation project in Spruce County.

Authors: Arthur Smith, wildlife biologist, and Jean Howard, landowner.

Location: DNR wildlife refuge and adjoining private lands, Spruce County.

Project Output: Two new breeding populations of blue grouse established within 24 months of project initiation; a videotape of project work, documenting success of the introduction, edited and available for viewing, within 30 months of project initiation; an analysis of how well the populations are doing in the new habitat, compared to a similar habitat close to an existing population of blue grouse. (*Note: This makes the project a passive adaptive management experiment; don't forget adaptive management!*)

Justification: We need this project in order to address our goal of restoring. . . . (*Note: A concise, readable, logical, and compelling analysis of why the project is needed.*)

Work Items: (*Note: This list defines all the actual work steps that will be necessary, from project start to finish.*)

1. Acquire permission to trap fledgling blue grouse from DNR.
2. Identify likely trapping locations.
3. Identify most promising unoccupied habitat patches, based on work from ongoing expert panel study.
4. Acquire permission from landowners to introduce blue grouse onto their property.
5. Select and begin monitoring a "reference area" near an existing blue grouse population.

⋮

15. Edit videotape into 15-minute program suitable for general viewing.
16. Present program to ten selected audiences.
17. Analyze populations in the areas stocked and the reference areas.

Time: 30 months from time of project authorization

Cost: $35,000

0.2 FTE of DNR Supervisory Ecologist. (*Note: An FTE is a common staffing term that represents one "full-time equivalent," or the time one worker puts in yearly, about 2080 hours.*)

0.3 FTE of DNR Field Technician.

0.2 FTE of volunteered videotape technician from local TV station.

One week each of 5 volunteer hunters, to locate birds.

One week each of 5 university graduate students, to trap and move birds.

---

12). Box 11.9 outlines a hypothetical project narrative that shows the link between mission and implementation.

We have presented an idealized form of strategic management, complete from mission to project. Although most individuals, groups, community organizations, and agencies do not act as formally as we have described, many are working hard to think and act strategically. As we have presented many times in this book, however, the particular process is much less important than the purpose—to learn and improve as we conduct our work. Strategic management is one way to make learning highly effective by making our intentions explicit and public. The next chapter takes strategic management to its rightful conclusion: evaluating what we have done and deciding how to do it better.

References and Suggested Readings

Committee of Scientists. 1999. *Sustaining the People's Lands: Recommendations for Stewardship of the National Forests and Grasslands into the Next Century.* U.S. Department of Agriculture, Washington, D.C.

Cortner, H.J., and M.A. Moote. 2000. *The Politics of Ecosystem Management.* Island Press, Washington, D.C.

Covey, S. 1990. *The Seven Habits of Highly Effective People.* Fireside Books, New York.

Crowe, D.M. 1983. *Comprehensive Planning for Wildlife Resources.* Wyoming Game and Fish Department, Laramie.

Drucker, P.F. 1974. *Management: Tasks, Responsibilities, Practices.* Harper & Row, New York.

Dye, T.R. 1984. *Understanding Public Policy,* 5th ed. Prentice-Hall, New York.

Ewing, D.W. 1968. *The Practice of Planning.* Harper & Row, New York.

Keystone Center. 1996. *The Keystone National Policy Dialogue on Ecosystem Management.* Keystone Center, Keystone, CO.

Kinsey, D.N. 1980. Organizing a public participation program: Lessons learned from the development of New Jersey's coastal management program. Coastal Zone Management Journal 8:85–102.

Kotter, J.P. 1996. *Leading Change.* Harvard Business School Press, Boston, MA.

Lee, K.N. 1993. *Compass and Gyroscope: Integrating Science and Politics for the Environment.* Island Press, Washington, D.C.

Lilienthal, D. 1944. *TVA: Democracy on the March.* Harper & Row, New York.

Nadler, G., and S. Hibino. 1990. *Breakthrough Thinking.* Prima Publishing & Communications, Rocklin, CA.

Nielsen, L.A. 1988. Improving planning in agencies and universities. Proceedings of the Organization of Wildlife Planners 10:7–14. (Available from Maryland Department of Natural Resources.)

Pajak, P. 2000. Sustainability, ecosystem management, and indicators: Thinking globally and acting locally in the twenty-first century. Fisheries 25(12):16–30.

Phenicie, C.K., and J.R. Lyons. 1973. *Tactical planning in fish and wildlife management and research.* U.S. Fish and Wildlife Service, Resource Publication 123. Washington, D.C.

Senge, P.M. 1990. *The Fifth Discipline: The Art and Practice of the Learning Organization.* Currency Doubleday, New York.

Steiner, G.A. 1979. *Strategic Planning: What Every Manager Must Know.* Free Press, New York.

Walter, S., and P. Choate. 1984. *Thinking Strategically: A Primer for Public Leaders.* Council of State Planning Agencies, Washington, D.C.

Wheatley, M.J. 1994. *Leadership and the New Science.* Berrett-Koehler Publishers, San Francisco, CA.

Experiences in Ecosystem Management:

If All It Took Was Money, Community-Based Conservation Would Be Easy

Heather A. L. Knight

In the late 1980s, The Nature Conservancy (TNC) became involved in a conservation project in northern Colorado. A historic ranch was threatened with development, and the family was looking for a way to keep developers at bay. In a final attempt to keep the land intact, the landowner negotiated with the State of Colorado to establish a state park. TNC acted as a consultant in this process, supporting protection of this site. Through TNC's strategic-planning process and biological inventory, conducted by the Colorado Natural Heritage Program, the ranch was identified as biologically significant. It was also one of the last remaining roadless river canyons along the Front Range of Colorado and lay at the crossroads between the short-grass prairie and the coniferous forest biomes of the southern Rocky Mountains. When it was evident that a solution would not be reached between the state and the family and that the only alternative was subdivision, TNC raised funds to protect as much of the canyon as possible.

By 1987, TNC had purchased in fee title 1120 acres, protecting approximately 4 miles of the canyon, called Phantom Canyon Preserve. At the same time TNC also acquired a 480-acre conservation easement protecting another 2 miles of the canyon.

From 1989, when the preserve was opened to the public, until 1995, Phantom Canyon Preserve was managed as an isolated site in a matrix of private land. During this period, TNC was going through a fundamental change in its approach to conservation. It was becoming increasingly obvious that to protect biodiversity on landscapes that were a blend of private and public lands, effective conservation strategies needed to incorporate the human communities into conservation efforts. TNC would never be able to buy enough land to ensure the existence of an area's natural heritage.

Through a strategic-planning process conducted by TNC and the state Natural Heritage Program located at Colorado State University, a 100,000-acre site was identified as having important ecological values for plant and animal communities. The site, called the Laramie Foothills, lay in a mountain valley situated on the east flank on Colorado's Front Range. The North Fork of the Cache la Poudre River winds its way through the area, forming spectacular granite canyons, including the Phantom Canyon Preserve. Stretching east to west, it connects the westernmost edge of the Great Plains to the beginning of the foothills of the Rocky Mountains, and north to south, it connects the southern end of the Laramie Plains to the northernmost edge of the Colorado High Plains. This valley is also a critical link of private land that, if protected, would reconnect the USFS Pawnee National Grasslands to the east with the coniferous forests and alpine tundra of the USFS Roosevelt National Forest to the west. This protection would once more allow traditional east-west migrations of species that had become increasingly fragmented by roads and housing developments.

At the same time, the TNC staff living and working in the community had begun to learn and appreciate the human history of the area. It was evident that this was a culturally rich community that valued land health. Since the 1800s, generations of families had stewarded this land, developing intimate relationships, caring for the water and soil upon which their livelihoods depended. Ranching was the sole remaining economic use of the land that had persisted for over a century; logging, mining, and farming had all boomed and gone bust. Ranching is not an economically lucrative land use, but it is a sustainable use if grass, soil, and water are husbanded. Indeed, what land use that generates great profits is sustainable? By definition a sustainable economy is one that lies on the economic margin of profit and loss.

But this ranching community was threatened. Like many western landscapes, fast-paced growth was rapidly converting former ranchlands at the valley's periphery into endlessly sprawling housing developments. Over 2000 ranchettes—small-acreage subdivisions—already rimmed the higher-elevation private lands to the north, south, and west, and abutted the Roosevelt National Forest. Land that was valued at a $50 an acre for ranching was worth one to two orders of magnitude more for houses. When ranches went on the market, ranchers could no longer afford to buy this land; developers had the trump card. Land use was rapidly changing from a once agriculturally dominated landscape to a commuter landscape of city people "living country." Water, essential for hay production, was gradually being shunted to the cities for lawns and swimming pools. In only 3 years, a share of water from the North Fork of the Cache la Poudre River had gone from $5000 to $40,000.

The biological communities, too, were under threat. Fire suppression, altered grazing patterns, water diversions, invasions of exotic plant species, new roads and subdivisions, and increased recreational activities were all fragmenting and disrupting the landscape at a larger and more intense scale than ever before.

Along with all of these changes, we were witnessing the loss of a generation of land stewards. People whose livelihoods were tied to the land through animal husbandry were being replaced by others who appreciated the land for its beauty but who had not taken the time to learn about its human or natural communities. Where conversations once centered on the weather, now the weather was merely background noise. As these families of stewards were lost, so was their knowledge of land, water, grass condition, animal husbandry, wildlife, and stories of preceding generations. The "newcomers" among us, although caring, lacked the skills to learn and often did not even know the right questions to ask of the "oldtimers."

Through listening and learning from neighbors, it became apparent that this was a place worth saving not only because of its biological diversity, but also because of its cultural heritage. In fact it was clear that we, TNC, would always be one of multiple newcomers and that the best way to help the Laramie Foothills would be to enable those landowners who stewarded the land to find sustainable ways to stay on the land. After all, they were the individuals who knew the place best and who were bound to the land by family and time. In such a rapidly changing landscape, could TNC work in this community to build partnerships that would achieve the conservation both of land health and of culture? It was essentially up to the community to decide; TNC was willing to try.

Slowly TNC, unknowingly at first, had embarked on what has come to be called community-based conservation. The Phantom Canyon Preserve Steward found herself spending time at the kitchen tables of ranchers listening, drinking coffee, and discussing the issues around land-use change. Likewise, time was spent participating in community events, riding and mending fences, branding and checking cattle. The TNC steward, who now lived in the valley and was also a landowner, had begun to ask questions of private landowners and public land managers, exploring issues and visions of land management that each held in common, and that separated them (Figure A).

These activities were taking place against a backdrop where, at the county level, anxieties were being fueled by rapid growth as Colorado was losing over 270,000 acres of open space each year to residential and commercial development. It was "boom time" once more in the Rockies, and people were taking a stand as either pro-growth or

Figure A. *A typical community meeting in the Laramie Foothills region. Can you spot the TNC representative in this group? If not, why not?*

anti-growth. At the county level, community meetings flared around private property rights laced with heated arguments and threats. In this atmosphere, TNC felt that a more positive approach would be to work on less-threatening projects that would bring folks together over issues that affected land and human health.

In such an environment, how could the Laramie Foothills community find a safe and cohesive place to work together cooperatively? Through community meetings and time spent walking the land and listening to the water, it became obvious that there was a pervasive and noncontroversial issue that threatened everyone: the invasion of weeds. TNC staff quietly took every opportunity to pass on information about weed identification and management. When asked, TNC responded to neighbors, and continued to work diligently on weed control on its own lands. Over time, it found itself being invited to work on neighboring lands and waited for other landowners to respond.

In the winter of 1998, neighbors came to TNC and asked if it would join forces to work on weeds. A small group was established, and a brainstorming meeting was held. The group recognized a common concern and committed to a long-term weed project. A vision was defined, goals set, and tasks assigned. A larger community meeting was organized, inviting everyone in the watershed and purposefully including key players such as the Western Governors Association and the U.S. Forest Service. The group decided to start small and work on projects that ensured success. During the summer of 1999, four "weed tours" were hosted, visiting landowner's properties, identifying weeds, and discussing how to control them. A local newspaper reporter became involved and ran a story on weeds.

By the end of the first season, we had over 30,000 acres enrolled in the project, including private, state, and federal lands. Unexpected partnerships developed, several of which involved individuals and groups who were at loggerheads on other issues. Who would have imagined an irrigation company board member, a nun, and a rancher working together? The group expanded their vision and set out goals for 3 years. The project was so compelling that some landowners quadrupled their monetary contributions. The Colorado Division of Wildlife (CDOW) found money to be used for weed control across their administrative boundaries on private lands, anticipating further cooperative stewardship ventures. The CDOW's monies were leveraged with other groups and individuals, and TNC helped create a community-based weed coordinator. The group named itself the North Fork Weed Cooperative and began publishing a newsletter, *The Weed Roundup*; over 300 neighbors received the first mailing. Training workshops were held, and landowners prepared weed management plans for their properties. These conversations and actions—based on a cooperative integrated weed management area—led to the idea of a cooperative stewardship area, including weed control, fire, and grazing. Although the group's challenges remain great, and weeds are still a threat, the community had demonstrated it could collaborate on issues that dealt with land health (Figure B).

Figure B. *Two community projects in the Laramie Foothills region, using cooperative and largely volunteer help: controlled burning (top) and planting of native vegetation (bottom).*

During this time, the listing of Colorado's first threatened subspecies of a mammal, the Preble's meadow jumping mouse, landed on the community's doorstep. Exurban development along the Colorado Front Range had destroyed most of this species' habitat, restricting it to a few pockets of land still devoted to ranching. The largest and most intact population in the state was found in the Livermore Valley, our community. Ironically, after generations of good stewardship that allowed the mouse to persist, landowners were now faced with the Endangered Species Act and restrictions on the very activities that had protected the subspecies.

The challenge was to respond positively. Through the determination and leadership of community members, we were given permission to embark on our own Habitat Conservation Plan (HCP). Landowners, realizing that their community was under threat, took the initiative to attend state meetings on the mouse to learn how to prepare an HCP. Local meetings were organized, and neighbors struggled with issues regarding the historic uses of their lands and what was required to help save the mouse. The U.S. Fish and Wildlife Service (USFWS) was invited to tour the valley with the landowners and discuss their concerns. Slowly a partnership was formed between private landowners, TNC, CDOW, and USFWS. Landowners worked hard to draft the first version of the HCP. They identified a core conservation zone along 216 miles of streams in the valley, totaling 3540 acres of land.

TNC continues to work in this community, viewing itself as only a partner with much to learn. It is clear that TNC's success in conservation actions is mostly due to a community willing to allow its participation. Slowly TNC has built trust, working hard to demonstrate its commitment to healthy and robust human and natural communities. Landowners continue to view TNC as a resource and a partner on land management and protection projects. As of spring 2000, TNC held conservation easements on 7350 acres and held title to 1660 acres—a far cry from when it owned a tiny preserve surrounded by immense lands whose owners it did not know. TNC's management responsibilities and role have vastly expanded. More time and resources are spent each year on properties across the fence from TNC's boundaries. New ideas are no longer unusual in the valley, nor are diverse partnerships. TNC personnel are reminded that to be successful in protecting the area's natural communities, the human community must assume responsibility.

In an effort to strengthen community commitment to each other and to the land, an education program has been piloted. The mission of the Poudre River Ecology Project (PREP) is to implement a place-based river ecological curriculum for kindergarten through sixth-grade schoolchildren focused on conservation in the watershed. Three mountain schools have formed a partnership with public land managers, private landowners, conservation groups, parents, and community volunteers to implement an interdisciplinary river ecology project in the local community. Partners act as instructors and mentors as students undertake conservation projects at sites located within the watershed. Curricula are designed to (1) build a strong sense of place and a land ethic through experiential learning in the student's local community, while interacting with local people; (2) meet school district learning needs; (3) meet state curriculum standards; (4) enhance the current curricula by providing extended learning opportunities in the school's local community; and (5) provide meaningful scientific data to local land managers and landowners who are addressing conservation issues. Currently, 60 children once a month go out onto neighbor's lands, write of their experiences, ask questions, and discover the wonder of inquiry.

Although these successes have been important and have made a difference, not all projects have come to fruition, try as we might. We continue to be humbled by our limitations and realize that the challenges ahead of us will be more complex and require more creativity, patience, and compromise. The conservation of one ranch in particular exemplifies this.

When a fifth-generation ranching family asked TNC to help them protect their 16,000-acre ranch, it jumped at the chance. The family had worked for over 30 years to ensure that the integrity of the ranch would be protected from development. The ranch had been identified as a critical component in the valley, being both culturally and biologically important. TNC believed it had "protected" the ranch because the landowner had approached it, the site was biologically important, the money was available, and the landowner was willing. Unfortunately, TNC underestimated two critical factors that can prevent successful conservation: the power and disruptive influence of minority voices in the community and the effects of a disconnected family.

Misinformation had been spread through the community via "neighbors" and members of the media that wanted sensationalism rather than facts. Indeed, in some newspapers TNC was labeled as "spies" sent to "seduce" the family. Hopes were dashed. Attorneys were paid huge amounts of money. Grants to buy the ranch were withdrawn. Most importantly, the family that owned the ranch carried a huge burden of worry, and TNC appeared to fail to keep a promise. After all that TNC had learned about community-based conservation, it had reverted to the "old way of doing business"—that is, it treated the project as just another real estate transaction rather than a complex family issue. It took the quickest, easiest route, rather than show empathy and concern for the family problems. It hired renowned but distant lawyers who had yet to visit the valley. After a quiet time, thanks to a neighbor's efforts, the family is starting once again to talk about finding ways to reconcile their differences, which may ultimately result in conservation of the ranch.

After working and living in this small valley, TNC has learned that successful community-based conservation is not just a matter of money, it is about successful relationships. In order to work, community-based conservation requires a long-term investment in building relationships, listening to neighbors, developing trust, promoting honest conversations, getting to know the human and natural histories of a place, a willingness to get dirt under your fingernails, an ability to make mistakes and ask for help, creative thinking, and a shared community vision. Community-based conservation is about becoming part of a place. Although it may require more time and effort than traditional conservation approaches, its results may last longer.

Discussion Questions

The Scenario: You have just moved to a new place, a small, rural, agricultural-based community. As you drive into town, you notice "for sale" signs, new roads, and utilities being placed across the first piece of farmland you see. The majority of the land, however, remains in large private agricultural ownership. To the west lies an extensive tract of public land, albeit increasingly fragmented with private in-holdings. You have taken a position with a federal natural resource agency as the newly hired conservation biologist, the first on staff. You do not have a rural background as such, but are willing to work hard; you care about people and the land, and you are ambitious. Upon your arrival, experienced and established colleagues advise you "not to try anything new because you will fail, as this community has tried everything and does not like newcomers."

1. What is your initial course of action (a) with your co-workers and (b) with the community?

The list of tasks handed to you by your supervisor focuses on inventory and monitoring of species on your agency's lands. Information in files identifies issues of concern in the community that are associated with economic sustainability and land-use change.

2. What should you do with this information?

You hear about a local community building project. Community members will be volunteering their time on Friday afternoons and Saturday mornings to renovate the Community Hall. No one has di-

rectly invited you, but there are posters in town inviting community participation. You ask your peers, and they say not to go; besides, part of it is during work time.

3. What do you decide to do and why?

As you are working on the project, people ask you why you would work for such a federal agency, and they relate stories of all the bad things your agency has done in the past.

4. How do you respond? Do you identify yourself as an agency person or as a community member while you participate in the project? (In other words, what "hat" will you be wearing, and how will you convey that to community members and your colleagues?)

You have been in the community now for about 6 months. Your agency at the national level is sued, and newspaper headlines read "X Species Listed as Endangered!" The largest potential habitat is identified in your watershed. A community meeting is called, and your agency is not invited.

5. What is your strategy for dealing with this issue?

Two years later, you have been making headway in the community. You now hear rumors of your transfer, and your colleagues kid you that you have "gone native."

6. Evaluate how becoming part of the community has been a strength and a challenge for you as (a) a federal employee and (b) a new community member.

Evaluation

EVALUATION IS PERHAPS THE MOST AMBIVALENT FEATURE of modern organizational life. We all know that we *should* evaluate our actions and our performance, but avoid doing so because it is difficult, time-consuming, and often confrontational. Our interest in evaluation arises from the popularity of another concept—accountability. We expect government bureaucrats to be accountable for their actions to the electorate, corporate executives to their shareholders, local officials and community groups to their constituents, and school officials to parents, administrators, legislatures, accrediting agencies, employers—and sports fans! Therefore, evaluation has become the constant companion of people who work in public settings, whether in agencies, NGOs, or community groups.

The members of an ecosystem group also must be accountable. Because of the nature of ecosystem management—voluntary, cross-boundary, aspirational, and adaptive—the plans, processes, and products of ecosystem management must be open, reported, and evaluated. Earlier chapters presented the ecosystem approach as uncertain and evolutionary; if we are to improve our ecosystems under such conditions, then we must be eager to learn how we are doing. Therefore, evaluation is an essential tool of ecosystem management.

> EXERCISE 12.1
>
> **Think About It!**
>
> How do you feel when you are told that your work will be evaluated? What pleases or bothers you about being evaluated? How do you feel when you are asked to evaluate someone or something? Do you have different feelings when you are the evaluator or the one being evaluated? In a perfect world, how do you think an evaluation should be done? Why is evaluation not done that way now?

The Context for Evaluation

Evaluation is the feedback loop in strategic management, helping us learn as we go. The four-step model of strategic management presented in Chapter 11 implies that this feedback occurs after implementation, but that is only part of the story. As this chapter will illustrate, evaluation should occur before, during, and after each part of the strategic management process. The questions we ask and the data we use to answer them may vary substantially at each stage, but the premise is always the same: As adaptive members of an adaptive team, we want to learn from our previous decisions and actions.

Evaluation should also occur at all scales of management. A **policy** is the highest level of decision making; policies create groupwide rules, guidelines, priorities, and culture. Evaluation at this level helps guide the decisions of those who set the group's mission, strategies, and goals. A policy for an ecosystem management team might be that no land protection will be undertaken that causes harm to one person in order to benefit another. Evaluation would then seek out those who had been affected by a decision (e.g., landowners who agreed to the creation of conservation easements) and check their status.

A **program** is the main operational unit for a group; programs often define what gets done and who does it. Evaluation at this level helps guide decisions about goals, objectives, problems, and tactics. A program for an ecosystem team might be the creation of conservation easements to protect habitat; evaluation might track the numbers, costs, sizes, and level of participant satisfaction of creating and managing conservation easements over 1 year.

A **project** is an implementation action for a group. Evaluation at this level helps guide decisions about the specific conduct of the project—the who, what, where, when, and how of group action. A project for an ecosystem team might be the creation of the next conservation easement. Evaluation would track progress and check whether the stated intent was being accomplished as the easement was being created.

Notice our repetition of the phrase "evaluation helps guide decisions." This is an intentional reminder that formal evaluation, by itself, is only one input to decision making. Decision makers—whether elected officials, community leaders, or working group chairs—supplement formal evaluation with their own perspectives and personalities to make their decisions. Experience, training, intuition, political savvy, global perspective, and ambition all play a role. For example, a technical evaluation of optional sites for a new community ecosystem education center might show that one site is superior in cost, setting, accessibility, and proximity to natural features. However, the group's leaders might come to a different conclusion because they know the personalities of the potential sellers and the character of the neighborhoods.

Evaluations have three main purposes and three corresponding types or approaches:

EXERCISE 12.2

Collaborate on It!

Here are three decisions that might need to be made in an ecosystem management setting, one for each scenario:

- ROLE Model: The Nature Conservancy has asked the ROLE Steering Committee to select a name for its landholding on the shores of Little Lake.
- SnowPACT: The Semak Nation has received permission to introduce an experimental herd of elk (20 individuals) onto a small land tract and has requested the advice of the Community Circle to select the best possible site.
- PDQ Revival: The U.S. Fish and Wildlife Service and the state department of natural resources are collaborating on a priority-setting strategy for writing and implementing plans to recover species in the watershed that are in peril, and they've asked PDQ Revival participants to develop criteria for ranking species.

What objective data would you use to help make such a decision? Do those types of data include everything you need to know to make the decision, or would you want to supplement them with other subjective information or feelings? If so, what subjective kinds of data would you use?

Now, think about ecosystem management decision making in general. What sorts of decisions might you want to be made based mostly on objective data? Subjective data? Make a list of the kinds of decisions needed within your scenario, in order from those that should be based on all objective data to those that should be based on only subjective data.

1. Formative evaluation helps planners decide whether or not to initiate a policy, program, or project, and, if so, what resources to allocate.
2. Process evaluation helps planners decide whether or not to modify a policy, program, or project, in terms of resource allocation or performance expectations.
3. Summative evaluation helps planners decide to continue or terminate a policy, program, or project.

Most people focus on the third approach when they think about evaluation. This explains the general fear that evaluation is used primarily for punishment. But remember that evaluation is about learning, feedback, improvement, and adaptive management—not about punishment. This perspective makes the first two approaches as important as the third—perhaps even more important, because proper formative and process evaluations can make summative evaluations easy and welcome (Box 12.1).

BOX 12.1

Evaluating Conservation Easements

Conservation easements are popular ways for bringing private lands under management for natural resource values. Strategic management might produce the following approach to enhancing conservation easements within the mission of a nongovernmental conservation organization:

Goal: Spread the use of conservation easements as a primary strategy that private landowners will adopt on their lands.

Strategy: Provide technical assistance and partial funding for the creation and execution of conservation easements for large-tract private landowners.

Objective: With the support and involvement of private landowners, protect 60,000 acres of high-priority lands via conservation easements over the next 5 years (5000 acres in year 1; 10,000 in year 2; 15,000 each in years 3–5).

All three evaluation approaches help form, implement, and judge the utility of conservation easements. The following questions illustrate the different kinds of information that might be sought and judged during each phase of evaluation.

Formative Evaluation

- Which lands (location, size, habitats covered) ought to be targeted?
- What kinds of landowners (education, personal objectives, age, socioeconomic status) should be offered programs and assistance?
- What forms of conservation easements (permanent/temporary, donated/purchased, complete/partial) are most likely to be enacted?

Process Evaluation

- Have we met the targeted number of acres at each yearly interval?
- Have all the new conservation easements been on high-priority lands?
- Have all the new conservation easements been acquired in large tracts?
- Have all new easements been reviewed and approved by real estate attorneys to ensure that they will stand up to a court test?
- Has the objective been achieved using the anticipated amount of time, space, equipment, and funds?

Summative Evaluation

- How do landowners feel about their decisions to put conservation easements on their lands?
- How have the media responded to the concept of conservation easements?
- Has the rate of inquiries for assistance in establishing easements increased?
- What changes have occurred in the condition of the ecosystem on and around lands protected by easements?
- Have easements generated additional interest in donations of land to public natural resource agencies and NGOs?

Formative Evaluation

Formative evaluation helps planners decide whether or not to initiate a policy, program, or project. It is called "formative" because it helps form the plans for the group or organization; consequently, it is also known as a planning evaluation or a feasibility study. Formative evaluation focuses on two general areas of inquiry.

1. *Determining whether the rationale for a proposed action is valid.* The evaluation asks how well the proposed action matches the characteristics needed for success; such characteristics may be ecological, sociological, or institutional. For example, a goal to reintroduce an extirpated species into an ecosystem might be evaluated in terms of habitat quality—whether the habitat is present, extensive, well distributed, and connected; control of the habitat—whether it is on public or private lands and whether conservation plans are in place; or public opinion—whether species is liked, missed, or hated by landowners and the general public. A formative evaluation also asks how the proposed action fits into the group's higher purpose and aspirations. For example, if the stated strategy for a community ecosystem group is to pursue conservation by helping landowners, then a proposed action to set up a fund for purchasing land would not match the group's direction; however, a fund to share the costs of habitat improvements would match well.

2. *Determine whether the proposed action is likely to work.* Are the available resources (personnel, funds, land, equipment, expertise) adequate? Are the methods appropriate, in terms of efficiency and effectiveness? Has a logical cause-effect relationship been shown, linking the problem or opportunity to this particular approach? Does the proposal meet institutional needs—that is, is it legal, does the group have the authority to act in this way?

The U.S. National Environmental Policy Act (NEPA) mandates formative evaluations for all government actions that might affect the environment. The act calls for environmental impact statements (EISs) or environmental assessments (EAs) that compare the environmental consequences of different possible actions for the purpose of selecting the action that minimizes impacts while achieving the main purpose. Among thousands of EISs and EAs done annually, the recent Northwest Forest Plan (NWFP) is perhaps the most famous example at a landscape level (see Chapter 4).

The decision federal land managers faced was

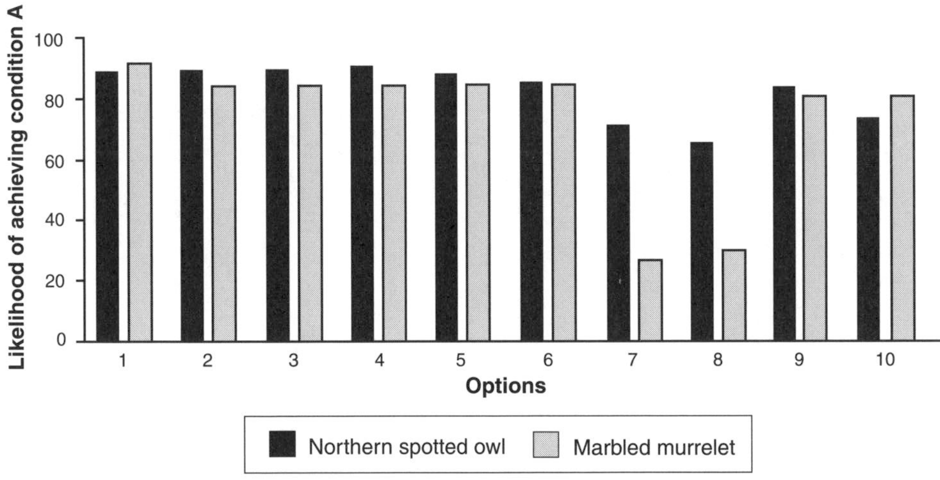

Figure 12.1. *Formative evaluation of ten options for managing federal lands in the Pacific Northwest. This graph shows the likelihood that two endangered species, the northern spotted owl and the marbled murrelet, would have suitable habitats for maintaining viable populations well-distributed across the federal lands (Condition A); an 80% or greater likelihood was considered a desirable outcome. (From Espy and Babbitt, 1994.)*

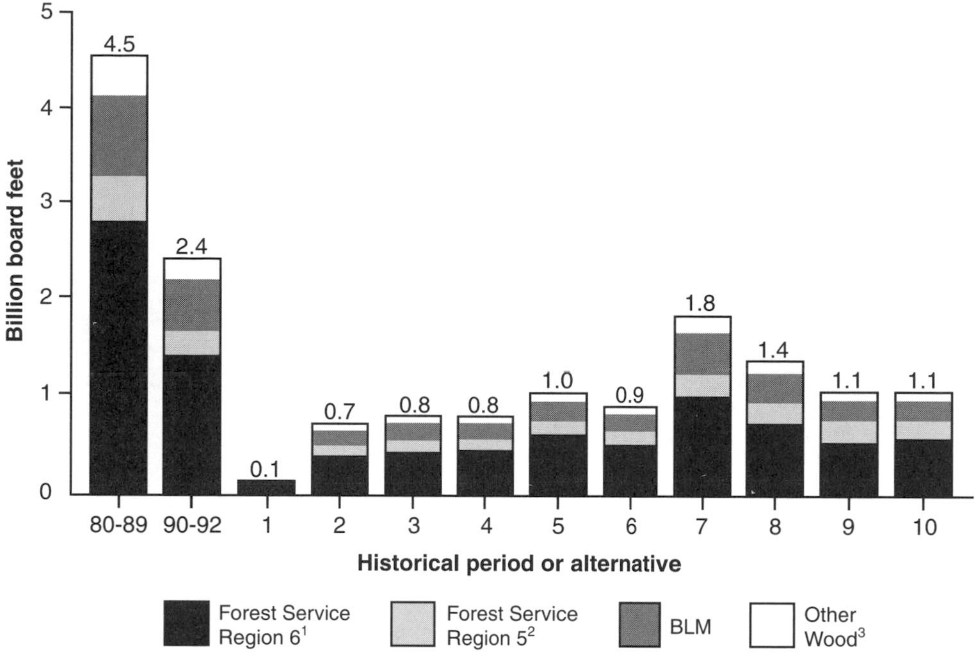

Figure 12.2. *Another part of the NWFP formative evaluation, providing assessments of the probable sale levels of timber annually over the first decade after enacting the plan. Note that this figure provides benchmarks from previous periods that can help decision makers understand the situation. (From Espy and Babbitt, 1994.)*

how to allocate federal lands in Washington, Oregon, and northern California into a series of management categories, ranging from reserves to lands managed actively for timber. They used a variety of criteria to help them evaluate the outcomes of ten different allocation schemes. There were five biological criteria: the viability of northern spotted owls, the viability of marbled murrelets, the viability of at-risk fish stocks, the viability of the community of other species closely associated with old-growth forests, and the likelihood of a properly functioning old-growth forest ecosystem (Figure 12.1). There were two socioeconomic criteria: the probable annual timber harvest and the anticipated timber industry jobs (Figure 12.2). Based on these analyses, the decision makers, Secretary of the Interior Babbitt and Secretary of Agriculture Espy, chose what they considered to be the best option.

CHARACTERISTICS OF FORMATIVE EVALUATION

Formative evaluations have several distinctive characteristics. They are *prospective*, meaning that they take place before a policy, program, or project is undertaken. Consequently, the evaluation is speculative, because the actual results of the proposed action cannot be known before it occurs.

Formative evaluations are largely *qualitative*, based on the judgment of decision makers. Some quantitative techniques may be used (e.g., a proposed time frame might be compared to times that were actually used for similar activities), but the most important judgments are seldom entirely explicit or data-rich. For example, a local committee selecting teachers to invite to a workshop on "teaching the ecosystem concept in elementary school classes" will probably make its judgment based on much more than what is written on the application forms.

Formative evaluations are *normative*, meaning that there is a model of the ideal against which a specific case can be examined. The Boy Scout Law, for example, states that a Boy Scout is trustworthy, loyal, helpful, friendly, courteous, kind, obedient, cheerful, thrifty, brave, clean, and reverent. That is one normative model of the ideal young man. Normative models take various forms, but they are often lists of criteria that the planned action should match (Box 12.2). In the NWFP, for example, the team developed a "norm" for population viability that it asked experts to use when evaluating alternatives. The norm was an 80% probability that the habitat for a species would be sufficient that populations would be numerically stable and well distributed across the species' range for 100 years.

Because they depend so heavily on experience, formative evaluations tend to be *conservative*. Proposals that are familiar and that fit a group's current culture are likely to be viewed positively; those that challenge the status quo may be rejected. Consequently, an explicit commitment to learning, perhaps via adaptive management, is often needed to stimulate risk taking.

Because formative evaluations cannot be totally quantitative, they are highly dependent on the judgment of an individual or team. Many techniques have been developed for formalizing group judgment, such as the Delphi methodology, which asks experts to predict probable outcomes through a series of iterative discussions; the nominal group technique, which assesses preferences through a series of listing, discussion, and voting steps; and expert systems, which attempt to model human judgments via a computerized algorithm. Whatever the method, the composition of the decision-making team is crucial. A diverse team, in terms of social demographics (e.g., ethnicity, gender, age, religion) and organizational characteristics (e.g., private/public, scientist/practitioner, supervisor/worker), always makes the best team (Figure 12.3).

BOX 12.2

The Ideal State Fish and Wildlife Agency

The Wildlife Management Institute occasionally issues summaries of the status of state fish and wildlife agencies in the United States. Along with the summary, the institute also lists the qualities of an ideal agency, including the following:

- An executive agency, responsible to the state governor.
- A decision-making commission composed of unpaid citizens appointed by the governor on a statewide basis.
- An executive director with professional qualifications in fisheries and wildlife who is chosen by and reports to the commission.
- Funding that comes from both license fees and general revenues.
- Five departments covering wildlife, fisheries, lands and engineering, information and education, and law enforcement.

This description of the characteristics of an agency is normative, defining the institute's view of how an effective agency should be structured. This structure, however, is based on the traditional work of a fish and wildlife agency. For an ecosystem approach to conservation, what normative characteristics should a state agency have? A federal agency? A nongovernmental organization? A community-based group?

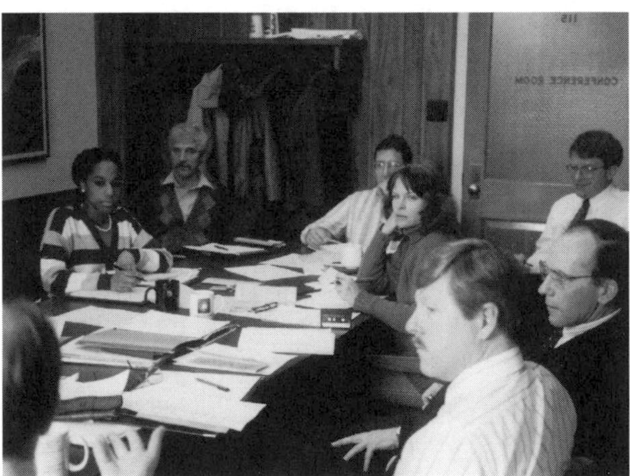

Figure 12.3. *A diverse team of people participating in making decisions helps ensure that the decisions are fair and comprehensive.*

EXERCISE 12.3

Collaborate on It!

Imagine that a large private foundation has created a grant program for helping community-based ecosystem management efforts and that the foundation has preselected the following project areas for support, one in each scenario:

- ROLE Model: Restoring unique or ecologically significant habitat patches (e.g., lilybush sites, old-growth forests).
- SnowPACT: Developing small-scale ecologically sustainable business ventures (e.g., native plant nurseries, guided tours of unique areas).
- PDQ Revival: Creating "living history" sites that teach children about their natural and cultural heritage (e.g., colonial agriculture, Native American fishing).

They have asked for a team of participants in each community to develop a formative evaluation process to help decide which specific ideas to support in the ecosystem. In small teams, and based on what you have learned in previous chapters, create a list of characteristics that should be required for projects to be supported and what their relative importance should be (i.e., the weight assigned to each characteristic). Include ecological, socioeconomic, and/or institutional factors.

Process Evaluation

Process evaluation tracks the implementation of a policy, program, or project, ensuring that it is being done correctly. Process evaluation goes by many names, including implementation evaluation, implementation monitoring, and monitoring. Confusion sometimes arises with the word "monitoring," because many people equate monitoring with the assessment of ecological conditions. Therefore, we prefer the term "process evaluation" to avoid confusion.

Process evaluation helps planners decide to adjust a policy, program, or project while it is progressing. This type of evaluation is a practical process, organized around getting the job done. Process evaluation involves two major aspects, one focused on assessing progress and the other on making adjustments based on the assessment.

ASSESSING PROGRESS

In examining how well an activity is progressing, evaluators typically ask four general types of questions:

1. *Is the budget being used as planned?* This question addresses how much money is being spent; the pace of spending (unspent funds often revert to the provider of the funds, so tracking the pace of spending is crucial); and what the money is spent on (equipment, materials, and services). If the planner intended to use most of the project's budget for paying workers but larger expenditures for equipment were needed, something may be wrong.

2. *Is the activity helping those for whom it was planned?* Decision makers should target a specific group to benefit from an activity. Questions need to be asked regarding the number, distribution, and kinds of clients served. This includes checking whether all intended clients, and any unintended clients, are being served. For example, if a program were intended to educate children about their ecosystems through classroom programs but 90% of educational activities to date have occurred through speeches to conservation or sporting organizations, the match may not be good.

3. *Are the number and pace of "units" produced adequate?* All policies, programs, and projects should have timelines, with milestones of productivity along the way. The leaders of the activity need to keep a regular scorecard that charts progress. For example, if a 5-year project for restoring patches of native prairies were intended to restore one patch per year for 5 years, but only one patch has been created after 3 years, progress may not be good. However, it might be quite reasonable to expect that restoring the first patch would take 3 years, given the logistical requirements. A timeline created in the formative evaluation that stated this in advance would allow proper process evaluation.

4. *Does the activity conform to the letter and spirit of the rules?* This question involves such things as

BOX 12.3

The Routine Questions of Process Evaluation

Implementing any project involves the use of real resources—time, money, equipment, supplies—and has effects on real people and places. Moreover, the community-based context of ecosystem management means working with stakeholders and partners to get the job done. To ensure that implementation is going as planned, process evaluation requires routine monitoring of several kinds of information.

Financial

- Is the activity within its budget, as planned for this stage of the work?
- Is spending following established general categories (e.g., personnel, equipment, operations)?
- Is spending following established procedures (e.g., grant eligibility, contracting protocol, procurement rules, accounting practices)?
- Will all the allocated funds be spent appropriately within the allowable time?

Clientele

- Are the clients who were targeted actually being served?
- Are clients who are not part of the target audience also being served, or being served instead of the intended ones?

Output Production

- Are the planned number and kinds of outputs being produced?
- Does the actual pace of output production match the scheduled pace?

Legal and Administrative

- Have all necessary permits and other permissions been acquired?
- Are all environmental laws, local ordinances, and other requirements being followed?
- Have all personnel been hired and evaluated according to accepted practices?
- Have all complaints by employees or stakeholders been addressed according to accepted practices?
- Have needed reports been written and transmitted as required?

Partners and Stakeholders

- Are all partners contributing what they said they would (support, involvement, materials, equipment, funding, and volunteers)?
- Have regular communications occurred as planned, with interested and potential stakeholders and partners?
- Have scheduled public meetings and other forums for public comments been appropriately announced, publicized, organized, and held?

safety, animal care, public disclosure, obtaining permits, accounting practices, personnel practices, record keeping, and quality control. Good intentions will not sustain ecosystem conservation if participants are suffering from injuries or animals and plants are dying because of poor treatment.

A final concern is particular to the partnering style of ecosystem management. Because ecosystem approaches will work best when they are community-based, the status of community participation must be checked continually. One aspect of this partnership is communication, from both a legal perspective (some forms of communication may be written into laws and funding agreements) and a trust-building and team-building perspective (Box 12.3).

As decision makers address these questions, process evaluation moves on to its second role of helping to adjust the activity. Seldom will an activity occur exactly as planned—after all, there are many unknowns to be encountered. Our commitment to adaptive management requires that we adjust our activities and our expectations as we learn.

MAKING ADJUSTMENTS

Adjustments usually focus on two general categories: resources and expectations. Changes to

BOX 12.4

Process Evaluation of a Conservation Education Program

The Conservation Leadership School (CLS) was a Pennsylvania tradition. For more than 50 years, high school students attended one of three 2-week sessions of a residential camp that taught natural resource management and leadership. Testimonials from former students were always glowing—the camp changed their lives, often leading them into careers in forestry or wildlife.

However, CLS was facing a crisis. Government funding had been eliminated and enrollments were dropping. In 1997, Pennsylvania State University's School of Forest Resources assumed management of CLS. A 5-year implementation plan called for a marketing strategy that would lead to full enrollments within 2 years, a programming strategy that would organize and codify lesson plans in 2 years, a development strategy that would create a private endowment within 5 years, and a financial strategy that would begin with 2 years of small deficits on the way to a self-supporting program in 5 years.

The programming strategy was implemented on time, with the help of a graduate student (a new resource assigned to the project). Likewise, the marketing strategy developed, printed, and distributed brochures on schedule, increasing enrollments over 2 years. But then enrollments began to fall again. The development strategy lagged with a few failed grant applications and then little attention over the following 2 years. The lack of progress on the development strategy also affected the financial strategy. Adjusting along the way, decision makers started raising tuition to cover more of the costs, and the CLS leaders did not receive a salary for 2 years.

However, funding continued to falter. With program deficits growing over the first 4 years of the strategy from $21,000 to $37,000, further formal adjustments were needed. A list of 16 possible adjustments was generated in the year 2000, ranging from program termination to a reduction to a 1-week program to tripling tuition costs.

When word of the possible changes to the CLS began to spread, Penn State received more than 75 letters of protest from former participants, legislators, counselors, and groups that sponsored scholarships each year. At the same time, a budget cut at Penn State eliminated the opportunity to continue to build a deficit into the program. Also, a variety of promising environmental education programs of other kinds, including 4-H clubs, day camps, and school teacher training courses, were also seeking funding and staff time.

Consequently, university leaders decided to close the Conservation Leadership School after the 2001 session. The process evaluation showed that sustaining the CLS would require funding that was needed elsewhere, and not enough of the intended clients were being served. Although the program had been very successful over a long period, the results of the process evaluation helped decision makers choose new priorities for environmental education.

resources frequently involve money, adjusting the budget up or down because the project is more or less expensive than anticipated.

Adjustments to other resource areas may be just as important and relevant to the success of a project. For example:

- An activity may have the wrong expertise. It needs an ornithologist, but a mammalogist was assigned.
- Land may be insufficient. The home range is 10 acres, but restorable habitat patches are only 5 acres each.
- The institutional climate may change. An easement policy was based on a state-wide, open-lands preservation law, but a court injunction has been filed, effectively stopping new easements.
- The timeline may be wrong. Designing and pretesting an amphibian species-richness survey was scheduled to take 1 month, but it has taken 3 months so far and is not completed.
- Necessary components of a strategy may change. A public information campaign was based on notices in a local weekly newspaper, but the newspaper went out of business.

In each of these cases, the group will need to adjust resources to get the project back on track.

Adjustments may also focus on the expectations of an activity. Rather than raising the budget, for example, the expected outputs could be lowered; if restoring native prairies is technically more difficult than anticipated, the project's output could be changed from five to three patches. If the project is running behind schedule and the reasons for the delays are justifiable, the timeline could be extended.

Adjusting resources or expectations, however, may have far-reaching consequences. Because resources are always limited, investing more in one activity will mean investing less in others. Consequently, the group's higher priorities (mission, goals, objectives) must guide resource reallocation. A process evaluation might cause decision makers to abandon an activity if the match between intended and real outcomes proved to be so poor that rescuing it would cripple other activities that have a higher priority. For example, a legal challenge stopping new easements might require abandoning that program, at least for the immediate future, because fighting the challenge would drain all available time, energy, and money from other projects.

Another possible decision is to concentrate resources into one project and cancel or modify others (Box 12.4). For example, the need to double efforts on a high-priority native prairie restoration may require canceling a lower-priority farm pond project, even if the farm pond work is popular and progressing as planned.

CHARACTERISTICS OF PROCESS EVALUATION

Process evaluation has several unique characteristics. Because the purpose is adjustment, process evaluation must be *concurrent* with the activities. Information must be collected and reported quickly and inexpensively—in real time. Also, because adjustments occur during an activity, the activity must have internal milestones. A 5-year project will need annual or semiannual targets against which progress can be judged. Waiting until the end voids the reason for performing a process evaluation.

Process evaluations are *quantitative*, heavily dependent on objective data. Budgets, time sheets, quarterly reports, and site inspections are the raw materials of process evaluation. For data to be available in real time, process evaluations need to be routine—standardized forms, cheaply available, and, if possible, electronically transmitted. Everyone who needs the data must be able to access them in a user-friendly form.

Process evaluations focus on efficiency. **Efficiency** means using the allocated resources appropriately; doing what the activity was intended to do, within budget, on time, and within the law. Consequently, process evaluation often attracts

EXERCISE 12.4

Talk About It!

Imagine you are chairperson of the state implementation committee of a nongovernmental organization that helps restore riparian areas as part of community-based ecosystem conservation programs. In this program, communities apply for small grants to install various riparian protection strategies along short stretches of degraded streams. Strategies may include a range of techniques, such as bank stabilization, channel narrowing and deepening, shoreline plantings, construction of in-stream devices, and the restoration of large woody debris. The projects are modest, ranging from $10,000 to $25,000, require one-to-one matching funding (either cash or in-kind contributions of time and materials), and must be completed within 2 years of approval. At any time, there are about ten ongoing projects, some in their first year and some in their second.

Create an outline for a process evaluation that the individual projects would need to perform, and report to the implementation committee. Indicate in the outline the general categories of evaluation criteria, the kinds of data collected under each, and the timetable for conducting the evaluations. Consider what sorts of decisions the statewide implementation committee might make based on the outcome of the process evaluation for each ongoing project.

BOX 12.5

Summative Evaluation of the Klamath River Basin Fisheries Restoration Program

The U.S. Congress created the Klamath River Basin Fisheries Restoration Program in 1986. The program, designed to help rebuild anadromous fish populations in the Klamath and Trinity Rivers of northern California and southern Oregon, was authorized for 20 years. In creating the program, Congress noted that floods, dams, mining, timber harvest, and road building had all contributed to reduced anadromous fish habitat in the Klamath-Trinity watershed. Included in the authorization was the need for a long-range strategic plan (completed in 1991) and a summative mid-term evaluation (completed in 1999).

The long-range plan included five goals: (1) restore biological productivity to provide viable commercial and recreational ocean fisheries and in-river tribal and recreational fisheries by 2006; (2) support harvest recommendations to provide viable fisheries and escapement; (3) recommend actions that federal, state, and local governments must take to protect fish and habitats; (4) inform the public about the value of anadromous fishes and gain support for the program; and (5) promote cooperative relationships among stakeholders.

The summative mid-term evaluation was designed to provide information about progress on all five goals, with specific tasks to assess the following:

1. How well the overall intent of the program and the five individual goals had been addressed.
2. How much the anadromous fish stocks had actually changed.
3. How plan policies and tasks had been implemented thus far.
4. How much leveraging of funds occurred because of program expenditures.
5. How fish habitat had changed since the program began, indicating what had happened because of the program and because of natural events.
6. How the organization and behavior of various groups had contributed to reaching program goals.
7. How much of the program's expenditures had been allocated among groups and organizations.
8. How the responsible party (the USFWS's Klamath River Fish and Wildlife Office) had performed.
9. How the knowledge level of watershed residents had changed.
10. How fish hatcheries in the river basin had contributed to reaching program goals.

Among a large number of specific findings, the contractors who performed the summative evaluation also noted several general findings:

- The program had established an effective organizational structure to pursue its goals.
- Nonetheless, fish stocks had continued to decline, and the program needed to confront the more contentious issues influencing anadromous fish restoration.
- The program needed to adopt an operating style that would allow participants to work through these contentious issues and find acceptable solutions.
- More efforts were needed to work with California state agencies to alter their land management practices and guidelines to improve stream protection.
- Pilot projects using stakeholder-based restoration planning processes should be expanded throughout the basin.
- The program should seek to have its plan be recognized and funded as the recovery plan for federally endangered anadromous fish stocks in the basin.

criticism and ridicule (ever heard of "bean counting?").

Process evaluation has consequences for adjustments *beyond the particular activity* under review. Differences between planned and actual outcomes in one activity may require higher-level decisions that affect goals, objectives, and assets allocated for other activities. A true ecosystem approach does not isolate actions within little boxes, but requires that the team views actions comprehensively.

Process evaluation often seems like harsh judgment. If evaluation leads to the modification or cancellation of an activity, how could it be viewed otherwise? This especially complicates community-based ecosystem work, in which volunteer participants have a deep personal commitment. All participants need to understand that adjustments help achieve the purpose they embraced when forming the team or joining the organization. Nevertheless, adjustments are hard—to make and to accept—and leaders must always be sensitive to the reactions of stakeholders, from employees to volunteers to neighbors to legislators.

Summative Evaluation

Summative evaluation, the traditional approach, examines how the policy, program, or project turned out. In other words, summative evaluation asks the question posed in the fourth step of the strategic management model: "Did we make it?" Some resource agencies call this effectiveness evaluation because it addresses an activity's **effectiveness**, how well it accomplishes it intended purpose.

Summative evaluation helps planners assess the outcomes and benefits of a policy, program, or project. Although summative evaluation can be directed at any level of strategic management, it typically aims at missions and goals—are we accomplishing, or even addressing, our stated mission?—and at strategic direction—is our approach having the impact that we intended?. Summative evaluation may also look beyond stated aspirations to assess the broader impacts, both positive and negative (Box 12.5).

CHARACTERISTICS OF SUMMATIVE EVALUATION

Summative evaluation occurs after an activity has been completed, sometimes long after. As a result, stakeholders, who often want answers quickly, need to be patient. Stakeholders must also understand and accept that summative evaluation can work only on a delayed schedule. It makes little sense to ask whether a native prairie has reestablished itself only a few weeks after planting, for example; several seasons may need to pass before we know whether the plants are healthy and self-sustaining and if the associated biotic community has developed as hoped (Figure 12.4).

Summative evaluations are both *prospective* and *retrospective*. Because this type of evaluation helps determine whether a change occurred because of an activity, data must be collected before and after the activity (this is part of adaptive management). The effectiveness of a program to increase groundwater infiltration, for example, can be judged only if the amount and timing of groundwater movements were measured both before and after the program was implemented. Consequently, summative evaluation must be designed at the very beginning of any activity, and data must be collected before the actions are initiated.

Summative evaluations use many types of data. Both objective and subjective data contribute to assessing outcomes and benefits. Objective data include ecological information (e.g., habitat changes, abundance, and distribution of organisms), as well as socioeconomic and institutional data (e.g., stakeholder surveys, economic analyses). Likewise, subjective data also cover ecological information (expert judgments of ecosystem condition, overall descriptions of a region) and socioeconomic and institutional information (endorsements by citizens and decision makers, reports of stakeholder panels).

Summative evaluation focuses on effectiveness—are we doing the right things? Because judging effectiveness is elusive, summative evaluation is a very difficult process. Did things turn out as expected? Would things have turned out this way anyway, without our actions? Would things have turned

Figure 12.4. *Prairie restoration is a complex process that may take many years to complete. Summative evaluation, for prairie restoration and almost any other ecosystem project, must wait until the system—ecological, socioeconomic, or institutional—has become suitable for study.*

out better if we had used a different strategy? And whatever way things turned out, was it worth it? Such questions cannot be answered with quantitative data only; answers require substantial thinking, discussion, and interpretation.

Each summative evaluation is unique. A summative evaluation must be individually designed for the policy, program, or project under review (Box 12.6). Because of these requirements, this type of evaluation looks much like research. In fact, summative evaluation is often called evaluation research, with hypothesis-like statements driving data collection and subsequent assessment. Remember that strategic thinking uses a step-down process to proceed from general goals (e.g., reestablish native plant communities) through specific objectives (e.g., double native prairie acreage in 5 years) to selected actions (e.g., burning, plowing, and replanting at ten sites per year). Rephrased as a hypothesis, this would read: "Burning, plowing, and replanting ten sites per year will double native prairie acreage within 5 years and reestablish native plant communities." This hypothesis can be tested, forming the basis of a summative evaluation.

Summative evaluations are characterized by a high rate of failure. The need to design a summative evaluation well from the beginning of an activity often gets overlooked. In the desire to get on with an activity, prospective data are not collected, and in the need to trim the budget, preactivity data collection is often the first item cut; thus, comparisons may not be available at the end. Data collection for summative evaluation at the end of an activity can be very expensive, and almost every project costs more than planned; consequently, the funds for summative evaluation often get used up along the way. Because summative

BOX 12.6

Design Criteria for a Summative Evaluation

Summative evaluations are like research projects: They require careful design so the evaluation issues will be useful and well received. Some of the questions that must be answered in the design are listed below, with different possible answers for the example of evaluating federal waterfowl management.

1. Domain: Will the evaluation focus internally or externally?
   - U.S. Fish and Wildlife Service (the primary agency responsible for setting waterfowl management policy).
   - State agencies, Native American tribes, Canada and Mexico, private landowners (the full set of stakeholders directly affected by waterfowl management policy).
2. Constituency: Who is the evaluation designed to inform?
   - Federal agency leaders in the U.S., Canada, and/or Mexico.
   - U.S. Congress.
   - International Association of Fish and Wildlife Agencies (a prominent lobby group for state agencies).
   - Ducks Unlimited and similar non-governmental organizations
3. Level: At what scale will the evaluations be done?
   - North American, or just the United States.
   - Nationally, or by individual flyways.
   - Species by species, or for all species combined.
4. Time Frame: Will the evaluation be medium or long-term?
   - Annual changes in waterfowl populations and harvest.
   - Five-year moving averages of populations and harvest.
   - On a life-span basis for the average individual of a population.
5. Data Type: Will the evaluation use objective and/or subjective data?
   - Waterfowl populations and harvest.
   - Habitat abundance and condition.
   - Expert opinions about the condition of populations and habitats.
   - Stakeholder views of trends in populations and habitats.
   - Stakeholder views of the acceptability of conditions and management decisions.
6. Frame of Reference: What expectations are used as a basis for the evaluation?
   - Objective-centered (e.g., did we hit or exceed established targets for population sizes and harvests?).
   - Comparative (e.g., were population trends better in the Central Flyway than in the Atlantic Flyway?).
   - Improvement (e.g., are mallard populations larger than when we last measured?).
   - Normative (e.g., does the population structure of each waterfowl species match the characteristics of a viable population?).

Table 12.1. *A Comparison of the Three Types of Evaluation*

	Formative Evaluation	Process Evaluation	Summative Evaluation
Purpose	Initiate	Monitor/adjust	Continue/terminate
Focus	Effectiveness and efficiency	Efficiency	Effectiveness
Timing	Before	During	Before and after
Data	Qualitative	Quantitative	Qualitative/quantitative
	Subjective	Objective	Subjective/objective
Basis for judgment	Normative	Achievement of stated targets	Case-specific
Cost	Medium	Low	High

evaluation comes at the end of an activity, much interest is already focused on the next activity rather than on the last one, so it is hard to keep attention (and resources) assigned to an evaluation.

Finally, and perhaps most importantly, many people involved in an activity do not really want an explicit, comprehensive summative evaluation. They may have invested years, even a career, and their professional reputation focusing on a single goal or strategy, and they are not interested in hearing that it was not the right thing to do.

The characteristics of the three types of evaluation are compared in Table 12.1. Evaluation is one of the most important tools in ecosystem management or in any aspect of life. Thoughtful consideration of our goals, our tools and techniques, and our performance will help us decide well about our future. Because ecosystem management is about our future—living and working in a sustainable world—evaluation is a fundamental activity. Evaluation becomes even more useful when conducted in such a way that the learning can be shared. And because ecosystem management is a group activity, involving many organizations and people, evaluation must be communicated and conducted as an open, explicit, and public process.

EXERCISE 12.5

Collaborate on It!

Cooperative agreements are ways for groups to become partners in achieving common objectives. Cooperative agreements are often formed among government agencies, NGOs, businesses, and community groups. Work with the cooperative agreement from your scenario (below) and generate a series of evaluation questions that might be asked to determine whether the project had been effective.

- ROLE Model: In the Round Lake ecosystem, the state DNR, Friends of Round Lake, and the American Tackle Company jointly create an aquatic biodiversity project to educate school-aged children about their ecosystem.
- SnowPACT: In the Snow River ecosystem, the U.S. Fish and Wildlife Service, ROCin', and The New Century Trust for Conservation link up to identify and post areas on the Red Cliff escarpment that are sensitive (and therefore off limits for climbing) and areas that are designated for climbing.
- PDQ Revival: In the PDQ watershed, Sonny Tymes, the state DNR, and Camp Fraser get together to fund and implement a program to conserve pine poccosins throughout the watershed.

References and Suggested Readings

Bozeman, B. 1979. *Public Management and Policy Analysis*. St. Martin's Press, New York.

Cameron, K. 1980. Critical questions in assessing organizational effectiveness. Organization Dynamics, Autumn 1980:66–80.

Dunn, W.N. 1981. *Public Policy Analysis*. Prentice Hall, Englewood Cliffs, NJ.

Espy, M., and B. Babbitt. 1994. *Record of Decision for Amendments to Forest Service and Bureau of Land Management Planning Documents Within the Range of the Northern Spotted Owl*. U.S. Department of Agriculture and U.S. Department of the Interior, Washington, D.C.

Kier Associates. 1999. *Mid-Term Evaluation: Klamath River Basin Fisheries Restoration Program*. Kier Associates, Sausalito and Arcata, CA.

Maguire, L.A. 1986. Using decision analysis to manage endangered species populations. Journal of Environmental Management 22:345–360.

Milbrath, L.W. 1983. Public decision-making with regard to managing major natural resources. Renewable Resources Journal 1(4):18–23.

Moore, C.M. 1987. *Group Techniques for Idea Building*. Sage Publications, Beverly Hills, CA.

Rorbaugh, J. 1979. Improving the quality of group judgment: Social judgment analysis and the Delphi technique. Organizational Behavior and Human Performance 24:73–92.

Rossi, P.H., and H.E. Freeman. 1982. *Evaluation: A Systematic Approach*. Sage Publications, Beverly Hills, CA.

Wildavsky, A. 1979. *Speaking Truth to Power: The Art and Craft of Policy Analysis*. Little, Brown, Boston, MA.

Zuboy, J.R. 1981. A new tool for fishery managers: The Delphi technique. North American Journal of Fisheries Management 1:55–59.

Experiences in Ecosystem Management:
Participation in Local Government Land-Use Decisions

George N. Wallace

By now you understand the importance of cross-boundary cooperation among public land managers to achieve ecosystem management objectives. You also know that community-based partnerships like Habitat Conservation Plans (HCPs) and watershed or special area planning groups are important for achieving these objectives in places that have multiple jurisdictions and a mix of private and public lands. Voluntary cooperation among stakeholders aimed at improving land health is increasing and shows much promise. One visible outcome of such partnerships has been better communication and tolerance among once-divided stakeholders and an improved understanding of ecosystems (Beatly, 1994).

There are limits, however, to the extent of the privately owned landscape that can be protected by partnership groups. Focusing on lands having endangered species or wetlands issues and prompted by strong federal regulations, many of these groups have had mixed results (Porter and Salvesen, 1995). Most private land in the United States remains in the hands of millions of individual landowners who are not stakeholders in a community-based partnership and may not be motivated by either ecosystem management goals or federal regulations. All these landowners, however, are subject to the plans and land-use codes of the jurisdictions where they reside. This is the "zone of regulatory or management authority" discussed in Chapter 2.

Working within this zone requires that land managers, nongovernmental organizations (NGOs), and citizens who care about land health regularly participate in the legally binding land-use decision process that takes place at the local government level. To do this, we must become familiar with how cities and counties make land-use decisions and understand the connection to ecosystem management. Let's start with an actual example of a land-use decision made in a typical community (names have been changed).

The Colter Slough: An Everyday Mini-Drama

The town of Maryville, Colorado, has one warm-water slough known as Colter Slough. During a cold snap, it does not ice over and provides important shallow-water habitat for a variety of species, especially the waterfowl that winter there. The 40 acres of land around the slough were farmed for the last 50 years, but recently they were annexed by the city, zoned for commercial use, and acquired by a developer for a planned-unit development with retail shops, an office complex, and space for light manufacturing. The city and county Open Space and Natural Area Program (OSNA) and the Colorado Division of Wildlife (CDOW) have this site listed as an important natural area. The City Master Plan calls for minimum lot-line setbacks of 150 feet from wetlands but provides no special guidelines for unique habitat areas like warmwater sloughs.

During the development review for this project, city planning staff sent the proposal to CDOW and

OSNA for their comments. Responses were slow, so planning staff contacted these agencies again and discussed the proposal. After this discussion, the planning staff recommended that any subdivision of land have a 250-foot lot-line setback from the high-water mark of the slough and that there should be an additional 100 feet between lot-lines and any building envelope. They also requested that parking areas be located farther from the slough to reduce human impacts, that lighting be directed away from the slough, and that vegetative filter strips and landscaping be used as buffers.

During the public hearing before the Planning and Zoning Board, the developer testified that these requirements would force him to "sacrifice" too much developable land. He noted that there was nothing in the City Master Plan or Code requiring such measures, and if the new conditions of approval were required by the board, it would constitute a "regulatory taking" of his property. The lawyer representing the applicant hinted at a possible lawsuit if these conditions of approval were added. Planning Board members were divided on the issue. Those opposing the development proposal were hoping for supporting testimony for the restrictions before closing the hearing to the public. Unfortunately, other than the planning staff's recommendations, no official letters from CDOW or OSNA were included in the application materials given to the board for their review. No representatives from those departments or local environmental groups were present to make a case for stricter protection that would go beyond informal conversations with the planning staff and become part of the public record.

Board members supporting the proposal as presented by the developer observed that protecting the slough "must not be very important, or wildlife and natural area representatives would be there to give testimony" and formally place statements in the record. Subcontractors for the development testified that other developments in the area next to ponds and wetlands had plenty of Canada geese and mallards that were seemingly unaffected by development. The board voted 5 to 4 in favor of the proposal requiring only the minimum setback and no restrictions on where buildings could be placed. Vegetative screening was required (without specifying native vegetation), but no restrictions were placed on lighting or parking areas. Vegetative strips to filter runoff were not required.

The next day there was an outcry in the press from environmentalists blaming wildlife officials for not defending wildlife. Staff at both the OSNA and the CDOW offices felt sheepish that nobody had been assigned to follow up on discussions with the city planning staff and admitted to one another that this one had fallen through the cracks. Others rationalized by complaining that their supervisor had never made the review of development proposals a priority. The district wildlife manager said "it was difficult to work a 50-hour week dealing with sportsmen, landowners, and problem wildlife and then be expected to attend evening meetings . . . and besides, no one had good data on what species were using the slough anyway." Across town, the biologist for the OSNA program lamented that she knew of no good studies that would support the request for a 250-foot setback request.

On any given day in most cities and counties across the United States, and in equivalent jurisdictions in many countries across the world, numerous proposals for developments are on the table. Some are for the subdivision of land into smaller parcels, others for the construction of buildings or communication towers, the extension of roads and sewer systems, or the rezoning of land from extensive to more intensive uses. Other applicants will come in for a "special review" of a proposed stone quarry, gravel pit, livestock feeding facility, or other land use that is not a "use by right" under the existing zoning for that location. Still others will come to request a "variance," which means they wish to be exempted from existing land-use or building code regulations.

On the same day, there are also plans to protect land and natural processes. These might be proposals to purchase development rights or transfer them away from sensitive lands, or to cluster development away from important habitat areas and put conservation easements on the remaining land (Figure A). There might be a "right to farm" ordinance passed in support of farmers and ranchers

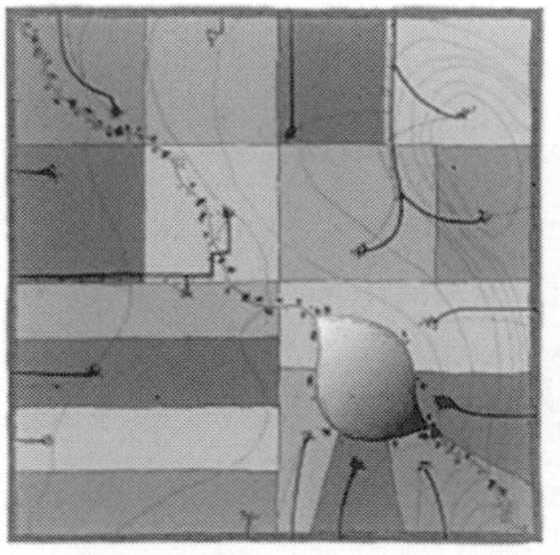

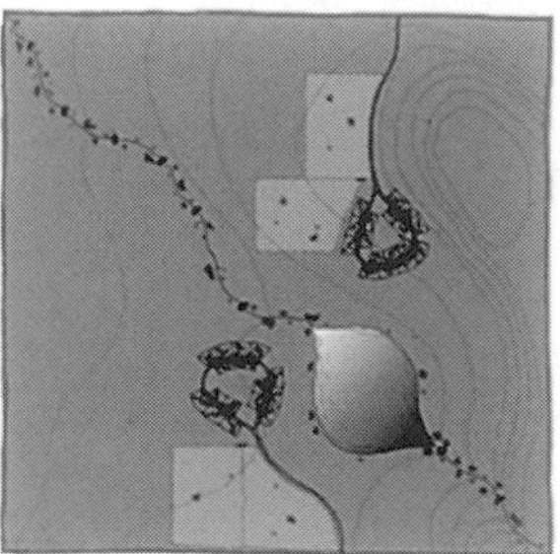

Figure A. *Clustered development can greatly reduce fragmentation effects; protect ridgelines, riparian areas, and lakeshores; and eliminate miles of fencing, roads, utility lines, and vehicle access points.*

who choose not to subdivide in spite of development pressure. There may be draft amendments to the land-use code that would remove the development potential from floodplains or fire-prone areas. Two cities might consider an intergovernmental agreement for the joint purchase and management of open space that would keep their communities from growing together and have the effect of creating a wildlife movement corridor.

Such land-use proposals, especially those for development, are relentless and cumulative. Each one slowly changes the face of the landscape in ways that will either increase or decrease threats to biodiversity and ecosystem health, and either block or allow for the continuation of natural processes. Obviously, many citizens and officials have not yet been exposed to basic concepts of conservation biology or ecosystem management. Consequently, few of the principles from these disciplines now appear in legally adopted planning documents, in the recommendations of city and county planners, or in the public testimony—all of which influence local land-use decisions. This situation is slowly changing, as cities and counties are prodded by federal and state environmental legislation regarding hazards, wetlands, threatened species, and air and water quality. Community groups concerned about rapid growth, sprawl, and open space indirectly motivate actions benefiting ecosystem management, even if it is not their main goal. Local land-use decisions that favor ecosystem integrity will require even more effort from ecosystem managers or advocates when they involve multiple jurisdictions.

How Land-Use Decisions Are Made by Local Governments

Becoming involved in local land-use decisions requires understanding of the legal framework and procedures used at that level. Except for shared national concerns about endangered species, air and water quality, or floodplains and other hazard areas, the federal government has given the responsibility for land-use decision making to the states. State constitutions, in turn, have passed most land-use decisions on to county and municipal governments. In doing so, they have enabled and instructed local governments to create the institutions and procedures necessary for making landuse decisions. Across the U.S., most important land-use decisions are legally structured to involve a combination of local elected officials, advisory boards and commissions, city or county planners, input from other agencies, and citizens.

At minimum, most states ask local governments to appoint planning commissions to develop master plans and a land-use code that includes zoning ordinances and regulations for subdividing land. Local government then hires planning professionals who help analyze and process proposals for the subdivision of land, development, and changes in the zoning (permitted uses) of land. A jurisdiction's planners provide technical assistance to those who wish to develop and to decision makers who will approve or deny a proposal. Decision makers are those members of the community appointed to sit on the planning commissions or who are elected to serve on town councils or county commissions. These appointed or elected officials will analyze information presented by planning staff, as well as public and agency input, to make final decisions about planning documents or proposed land-use changes.

Certain details about this process are useful to

understand before we discuss how protected area managers or other advocates of ecosystem health can influence outcomes. I say "influence" because, unlike consensus-driven partnerships, the local land-use process is "quasijudicial" and not truly collaborative.

Most towns have used state laws enabling them to "incorporate," that is, to develop a legal charter, elect officials, and levy taxes to pay for the services that their citizens come to expect. Rural people living outside an incorporated city or town are said to live in the "unincorporated" part of a county. Counties, by state law, always have some form of government, elect officials, appoint boards and commissions, and tax residents, but supply lower levels of service for things like roads, police, and fire protection. Nor are counties generally in the business of supplying water and sewer or other urban services such as libraries or recreation programs. Zoning inside incorporated urban areas (city limits) almost always permits higher densities, more intensive uses, and more commercial and industrial areas, and regulates other aspects of development to a greater degree than county zoning does.

Both incorporated (cities) and unincorporated (counties) areas usually have "master plans" or "comprehensive plans." These documents contain a jurisdiction's larger vision regarding land use, growth, development, and the quality of life. A master plan's guidelines and policies then set the stage for an accompanying "land-use code." The code is usually a separate document, with detailed zoning and subdivision regulations guiding what people can do with their land in different parts of the county and the procedures for reviewing development proposals. These documents can be very important vehicles for improving ecosystem health.

Counties typically elect from 3 to 5 county commissioners who oversee county government. Likewise, municipalities elect a city council or town board with from 5 to 9 members. The Board of County Commissioners typically appoints a planning commission of from 7 to 9 members (odd numbers are always used to avoid tie votes), and City Councils appoint an equivalent body that is often called the Planning and Zoning Board. Most development proposals go before the appointed body and then elected bodies in the jurisdiction where they are approved or denied. In many states, planning documents, such as master plans or area plans, receive final approval from planning commissions/boards instead of elected officials. This is an extension of our democratic system of checks and balances.

Proposals for subdivision or development (such as housing, manufacturing, recreational or agricultural facilities, towers, gravel pits, roads, schools) are received and examined by a jurisdiction's planning staff to ensure they meet the basic guidelines in the master plan and regulations in the land-use code. The proposal is then referred to other agencies for review. This means that specific agencies and specialists are requested to provide feedback about a proposed development's effect on roads, groundwater, wildlife habitat, wetlands, community services, public safety, and the rights of other property owners. In the past, many land managers, conservation organizations, and specialists have not taken the initiative required to be incorporated by local government into this review process.

After "referrals" come back to the planning department and suggestions from reviewers are discussed with the applicant, modifications are made and the application packet goes to the planning commission/board for public hearing. Hearings are part of the "due process" procedures and are held to provide an opportunity for all citizens and stakeholders to comment. Once the public comment portion of the hearing is closed, the proposal is discussed and voted on by the commission/board in its original form or as a board-modified version. It then goes on to the county commissioners or city council with the minutes of the meeting (the public record) summarizing the discussion, the votes of commission members, and their recommendation to elected officials for final approval or denial.

Commission/board decisions are considered "quasi judicial" in nature and must follow due process procedures. Local governments must provide proper notification of any changes to the master plans, codes, or any development proposal—especially to neighbors or other affected stakeholders—and they must see that other procedures are fair and timely. Public and stakeholder involvement is required at several specific points for nearly all land-use decisions. The failure to include stakeholders may precipitate litigation.

Integrating Ecosystem Management into Local Land-Use Decisions

With this background information in hand, how can resource specialists, protected area managers, NGO representatives, or citizens legitimately participate in local land-use decisions that determine the location, extent, type, and spatial patterns of development? Here are some basic steps that organizations and individuals can take.

Participate in the Development or Revision of Local Master Plans or Regional Plans. If they are well written, these documents can help overcome the tendency of local land-use decision making to become a type of public-sector review of individual private-sector proposals that does not address land-use or conservation concerns in an integrated fashion (Marsh and Lallas, 1995). A master plan should be a foundation document with a vision statement, goals, objectives, guidelines, and directions for achieving desired conditions. It is the legal basis for the subsequent zoning, subdivision, and other regulations contained in a jurisdiction's land-use code. It is likely to be referenced in any future court proceedings, such as regulatory takings lawsuits.

Goals related to protecting sensitive natural areas, wetlands, endangered or threatened species, wildlife habitat, and migration routes—if they are to exist for private lands—will be in the master or comprehensive planning documents for each county or municipality. It is important to include a description of how protected natural areas add to the quality of life and the economy, and the importance of mitigating impacts to them. It is here also that reference to any collaboration with other land managers, jurisdictions, or area plans has the effect of legitimizing and institutionalizing such partnerships (Porter and Salvesen, 1995). It is possible to amend a master plan between revisions.

Although not as prevalent, opportunities to participate in regional intergovernmental planning efforts including several jurisdictions may occur and are very important for ecosystem management. Notable examples include plans for the New Jersey Pine Barrens, Chesapeake Bay Critical Areas, and Adirondack State Park, which have legally tied local planning to the larger landscape.

Participate in the Development or Revision of the Land-Use Code for the Counties and Cities Adjacent to Important Protected Areas. The land-use code establishes the specific rules for any development and ensures implementation of the vision and principles of the master plan. The code includes zoning regulations that lay out the permitted uses and development densities at any given point on the landscape. The code also specifies procedures for the subdivision of land, review of development proposals, and rezoning. It specifies the performance standards or criteria by which all development proposals are judged during the review process. Criteria are set for access and grading permits, road standards, setbacks, building placement, fencing, landscaping, lighting, emergency access, handling of human waste and gray water, and storm drainage. Code-driven decisions for thousands of development proposals have a significant cumulative effect. As the Coulter Slough example revealed, this is where a number of negative effects on biological diversity can be directly addressed and mitigated. When accurate and timely ecological information is presented by agencies, resource specialists, or conservation organizations during code refinement, it may well influence how these regulations are written and applied.

Participate in the Review of Development Proposals. By now you are probably aware that the Colter Slough was not well protected because the right resource specialists did not fully participate or provide good information during development review. To be effective, advocates for ecosystem health must fully understand the review process and be aware of when their participation is most needed. They should understand the type of

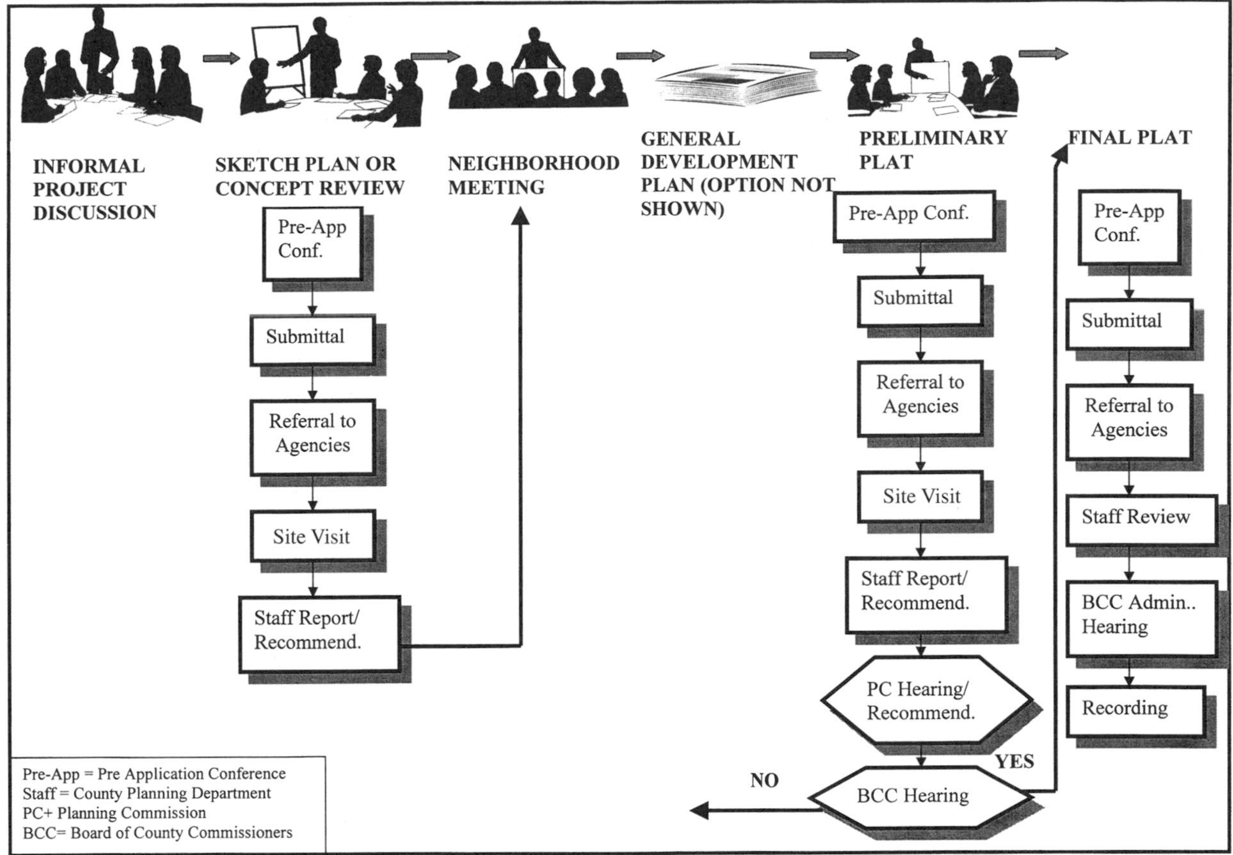

Figure B. *A typical development review flow chart. The development review sequence has several points where ecological information is formally requested and reviewed by local government staff and decision makers.*

information that must be presented and how to tie that information back to elements in the master plan and the code on which final decisions are based. Once a development proposal is brought forth, there are several formal opportunities for agencies, organizations, and specialists to provide input, and advocates must actively participate in them to be effective. Figure B illustrates the sequence of events followed during development review in one county. Sketch plan reviews, site visits, summary letters, and hearings are all important places for airing ecosystem issues.

Collaborate with Local Open Space Programs and Efforts to Protect Agricultural Lands. There will be times when land-use regulations and the review process itself may not offer the level of protection that critical areas such as corridors or unique habitat patches really need, even if voluntary land conservation techniques like transferable development rights are in place. In such cases, it may be possible to work with local government open space and agricultural land protection programs to protect these parcels. These programs are increasing in number and typically use public funds (sales taxes, real estate transfer taxes, state lottery funds) to purchase land or transfer or purchase development rights from willing sellers. The boards that prioritize such lands also require stakeholder input, follow due process, and are more readily influenced by good information about a parcel's ecosystem or biodiversity values than other boards and commissions. Open space plans are usually adopted as an element of a jurisdiction's master plan and provide an ideal opportunity to include ecosystem-related selection criteria. Open space and agricultural lands can provide important buffers for other protected lands, precluding many of the problems created by more intensive forms of adjacent land use.

Designate a Land-Use Specialist. This person should be assigned to work with local government officials, landowners, homeowners' associations, and nonprofit organizations such as land trusts to address land-use issues, especially land adjacent to existing protected areas. This is typically a full-time job, and the person needs to have the support of his or her supervisor, access to resources for managing spatial information, and the time to carefully establish a presence in surrounding communities.

Help Develop a Cross-Boundary Spatial Information Database. Many public and private protected areas or natural heritage programs have their own natural resource data, often displayed in a GIS format. Some now go outside their boundaries to obtain or share data. It is necessary to analyze and predict the cross-boundary effects of various land uses and describe them for others. The GIS map themes that have proven useful for such an analysis include a base map showing topography, hydrology, jurisdictional and ownership boundaries, and infrastructure; county and municipal zoning, as well as the management prescriptions and zoning on the protected area side; vegetation and unique ecosystem components and corridors that extend across boundaries; and a separate theme showing species-specific wildlife habitat and the locations used by threatened or indicator species. Specialized themes such as the location of special utility districts, sewer and water lines, and active development proposals help predict future development pressures.

Perform a Boundary Analysis. Using the spatial information described above, joint work sessions with land managers and local officials can be used to identify specific cross-boundary issues and opportunities along the interface between private and protected lands. A boundary analysis can include a "buildout" analysis, whereby a map is created that depicts what a given boundary segment would look like if it were subdivided and developed at the level allowed by the underlying zoning for that area. Recent studies indicate that areas next to public lands are proceeding toward buildout at a rate much faster than anyone had anticipated and that land values near protected areas are well above those of equivalent lands farther away. Because commercial and residential development out-bids almost all other extensive uses of land, a safe assumption for ecosystem managers is that adjacent land uses will proceed toward buildout unless some combination of land conservation and mitigation techniques are used.

Propose the Creation of an Overlay Zone Along Protected Area/Private Land Boundaries. One way to institutionalize cross-boundary collaboration might be an agreement by local governments and protected area managers to create a land-use "overlay zone," a district that is superimposed over existing county zoning and protected area management prescriptions. Such a zone would be placed on lands within a certain distance of the boundary in both directions. It is important that this be seen as a zone of two-way collaboration, to avoid the stigma of trying to create a buffer zone that would only benefit the protected area. Within this overlay zone, officials on both sides of the boundary would formally agree to include each other in the land-use decision-making process. Along a boundary segment where sensitive ecosystem elements occur, both parties might agree to adjust allowable densities, permitted uses, and the performance standards for development. Areas for transferable development rights might be created within the zone, or participants could target land that included a unique vegetative community for open space acquisition.

Develop Memoranda of Understanding. To legitimatize interjurisdictional collaboration where intergovernmental agreements are difficult (i.e., between federal and municipal government), it is advisable to summarize them in a memorandum of understanding (MOU). Long-standing legal and territorial differences can be partially overcome using well-worded MOUs that outline matters of mutual concern, references to supporting policies, descriptions of collaboration mechanisms, and references to particular activities. An MOU makes it clear that neither jurisdiction is being asked to abandon its legal responsibilities.

Use These Opportunities to Be Advocates of Land and Community Health. Involvement in the legally binding land-use decision-making process used by local government provides an institutionalized and ongoing forum for advocates of ecosystem health. It affords an important opportunity within a quasi-judicial setting to carefully explain concepts such as viable populations, the role of natural disturbance, and landscape structure to community members who are not natural resource professionals. The long-term implications of the loss of biodiversity, the rationale behind ecosystem management, and the threats to public land protected areas from adjacent development, for example, are poorly understood by the general public. It is important to take the dialogue about such matters out of the classroom, academic journals, and professional meetings and into the legal arena where most secular land-use decisions are made. Hearings related to land use are now routinely televised in many communities, and land-use proceedings and decisions are almost always summarized by the print media. Testimony given during the planning process and development review is high profile, begins to shape public opinion, and becomes part of the public record. "Decisions of record," as final approvals or denials for proposed land division, rezoning, or development are called, are like case law in that they slowly shape our collective attitude and norms about land use, land health, and their relation to the quality of life in a community.

For 6 years I have been a county planning commissioner. I have visited the sites, read the referrals and recommendations, listened to citizens and stakeholders, and voted on many development proposals. Each one changes the land around us to some degree and more often than not reduces our remaining supply of natural places and processes. It is at once frustrating and fascinating. There have been scores of mini-dramas like Coulter Slough—some more far-reaching and emotional. Although we have improved our county master plan and land-use code, thereby reducing some of the effects of development, many decisions have not gone like they might have if more people concerned with ecosystem health had been involved.

We tend to get excited about old-growth forests and grizzly bears, field inventories, and concepts like "rewilding," but we have less patience with land-use codes, long evening meetings, and the details of so many proposed developments. We are good at identifying unique vegetative communities but are not yet adept at explaining at public meetings why they are unique. Developers and representatives of other special interests, however, are always well prepared and at the table. This situation will be countered only when many more of us include participation in the local land use decision-making process as an important part of ecosystem management and an interesting part of our civic lives.

Discussion Questions

1. Imagine you are present at the Colter Slough public hearing on behalf of an open space program. You are looking for opportunities to explain the ecological importance of the slough. How would you respond to the person who said that there were other wetlands with development close to them that had plenty of Canada geese and mallards, so why worry about additional protection for this slough?
2. If possible, attend or watch on TV a planning commission meeting for your county. Agendas are posted on Web sites in many counties. Your instructor can help you watch for a hearing having proposals that would affect biological diversity or ecosystem health in some way. Discuss the issues, procedures, and participants at the hearing. Were you able to distinguish between applicants, staff, commissioners, agencies, and

other stakeholders on both sides of the selected issue or proposal? Was information presented or testimony given on behalf of the "land organism" by land managers, specialists, non-governmental groups, or citizens that contributed toward a decision favoring ecosystem health? Where might you have provided such information?

3. What provisions exist now in your city or county master plan for protecting natural systems, habitat, wetlands, wildlife, rural connectivity, biodiversity, and related concerns? Is there any reference to collaboration with other entities or jurisdiction to do this? What would you add to the plan to set the stage for land-use decisions that favor ecosystem health? Who is likely to support or oppose such inclusions or additions?

References

Beady, T. 1994. *Habitat Conservation Planning: Endangered Species and Urban Growth.* University of Texas Press, Austin.

Marsh, L. L., and P. L. Lallas. 1995. Focused, special-area conservation planning: An approach to reconciling developmental and environmental protection. Pp. 7–33 in D.R. Porter and D.A. Salvesen (eds.). *Collaborative Planning for Wetlands and Wildlife: Issues and Examples.* Island Press, Washington, D.C.

Porter, D.R., and D.A. Salvesen (eds.). 1995. *Collaborative Planning for Wetlands and Wildlife: Issues and Examples.* Island Press, Washington, D.C.

A Final Word

We have taken quite a long journey in this book, yet your individual professional journeys have barely begun. You have an exciting and challenging career ahead of you which may be spent in the service of protecting and managing natural resources. Natural resource management has never been easy, and it will only become more difficult in coming years and decades as more people come to depend on the Earth's shrinking resources. We hope that the information, situations, and ideas presented in this book will begin to guide you on your journey and provide a basis for the many challenges you will face.

We stand at the brink of a different approach to and relationship with the natural world. Ecosystem management, community-based conservation, or whatever we wish to call it is still in its early stages of development, but we are convinced it will increasingly be the way we think about the land, its people, and their interconnectedness. As it evolves and we gain more experience and understanding (through the future efforts and hard work of people like you), ecosystem management will help us overcome many of the obstacles, dead ends, and frustrations that haunt natural resource management. It is not perfect, it is certainly not a magic solution to our problems, and there is much to learn, but it is the best hope, we think, for a new relationship between people and landscapes, and people and people. It is perhaps our best guide at this time to learning to live on a piece of land without spoiling it.

We have tried in this book to give you a realistic feel for what it might be like to actually participate in difficult decisions that will determine the fate of landscapes and their human inhabitants. Rather than simply learn the theoretical and scientific basis of ecology and conservation, it is critical that we all embrace the realities of working in complex socioeconomic, institutional, and ecological landscapes, and the countless and diverse factors that influence society's decision making. Never again can we naively think that, as long as we have the science correct, we can make the proper decisions and all will be well. Science is only one part of a complex picture.

An important component of ecosystem management is recognizing that landscapes are shaped by ecological processes as well as by complicated administrative boundaries. It is critical that we not restrict our actions to protected lands; we have to find ways to work creatively across all land ownerships, from federal, state, and local jurisdictions, to first nations and private lands. Remember that most countries and many states in the U.S. consist of more privately owned lands than lands in the public domain. These private lands may often be the most productive for and critical to the persistence of biological diversity. Managing natural resources at an ecosystem level requires forging creative partnerships with diverse constituencies and landowners. This approach obligates all of us to think and act differently from how we have behaved in the past. What better time than now to begin viewing ecosystems and human communities differently, acknowledging their complexities, and eagerly devising approaches that reconnect both human and natural communities?

As you embark on your career in natural resource management, you are likely to encounter many frustrating failures and a few great successes. There will be times when it all seems overwhelming and you will wonder whether your efforts will pay off. You may view the many challenges and ask how you can possibly make a difference; it may seem onerous and exhausting. This is a natural response, and most professionals ask those same questions of themselves frequently. If it were easy, it would have been done already; it is not, and the world needs bright and talented people ready to accept a challenge. Here is a little story that may provide some guidance and inspiration on those discouraging days that undoubtedly await you.

Picture the following scene. It is early morning, barely light, with a heavy, damp mist hanging in the air. You are walking on a sandy beach; the tide is receding, but the heavy surf is a reminder of last night's storm. In the distance you see a tiny figure that seems to be moving in and out with the waves, almost in a fluid dance. As you get closer you see it is a young woman, bending down to pick up objects as the waves retreat, and then throwing them into the water, beyond the breakers. When you finally approach her you ask what she is doing. She says, somewhat out of breath, "The storm surge last night brought in all these starfish onto the beach. The tide is receding, and when the sun comes out, they'll dry out and die. I'm saving them by throwing them back." You look around you for a moment and then respond, "You must be kidding! I can see dozens of starfish right here, and there are hundreds of miles of beach. The job is too big. You can't possibly make a difference!" And as she throws back one more starfish she looks you squarely in the eye and responds, "Wanna bet? I just made a difference for *that* one! Now how about *you* picking up a few and making a difference for *them*."

It may sometimes seem that we are working against a receding tide and a rising sun. We cannot solve all the problems in the world, or even in our small part of it. But individually and together, as professionals and plain citizens and members of the biotic community, we *can* make differences in key places. So go forth in your career, find some starfish to throw back, and see what a difference you can make. Hand a starfish to someone else, and teach that person to make a difference. Proceed with enthusiasm and dare to do great things—one starfish and one person at a time.

Glossary

active adaptive management The form of adaptive management that is most like a scientific experiment, with random assignment of treatments and a full range of experimental treatments and controls.

adaptive management The process of treating the work of managing natural resources as an experiment, making observations and recording them, so the manager can learn from the experience.

alpha richness (α) The number of species within small areas of fairly uniform habitat.

area-sensitive species A species that requires a large area to persist, because of body size, movement requirements, or specialized needs.

beta richness (β) The amount of change or turnover in species (i.e., species gained and lost) in going from one habitat type to another (i.e., the species *difference* between two habitats).

biodiversity The variety of life and its processes; also, the composition, structure, and function of life considered from genetic to landscape levels of organization.

biological species concept A species concept based on reproductive (genetic) isolation; it defines a species as groups of actually or potentially interbreeding populations, which are reproductively isolated from other such groups.

biophilia The biological dependence and innate attraction that humans feel toward other living things and the natural world in general.

census population size (N_c) The estimated number of total individuals in a population, regardless of their reproductive states or genetic contributions to the gene pool. *See also* genetically effective population size.

charrette An intense, long, and continuous meeting (sometimes lasting for days) in which highly motivated stakeholders work together to develop a plan or position.

coarse-filter approach An approach to protecting biodiversity that focuses on managing at appropriate landscape scales. It assumes that biodiversity, from plants to insects to vertebrates, will be maintained if the correct mix of ecological conditions are provided. *See also* fine-filter approach.

command and control An approach to problem solving in which precise control over events and their outcomes is both desirable and possible.

consultation A type of stakeholder involvement in which decision makers ask stakeholders to comment on proposed decisions or actions.

demographic bottleneck A significant, usually temporary, reduction in genetically effective population size, either from a population crash or a colonization event by a few founders. *See also* founder effect.

density-dependent dispersal The movement of wildlife from areas of high density and intraspecific competition to areas where density is lower and access to critical resources is more attainable. *See also* natal dispersal.

desired future condition The qualities of an ecosystem or its components that an organization seeks to develop through its decisions and actions.

deterministic force A factor in the cause of extinctions that covers wide areas and results in reduced and isolated populations that are then susceptible to stochastic forces. An example is the conversion of rural landscapes to residential developments. *See also* stochastic force.

dispersal-sensitive species A species whose fitness (ability to survive and reproduce) decreases in fragmented landscapes, due to physical, behavioral, or physiological limitations or that experience elevated mortality rates from having to cross human-dominated landscapes.

documented trial and error A form of trial-and-error learning in which experiences are carefully collected, analyzed, and shared with other people.

ecological process approach An approach to managing for species communities that manages for ecological processes (e.g., flooding, fire, herbivory, predator-prey dynamics) within the natural range of historic variability. This approach assumes that if ecological processes are occurring within their historic range of spatial and temporal variability, then the naturally occurring biological diversity will benefit. *See also* landscape approach; species approach.

economically important species A species that has positive or negative consequences for the local, regional, or national economy.

ecosystem A dynamic complex of plant, animal, fungal, and microorganism communities and their associated nonliving environment interacting as an ecological unit.

ecosystem management An approach to maintaining or restoring the composition, structure, and function of natural and modified ecosystems for the goal of long-term sustainability. It is based on a collaboratively developed vision of desired future conditions that integrates ecological, socioeconomic, and institutional perspectives, applied within a geographic framework defined primarily by natural ecological boundaries.

edge effect The phenomenon whereby edge-sensitive species are negatively affected near edges by factors that include edge-generalist species, human influences, and abiotic factors associated with habitat edges. Edge effects are site-specific and factor-specific and have variable depth effects into habitat fragments.

edge-generalist species A species that experiences enhanced fitness near habitat edges.

edge-sensitive species A species that experiences reduced fitness near habitat edges.

effectiveness The extent to which an action or organization is accomplishing its stated purpose.

efficiency The extent to which an action or organization is properly using its allocated resources.

evaluation The fourth and last step in strategic management: examination of how an organization's plans and actions have turned out—and adjusting them for the future.

evolutionary-ecological land ethic A conservation philosophy derived from evolutionary and ecological perspectives and advanced by Aldo Leopold. Nature is viewed as an integrated system of interdependent processes and components, all of which are or can be important to the functional whole.

experimental approach An approach for estimating viable population sizes that experimentally isolates different-sized patches of suitable habitat containing the species of interest. Populations living in the patches are monitored over time, and an empirical estimate of the minimum viable population size is made based on how long the different populations persist. *See also* modeling approach; observational approach.

fine-filter approach An approach to protecting biodiversity that focuses on providing suitable habitat conditions for individual species, guilds (species that exploit a similar resource a similar way, such as scavengers), or other groupings of species. *See also* coarse-filter approach.

flagship or charismatic species A species that elicits emotional feelings from individuals, including a willingness to contribute financially to the species' well-being or otherwise support their protection.

focus group An organized meeting in which a small number of similar individuals respond qualitatively to a series of open-ended, opinion-seeking questions.

formative evaluation Analysis that helps planners decide whether or not to initiate an action and, if so, what resources to allocate, based on previous experience with similar actions.

founder effect The principle that the founders of a new population carry only a random subset of the genetic diversity found in their larger, parent population.

gamma richness (γ) The number of species within a defined region (i.e., the cumulative number of species observed in all habitats of a region).

gene pool The sum total of genes in a sexually reproducing population.

genetically effective population size (N_e) The functional size of a population, in a genetic sense, based on numbers of actual breeding individuals and the distribution of offspring among females. *See also* census population size.

genetic drift Random changes in gene frequency in a small population due to chance alone.

goal A broad general statement that further defines an organization's mission or mandate. *See also* objective.

Habitat Conservation Plan (HCP) A management agreement developed by private landowners whose properties support species listed by the Endangered Species Act and who may incidentally "take" individuals of those species while developing or managing their property. (A "taking" is an activity that harasses, harms, or kills a listed species.) HCPs are documents that share the responsibility of protecting listed species between the government and private landowners.

habitat fragmentation The process by which a natural landscape is broken up into small parcels of natural ecosystems, isolated from one another in a matrix of lands dominated by human activities.

implementation The third step in strategic management: converting plans into action by assigning assets and performing real tasks.

inbreeding The mating of individuals that are more closely related than by chance alone.

indicator species A species that is indicative of particular conditions in a system (ranging from natural to degraded) and used as a surrogate measure for other species or particular conditions.

internal fragmentation A process that occurs when linear or curvilinear corridors (e.g., roads, power lines, trails) dissect an area.

inventory A list of all the assets and liabilities of an organization, including physical, financial, personnel, and procedural aspects.

keystone species A species whose effect on the structure of a biological community is well out of proportion to its relative biomass. The addition or removal of a keystone species has large effects on the richness and relative abundance of many other species.

landscape approach An approach to managing for species communities that focuses on landscape patterns rather than processes and manages landscape elements to collectively influence groups of species in a desired direction. This approach assumes that by managing a landscape for its components, the naturally occurring species will persist. *See also* ecological process approach; species approach.

limited partnership A type of stakeholder involvement in which the stakeholders themselves become the decision makers, usually with some limit on the range of their authority.

mandate The highest-level stated purpose for an organization's existence; usually synonymous with mission.

matrix The most connected and extensive landscape element type. It can include both human land-use and vegetation communities. The matrix is important because it can often influence ecological processes that may affect biodiversity.

metapopulation A regional population consisting as a number of spatially discrete subpopulations distributed among habitat fragments and connected via dispersal.

minimum viable population (MVP) The smallest spatially discrete population having a certain probability (e.g., 99%) of remaining extant (not going extinct) for a certain period of time (e.g., 1000 years), despite the effects of demographic, environmental, genetic, and catastrophic events.

mission The highest-level stated purpose for an organization's existence; usually synonymous with mandate.

modeling approach An approach to population viability analysis that uses formal models to assess the extinction risks of species. *See also* experimental approach; observational approach.

mosaic The spatial characteristics of all the natural and human-created aspects of a landscape.

movement corridor A linear strip of a natural ecosystem that connects areas with conservation value to increase the likelihood of successful movement and reduce the extinction rates of isolated populations. Movement corridors may facilitate the daily movement of individuals from one habitat patch to another, or they may be much longer strips that allow the periodic long-distance dispersal of individuals.

natal dispersal A primarily innate behavior in young animals who leave their site of birth and move in search of a mate and suitable breeding habitat. *See also* density-dependent dispersal.

nominal group technique A formal process for acquiring group views by generating and then ranking views expressed in a round-robin listing by the group's members.

notification A type of stakeholder involvement in which stakeholders are informed of planned activities with no expectation of response or input.

objective A specific, quantitative statement that defines exactly what an organization seeks to accomplish or achieve. *See also* goal.

observational approach An approach to population viability analysis that examines populations of a species over time in natural habitat patches of different sizes. *See also* experimental approach; modeling approach.

passive adaptive management A form of adaptive management in which many of the requirements of a scientific experiment are not met, but the overall process is still approached with learning as a major objective.

perforation Habitat alteration in which human uses (e.g., houses, oil wells, campgrounds) alter small areas within a larger area of natural vegetation.

phylogenetic species concept A species concept based on branching, or cladistic relationships among species or higher taxa. This concept hypothesizes the true genealogical relationships among species, based on shared, derived characteristics.

policy The highest level of decision making, in which an organization creates rules, guidelines, priorities, and its culture.

populational view A philosophical perspective of the world that embraces and recognizes variation in classes of objects, including species; variation is seen as critically important.

population viability analysis (PVA) The science of model development to estimate extinction risk and closely related parameters.

problem In the context of strategic management, an explicit obstacle that stands in the way of achieving an objective.

process In regard to decision making and the success triangle, the formal steps or stages the decision makers promise to follow.

process evaluation Analysis that helps planners decide whether or not to modify an ongoing action, based on its progress to date.

program An organization's main operational unit, defining what will get done and who will do it.

project In the context of strategic management, an interrelated group of activities performed to achieve a planned objective.

proximate factor The immediate cause for the decline of a population, usually a decreased birth rate, an increased death rate, or both.

public meeting A formalized session, usually required by law or regulation, in which decision makers explain an issue and seek comments from the general public.

relationships In regard to decision making and the success triangle, the trust that develops among people because they have worked together honorably and successfully over time.

rescue effect The phenomenon whereby source populations bolster or recolonize sink populations through dispersing individuals.

resource conservation ethic A conservation philosophy derived from the views of forester Gifford Pinchot, based on the utilitarian philosophy of John Stuart Mill, in which nature is seen primarily as providing goods for humans.

review and comment A type of stakeholder involvement in which stakeholder reactions to a proposed activity are sought.

romantic-transcendental conservation ethic A conservation philosophy derived from the writings of Emerson, Thoreau, and Muir in which nature is viewed in a quasi-religious sense and emphasis is placed on keeping nature in a pristine and wild state.

scientific experiment A formal way to learn by setting up an explicit, logical test of an idea, with all the conditions for judging the outcome clearly established and written before the test.

sink population A discrete population where productivity is less than mortality; it is not sustainable without the immigration of individuals from other populations.

source population A discrete population where productivity exceeds mortality; a self-sustaining population.

spatially explicit model A population model that incorporates the actual locations of organisms and suitable patches of habitat and explicitly considers the movement of organisms among such patches across real landscapes.

species approach An approach to managing for species communities that focuses on manipulating a single species to affect many other species. Species chosen might be an invasive species, a keystone species, or any species that has important ecological interactions that other species depend on or are affected by. *See also* ecological process approach; landscape approach.

species-area relationship A well-known ecological relationship in which the number of species increases with area of an ecosystem.

stakeholder A person who wants to participate in a decision because the decision is important to his or her interests.

stakeholder orbits A visual analogy for the intensity with which individuals care about a decision or action; the lower the orbit, the more intense the interest.

stochastic force A factor in the cause of extinctions that results from random events such as demographic changes (birth and death rates, age and sex cohorts), the loss of genetic diversity, or unusual environmental factors, including an extremely cold winter, a wet spring, or a dry summer. Stochastic extinctions usually occur in populations that are small or already reduced by deterministic forces. *See also* deterministic force.

strategic management The continual process of inventorying, choosing, implementing, and evaluating what an organization should be doing.

strategic thinking The second step in strategic management: identifying and selecting the mission, strategies, goals, objectives, tactics, and projects for an organization.

strategy A general approach for accomplishing an organization's mission or mandate.

substance In regard to decision making and the success triangle, the facts relating to the decisions; items of information, such as acres of land in a certain use category, that are not subject to debate.

summative evaluation Analysis that helps planners decide to continue or terminate an action, based on the outcomes with regard to goals and objectives.

tactic In the context of strategic management, a specific method chosen to overcome a problem.

three-context model of ecosystem management The major conceptual model used throughout this book in which ecological, socioeconomic, and institutional perspectives, abilities, and constraints are all considered in seeking solutions to complex problems.

town meeting A method of stakeholder involvement consisting of a large-meeting format in which highly active participation by the attendees is expected.

tradition A way of learning, based on standards passed from person to person over time.

trial and error An informal way to learn by experiencing a single event (or series of events) and changing future decisions based on that experience.

trust responsibility In the federal government, a special duty required of agencies to hold and manage lands, resources, and funds on behalf of Native American tribes.

typological view A philosophical perspective of the world that embraces the existence of a "type" or perfect form of objects, including species; individual variation is seen merely as imperfections.

ultimate factor The underlying factor that drives changes in birth or death rates in a population.

umbrella species A species that, if secure or flourishing, would protect many other species because of its demand for large expanses of habitat.

vulnerable species A species (or population) that is particularly susceptible to extinction.

workshop A small meeting, usually of several hours, that is intended to generate a specific output.

Index

Island Press Board of Directors